AF400694

INSIDE THE LOST MUSEUM

INSIDE
THE LOST MUSEUM

<Curating, Past and Present>

STEVEN LUBAR

Harvard University Press

Cambridge, Massachusetts

London, England

2017

Library of Congress Cataloging-in-Publication Data
Names: Lubar, Steven D., author.
Title: Inside the lost museum : curating, past and present /
Steven Lubar.
Description: Cambridge, Massachusetts : Harvard University Press,
2017. | Includes bibliographical references and index.
Identifiers: LCCN 2017004601 | ISBN 9780674971042 (alk. paper)
Subjects: LCSH: Museum techniques. | Curatorship. | Jenks
Museum of Natural History and Anthropology.
Classification: LCC AM111 .L83 2017 | DDC 069/.4—dc23
LC record available at https://lccn.loc.gov/2017004601

For Lisa

Contents

INTRODUCTION

Explore

ON SEPTEMBER 26, 1894, John Whipple Potter Jenks died on the steps
of his museum. "He was in his seventy-sixth year, apparently hale and
hearty, his youthful enthusiasm not abated. He had lunched, perhaps
too heavily, with some dear friends with one of whom he walked up College Hall, stopped for a minute in conversation on the steps of Rhode
Island Hall, started to go upstairs to the Museum, sank down and expired
without a moment's sickness or suffering." "A martyr to science," one of
his students would write.[1]

Jenks was a naturalist, taxidermist, popular science writer, and beloved
professor, as well as the founder, director, and curator of Brown University's
museum of natural history. The Jenks Museum offered students and local
visitors glass cases packed with taxidermied animals, ethnographic items
from around the world, and other museum-worthy "curiosities"—some
50,000 items. Jenks founded the museum in 1871 and devoted the last
decades of his life to it. But even before his death, the museum had come
to seem old-fashioned. Like its director, it had not kept up with new ideas
in science and teaching. Brown closed the museum in 1915 and discarded
most of its collections in the university dump in 1945.[2]

In 2014, the Jenks Society for Lost Museums—a group of students and
faculty at Brown and the Rhode Island School of Design, along with artist-
in-residence Mark Dion—opened an installation called *The Lost Museum*
to celebrate the long-lost museum. The installation, located in Rhode

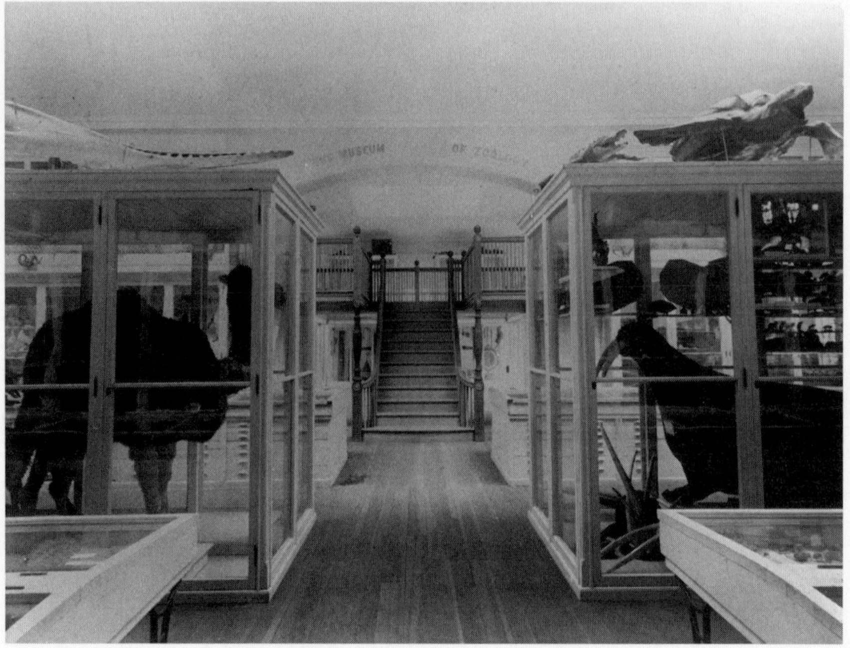

The Jenks Museum around the time of Jenks's death in 1894.

Island Hall, the building that had once housed the museum, told three stories. Visitors could peer into Jenks's workshop, seeing it as it might have looked before Jenks left for lunch that fateful day. Traps and nets lean up against the wall. Birds' nests and fossils cover tables. A taxidermy project is underway. Correspondence covers the desk. A Bible is near at hand. A copy of Darwin's *Origin of Species* is on the floor.

Another room, labeled "Museum Storeroom," contained ghostly reminders of some of the objects that had been in the museum. Animals, tools, weapons, and clothing of "primitive" peoples, artifacts of local history, and what nineteenth-century visitors would have called "curiosities" had been reimagined by eighty artists, based on Jenks's descriptions of them in museum reports. All were white, ghosts of objects long lost. A register provided the name of the artist, date of acquisition, and accession number for each of the objects.

A third display looked more like a traditional museum exhibit. A long case contained about one hundred Jenks Museum objects that had escaped the trip to the dump. But unlike in a standard museum exhibition, these were organized not by type or chronology, but by degree of decay. At the left end were objects that had been found useful and preserved in other university collections. At the far right lay dirt from the dump where the collections had been disposed of, and Jenks's handwritten labels for objects that are forever lost.[3]

The Jenks Society project told the story of the Jenks Museum, but it also revealed some truths about museums and museum artifacts more generally. The short history of the Jenks Museum and the laconic *Lost Museum* installation cast light on elements common to every museum. In their stories we can begin to understand collecting, cataloging, and categories; how museums preserve objects, learn from them, and tell stories with them; and how visitors interact with and learn from exhibitions.

The Lost Museum began with history, which led to reflection, and then to revelation. This book follows the same path. As the Jenks Society teased out the history of the Jenks Museum from annual reports, newspaper and magazine articles, archives, and surviving collections, I realized that the story of this one small museum provided a path into museum history, and from history to the present, that gets beyond the usual success stories that shape our understanding of museums. There was, and is, so much more going on, both in the history of museums, and in museums today.

In this book, I give you a look behind the scenes, revealing the debates and disagreements that shape museums. I do this not as a writer of an exposé, or as an outside critic, but as a former museum curator and director, and a teacher of students interested in museum work. The front rooms of museums, their galleries, are open, public spaces. But the rest of the iceberg that is the museum is mostly hidden. Locked storerooms hold collections. Closed-door conference rooms conceal passionate debates about what to collect, what to put on display, and what to say about it. What happens in the storerooms and conference rooms, and in curators' and educators' offices, shapes the stories told in the galleries. I will open the door to those silent storerooms and noisy meetings and accompany you inside, to see museums at work.

In *The Lost Museum* installation, we used the story of one museum to consider the fundamental nature of the museum. That's what I do in this book, too. In each chapter, I start by considering the Jenks Museum's collecting, preservation, exhibiting, or teaching, and move on from those stories to consider the work of museums in a more general way.

What makes museums unique is their collections, and so this book starts with the process of *collecting*. Multimillion-dollar art auctions and unexpected discoveries may get all the publicity, but most museum collecting happens quietly: gifts, purchases from specialized dealers or on eBay, even objects taken from the trash. The curator's challenge is what to collect. What will be useful for display, or for documenting an era, an artist, or a species? What makes something worth saving?

The focus of Part 2 is the *preservation* of collections. There are more than a billion objects in American museum collections. Museums need to take care of them, know what they are, and know where to find them. Museum storerooms are more than warehouses. They're also laboratories, places of discovery. Organizing and making collections available are essential parts of museum work.

Display, the subject of Part 3, is the public face of the museum, but here too, there's more going on than a visitor sees. Exhibitions are rarely simple displays of things, easily agreed on. Some are years in the making and cost millions of dollars. Even a small, simple exhibition can engender endless discussion about what story to tell and how to tell it, what objects to show and how to show them.

Part 4 considers the *usefulness* of museums, beyond collections and exhibitions. Museums can build and strengthen communities—though which communities, and how, is a matter of continuing debate. The objects in museums present research possibilities, from old-fashioned hands-on understanding to scientific analysis using the latest technologies. Visitors to museums learn from looking, especially from guided looking. Museums help us understand the past and present, nature and art, ourselves and others. They offer connection and empathy. They can change lives.

The last part of the book, the Coda, looks back at all of these topics through the eyes of contemporary artists.

Arranging the book this way cuts across the bureaucracy of museums, combining work that's often kept separate. How to organize a museum's hierarchy is more contentious than someone outside the museum world might imagine. A traditional structure might include three groups. "Historical (or artistic or scientific) resources" covers the work of registrars, conservators, collections managers, and curators. "Visitor experience" includes education, exhibition design, the shop, and docents. A third group deals with development, public affairs, human resources, and similar departments. In recent years, some museums have moved curators to the "visitor experience" category, separating them from "historical resources," which focuses only on collections. New organizational structures reflect new ideas about the role that collections and curatorship play in museums.

You can see these changes in another way that museums chart their work. Some put collections at the center, and around them, rings for conservation and preservation, curatorial expertise and research, then visitor learning, and community engagement. Other museums reverse this. They make a point of putting visitor experience or community engagement at the center, and of proposing a different set of rings: engagement, the museum building, objects and information about them. Yet others might list the audiences they focus on first, and build the museum's resources around them.

All of these systems offer a way of describing the work of the museum. They are also ways for me to circumscribe the parts of the museum I cover in this book. I am interested in objects and visitors, and the way the museum directs their interactions. This is curatorial work, very broadly defined, and the organizational chart I would draw would put objects and visitors at the center, two core activities connected by curators, educators, and exhibition developers. Those are the activities and people I focus on in this book.

There are parts of the organizational chart I don't consider. At the top of many museum charts are the boards of directors or trustees— community members who are interested in the topic, or in the idea of the museum, or in its value to the community, sometimes chosen for their ability to provide financial support for the museum, sometimes collectors. They have fiduciary responsibility for the museum, make strategic decisions, raise money, and hire the director. Next on the chart is the director, and under him or her the divisions of the museum, not just curators and educators, but also people responsible for fundraising, commercial operations, building design and maintenance, event management, business development, and more. Governance and management is essential to the museum, but that is a topic for a different book. I focus on the things that make museums unique: objects, visitors, and expertise.

I also focus on American museums. From their beginning, though, American museums have drawn on European precedents, and so I've included some of that early history. When museum work became professionalized, in the late nineteenth century, ties between museums in Europe and the United States were close. American curators and directors read British and German publications, and many of them took a grand tour of European museums. In the past few decades, England, Australia, and Canada have led the way in many aspects of museum work, and precedents and publications from those countries, and others, are an essential part of modern American museum philosophy and practice. They're part of the story I tell.

Museums are wonderful places, with a long history. It's a tangled history, more varied and more interesting than most museum visitors or even most museum workers know. The museum we accept today, with a particular vision of "museum quality" art and "significant" history, a narrow range of visitors, trained museum professionals, a scholarly standard of curation, and displays meticulously designed to be educational and attractive, represents only one way that museums have framed their work.

Inside the Lost Museum argues that museums' past and present are more connected than we realize. I focus on these continuities, considering the history of museums not only to understand the past but as a source of new ideas for the future. Over the course of the last century, the world of museums has moved toward agreement on a wide range of topics, from the untouchable preciousness of artifacts to the necessity of interpretation. In the process, it has abandoned other ways in which museums have structured the relationship of art and artifact to history, communities, expertise, and the public. This book is not a history of museums, but there's a lot of history in it, building on the explosion of scholarly research on museum history and theory over the past few decades. I explore some of the possibilities that were cast aside, the paths not taken, the history forgotten, as museum professionals forged the museums of today. I use history to get at what makes a museum, a museum. Much recent writing about museum history focuses on change.[4] I keep those changes in mind, but also search for a usable past. I consider not only what museums have been but also what they might be. I look inside those lost museums to find what's useful for today, and tomorrow.

The historian Neil Harris wrote that "repudiation of the immediate museum past as dusty, remote, lifeless, and unimaginative became an expressive ritual for each generation of museum professionals since the late nineteenth century."[5] As a curator, I've seen that repudiation in action. Indeed, I was one of those who disparaged the previous generation of curators when I worked at the Smithsonian. Why did they collect these things and not the things I need for my exhibition? Why didn't they take better care of them? It would have been so easy for them to find out more about the collections as they collected them. Why didn't they? Why were the exhibitions so dull? Why don't they address the important issues? Why are they so . . . old-fashioned? I'm certain that a new generation has asked the same questions about the work that I did.

This book answers those questions by giving voice to the museum workers who collected those artifacts, created those exhibitions, and made those decisions. It looks back further than just that previous generation to discover an astonishing diversity of ideas and activities in museum history. For institutions that are weighed down by collections, proud of and often tied to a long history, and too often poorly funded, museums have proven surprisingly flexible and remarkably resilient.

Museums are paradoxically always the same and always changing. Arguments from a century or more ago continue today. That is intrinsic to museum collections, to material culture. Art and artifact and specimen are at once grounded and unchanging, yet open to new interpretation. Museums can tell so many stories, to so many people. And those stories can coexist and change. We can build in many directions. Museums sometimes resist this diverse potential, looking for a narrow notion of truthfulness and value, but if they accept it, acknowledge it, perhaps even play it up, it becomes a source of strength.

I am fascinated by museums' peculiar capacity for stasis and transformation, the way they embody both stability and flexibility. Museums supply both the authentic presence of the artifacts from the past and the possibility of alternative ways to understand that past. Objects don't have fixed meanings. Collected for one purpose, they are used for many purposes. They are open to endless reinterpretation, a resilience invaluable both for individuals and communities. They bridge the past and the present. Museum collections offer both a canon of what was great and important and interesting, and a reserve of what might later seem great and important and interesting.

Museums are a site for informal education, filling the middle space between the formal education provided by the schools and that of friends and family. The range of exhibition styles and other kinds of object-learning is limited only by the imagination—not only of museum workers but also of visitors. Museums are a "third space," a social space between work and home, a place where diverse people can gather to learn new things.

Museums connect the creators of culture with the consumers of culture. They haven't always done that well. All too often they silenced the voices of communities regarded as primitive or uninformed, privileged the voices of the expert and powerful, and excluded certain visitors. But many communities who once had no say in how their objects were collected and displayed now play a role in telling their stories. Some have found museums a useful place to explore their past, to have their say. And the public too has gained power, as museums ask how they can be of service to new audiences and how they might break down once solid lines between expert and audience.

Museums provide society with a useful place to think through important questions. To do that well, museums need to consider their work in new ways. Are they about questions or answers? Are they about the past or about the present and future? How might museums best use new technologies while at the same time take advantage of the collections that make them unique? How might they not be weighed down by those collections but inspired by them?

The future of museums is in part about those new technologies of display, increased community connection, and a new focus on the public. But it is also about something more profound: the role of the real in an increasingly virtual world. Science fiction writer William Gibson predicted in 2007 that "one of the things our grandchildren will find quaintest about us is that we distinguish the digital from the real, the virtual from the real."[6] Museums, so heavily invested in the real, will have to consider the interface of the real and the virtual. They have a good head start on this, with centuries of understanding the complexity of the apparently simple notion of "the real thing," of authenticity. How does that empower new ways of using the virtual?

At the turn of the twenty-first century, the museum philosopher Stephen Weil defined the transformation of the American museum as "from being about something to being for somebody."[7] I suggest here that the museum needs to be both about something *and* for somebody. A museum will only survive as long as it has something to offer and provides a valuable service. To do that it needs to be alert to society's needs, and ready and willing to change to meet those needs.

The best museum people have always known this. In 1895 the first director of the United States National Museum, G. Brown Goode, declared (in all caps) that "A FINISHED MUSEUM IS A DEAD MUSEUM, AND A DEAD MUSEUM IS A USELESS MUSEUM."[8] Sir William Flower, first director of the Natural History Museum of London, wrote a few years later that a museum is "a living organism—it requires continual and tender care."[9]

When museums don't change, they fail to live up to their potential. Sometimes they fail outright. And that, perhaps, is the most important lesson of the Jenks Museum. When it was no longer useful, when it didn't change with the times, when it couldn't make its case for support—it failed. Its founding director didn't keep up with science, and so scientists stopped supporting the museum. The university, turning inward, no longer cared about the local audiences that had once come to the campus to see nature and culture on display. Professors weren't convinced that the museum was useful for teaching. Deprived of new leadership, underfunded, and committed to outdated notions of community, science, scholarship, and learning, the museum disappeared.

But in failing, it left a legacy. Some of the museum's collections survived to be used in new ways, for new audiences. Providence's Museum of Natural History, founded as "the people's university" two years after Jenks's death, acquired some of the collections. Some went to the Museum of the Rhode Island School of Design to be appreciated as art, not exotica. A few hundred ethnographic artifacts, stored for decades in the attic of a university building and discovered as it was being torn down, broadened the collections of Brown's Haffenreffer Museum of Anthropology, founded long after the Jenks Museum had been forgotten. And objects from the collection were used in *The Lost Museum* installation in ways Jenks could never have imagined.

Even more important was the Jenks Museum's educational legacy. Students learned things at the museum—an appreciation of nature, close observation, the arts of categorization and classification—that they could have learned in no other way, and they used the knowledge and habits of mind they gained to do good work in many areas. Some went on to become scientists and teachers. Herman Carey Bumpus became director of the American Museum of Natural History, in New York, where he was responsible for a new emphasis on education and outreach that would shape museums for many decades. Dallas Lore Sharp became one of the best nature writers of the early twentieth century. Benjamin Ide Wheeler, who become president of the University of California, vividly remembered Jenks's teaching with museum objects: "In that course of study in taxidermy I learned a great deal of pedagogy, the moderate and slow unfolding of a subject in connection with the use that it is to be put to."[10]

When Jenks went on a European tour in 1859, he asked Louis Agassiz, founder of Harvard's Museum of Comparative Zoology and an internationally famous naturalist, for a letter of introduction. Jenks had assisted Agassiz in collecting projects, and Agassiz was happy to help. The letter, Jenks recalled, was an "open sesame" at every museum he visited. One museum was closed, but Jenks showed the letter. He was admitted: "Oh come in, for we always honor his name."[11]

This book similarly honors Jenks's name. It continues the Jenks Museum legacy. The *Lost Museum* project helped me to understand museums in a new way. I hope that this book does the same for a wider audience, that it is an "open sesame" for its readers. Understand museums, appreciate them, and love them, not just for what they are but also for what they might be. Consider this book a letter of introduction from Professor Jenks, and me, to the world of museums.

I

❧ Collect ❧

❧ 1 ❧

WHY COLLECT?

John Whipple Potter Jenks was "positively ashamed" that Brown University, his alma mater, had no museum. In the 1860s it seemed that a university needed one. Alexis Caswell, Brown's president, explained why: "Without knowing something of the forms and laws of animal life which everywhere surround us . . . we seem to be walking blindfold in the midst of nature's richest treasures." A collection of natural history would help students develop "the powers of observation, comparison and generalization." Students too were concerned. Brown was falling behind other universities. One undergraduate wrote that "a cabinet of comparative anatomy is essential to any college. . . . Every plant and animal is an expressed thought of God, and cannot be presented through the medium of a professor. Brown has one ghastly skeleton and two or three small charts, and a few promiscuous bones!"[1]

Jenks volunteered to create a museum at Brown. In 1871 he retired from his job as headmaster of the Pierce Academy in Middleboro, Massachusetts, and moved his personal collection of taxidermied birds and mammals to Providence. He purchased collections and persuaded collectors to donate. He arranged for trades with the Smithsonian. Within a few months of returning to Brown he had filled several cases with minerals, fossils, shells, and animals. "The display," he wrote, "gave satisfaction to all."[2]

The founding of the Jenks Museum at Brown offers a snapshot of some of the reasons that museums build collections. Religion, love of nature, a feeling that artifacts are more true than words, their value in teaching, a

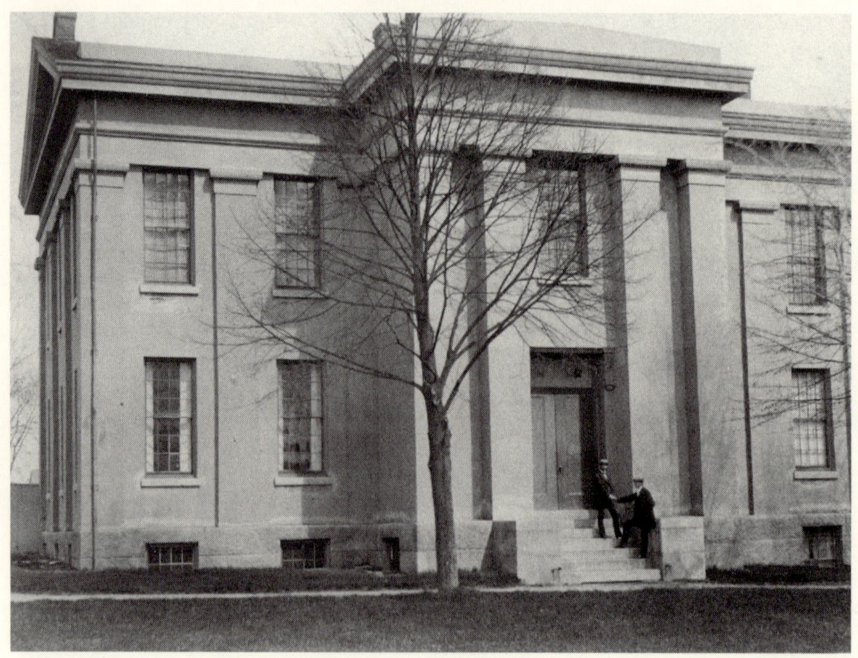

Brown University's Rhode Island Hall, home of the Jenks Museum, 1870s.

touch of envy; all of these were present in the formation of Brown's museum. Artifacts—real things—seemed essential for the university's work.

Museums Need Objects

These ideas about the importance of artifacts reflected widespread beliefs with a long history. Renaissance royalty and merchants collected and displayed rare and valuable objects to showcase their wealth, culture, and distinction. In the age of empire, nations built museum collections to flaunt their political and economic power. Artifacts were essential to science, technology, and agriculture, analyzed and explored in laboratories and workshops. And, of course, they were useful for teaching of all sorts. Objects properly displayed could reveal the hierarchies of nature and culture.

Objects in museums also served social purposes. The formation of museums with permanent collections, places where experts vouched for the authenticity of art and artifact, and arranged it to tell useful stories, reinforced the authority of political, economic, cultural, and scholarly elites. Museum displays made—and still make—powerful arguments for the stability of culture.

Collections are a good place to start our exploration of museums because museums are, at a fundamental level, about objects. There is endless discussion about whether objects are essential to the work of the museum. Museum historian Steven Conn raises the question in the title of his book *Do Museums Still Need Objects?*[3] I think they do. Collections—objects selected to provide evidence and to tell stories—are what make museums so interesting. They are what distinguishes museums from other educational institutions. Conn argues that the modern museum was created when objects were shown not for their own sake but as part of an object-based epistemology that made sense of them, put them in a larger story, and gave them an audience.[4] "Collections," argued a 1984 American Association of Museums report, "are the essence of the contribution museums make to society."[5]

Collections don't just happen. Acquisition and preservation are hard work. An object is in a museum because someone decided to collect it. Consider the 50,000 objects in the Jenks Museum. Jenks enjoyed telling the story of how they got there—about his expedition to Florida to explore and collect new specimens, the missionary who sent crates of ethnographic treasures from Burma, the local folks who would drop off strange things they thought should be preserved. Every curator has war stories like these: the thrill of the chase, the rivals outwitted, the bargains no one else appreciated.

But those personal stories are important only because of the larger story: museum collecting is disciplined collecting, for a larger purpose. It's different from personal collecting, different from hoarding. To build a museum collection one fits together the pieces of a complicated, multidimensional jigsaw puzzle, working with other curators, past, present, and future, over decades or centuries. It is a challenging intellectual exercise. The curator's job, as collector, is to determine what's worth saving and convince the museum it's worth saving. Then to move the object, and its story, into the museum, so that it can be put to use.

I say here, "the curator's job," but that's shorthand. Curators work with many others, both inside the museum and out: registrars, collections managers, educators, outside experts, dealers and donors, and, of course, the public. They work within cultural notions of what seems appropriate to collect, and practical, ethical, and legal constraints of what can be collected.

The objects Jenks collected represented more than just adventures, the museum more than a storehouse for exotica. Jenks believed that his taxidermied animals and African spears and ancient coins were indispensable teaching tools. They served a higher purpose as well. The objects he collected reflected his belief in natural theology: the glories of nature reflected the glory of God. A visit to the museum was, Jenks believed, an opportunity for a religious experience.

That bigger picture is different for different museums. It might be, as it was for Jenks, a religious story. It might be political, social, or cultural. It might be the story of a town or a nation, an art movement, or the natural world. Or it might be a grand vision like the wonderfully ambitious mission that Brooklyn Museum director Franklin Hooper announced in 1895: "all known human history, the infinite capacity of man to act, to think and to love, and the many departments of science and of art which he has developed. Through its collections in the arts and sciences, and through its libraries, it should be possible to read the history of the world."[6]

Not every museum aims this high, but all museum collecting has this in common: to use the artifact to tell a bigger story. The way museums do that is to move things out of their natural habitat and put them in a place where both the things themselves and information about them can be preserved and used in a new way. Museums give artifacts a new life, a new meaning. The new life is as part of a collection that tells that larger story.

I tell the collecting story in three stages. In this chapter and the next: What makes something museum-worthy? Then, in Chapters 3 and 4: How do museums build their collections, through gifts, purchases, and field collecting? And finally, in Chapter 5, I look more closely at what curators need to know to be good collectors.

What to Collect

Jenks collected thousands of objects each year the museum was open. He was acquisitive, even rapacious. His annual reports to the university's president tell some of those stories. How could he resist the offer of a dead giraffe from the Central Park Zoo? His collecting reflected his energy and his enthusiasm for the museum's mission. Not only could Jenks not control his own collecting, much came in unsolicited. What to do with the "valuable gifts, most of which came unexpectedly by mail or express"?[7] The basement of the museum building was full of unopened crates.

The Jenks Museum collected too much. In part, the problem was that Jenks loved to collect. In part, it was that the intellectual justification of his work, natural theology, meant that almost everything was useful. Jenks's always-uncertain funding at Brown compounded the problem: his desire to prove that the museum would be useful to "every department of instruction" meant that any new gift might bring the museum a new ally. The Jenks Museum lacked a clear collecting agenda.

It's hard for a curator to say no, to let an object go to another museum, to a private collector, or into the trash. It's a challenging problem for museum directors who manage curators. How can someone who is not an expert evaluate the collecting of a specialist? What do you say to the scientist who is the world's authority on fleas and wants to collect more? (They take up so little room.) How to judge the instincts of a decorative arts curator intent on building the museum's chair collection? When is a collection good enough? How do you balance the resources for collecting with those for taking care of and making use of the collections?

Frederic Lucas, describing his time as director of the Brooklyn Museum in the early twentieth century, acknowledged these difficulties: "Museums, so far as their collections are concerned, are seldom, or never, formed according to a definite plan. Their growth is sometimes influenced by circumstances, a bequest, the gift of a collection, or of money for the purchase of a collection. More often it reflects the interest of some . . . able or energetic curators of great persuasive powers, wealthy friends of the museum, or even collectors. . . . The result is that museums are more or less unbalanced, overdeveloped in some directions, atrophied in others."[8]

The Smithsonian Institution has worried about this problem since its founding. The first secretary, Joseph Henry, didn't want to collect at all, worried about the need for continued government support for collections. When the institution did start building collections, it was inundated with them, in part because of a policy that let curators decide what they needed and let them collect until they ran out of storage space and demanded more. Curators were professionals; they knew what they needed for their work. A 1977 report summed up the policy: "Each curator had pretty much first and final say regarding the collection process." For that report, the Smithsonian hired mathematicians to calculate the expected size of the collections fifty years in the future, were that policy to continue. Their predictions were scary. Depending on the assumptions—linear or compound growth—there would be either 161,000,000 objects in the collection by 2015, or 255,000,000.[9] The actual number as of 2016 is 156,000,000.[10]

Over-collecting is still a common problem. Collecting is exciting, fulfilling work. Some museums connect pay and promotion to collecting prowess. Museums are judged by their collections; museums compete with each other for collections, and curators like to win. Collecting is an easy-to-measure way of comparing museums. Museum director Steven Miller made the connection directly: "The better the collections the more highly respected the museum, and new acquisitions are the spiritual manna that prove an institution's worth."[11]

But curators, and museums, shouldn't be judged by the quantity of their collecting. Indeed, the first thing a curator needs to learn is how to say no. Collecting more things means less time to understand each object, less time to care for it and put it to use, less time to do the other work on the curator's agenda. It often means that the work of other departments of the museum is short-changed. Collections are expensive to maintain. John Cotton Dana, the plainspoken director of the Newark Museum, wrote in 1917, "It is better to spend money in making good use of one thing than in acquiring another thing. The worth of a museum object is in its use."[12] Dana came to museums from a long career in libraries, where he had fought for public education and access. He was a voice for making museums useful, writing a series of books with titles like *A Plan for a New Museum, the Kind of Museum it will Profit a City to Maintain*. Not everyone agreed with his opinions, and he knew it, writing that he was opposed to "most of the accepted museum conventions."[13]

A museum needs a good reason to accept an offered gift, and many turn down much more than they accept. As far back as 1884, a curator noted that "the paradox has been propounded that it is the chief duty of the managing committee to keep things out of their museum."[14] Nearly a hundred years later, in a 1976 survey, the aeronautics department of the National Air and Space Museum proudly reported that it had accepted only three of the nineteen aircraft it was offered. The anthropology department of the National Museum of Natural History reported in the same year that curators turned down about three-quarters of what they were offered, and that its collections advisory committee turned down about 15 percent of what the curators had approved.[15] Museum curators deflect inappropriate donations, often most of what's offered, politely suggesting other, more appropriate museums.

How much is enough? An old joke among curators of natural history museums had it that a series—a group of specimens useful for scientific work—was "one more specimen than any one has."[16] Funny, but true. Systematic biology, after all, depends on large collections. The largest single collection in a museum—the over 7.5 million gall wasps Alfred Kinsey collected for the American Museum of Natural History— offers both an example and a warning. Large series show the variation within species and population changes over time and place. But curators can get carried away. Kinsey, who went on to fame as a sex researcher, wanted to outdo his advisor, who had collected a million ants for Harvard, and he did, earning the nickname "Get a million Kinsey." In the high school biology textbook he authored he urged students to collect: "If your collection is larger, even a shade larger, than any other like it in the world, that greatly increases your happiness. It shows how complete a work you can accomplish, in what good order you can arrange the specimens, with what surpassing wisdom you can exhibit them, and with what authority you can speak on your subject."[17] Not good advice for a museum curator.

Objects are important to museums, but they need to be the right objects, collected thoughtfully, documented thoroughly—and not too many. The institution is committing to support them, now and for the future.

Museum-Worthy

Some types of objects seem museum-worthy, and some don't. The answer to the question of what's worth collecting, what's worthy of preservation, reflects the complex relationship between value and meaning.

Consider "manufactured collectables," the things made solely to be collected, like Beanie Babies, or Franklin Mint plates. These can bring great pleasure to their owners, but with very few exceptions they find no place in a museum. Museums do collect other kinds of collectables, but often with a different purpose than they have for collectors. Consider baseball cards. The Metropolitan Museum of Art has some 30,000 of them. While collectors would likely have the cards as a way to connect with their baseball heroes or to reminisce about their youth, at the Met the cards are examples of commercial printing, so-called low culture. The National Baseball Hall of Fame has even more baseball cards, to document the game and its culture. The National Museum of American History's card collection serves to document the role of the sport in American culture.

Art museums are perpetually the site of a battle over what art is and what belongs in a museum. Collecting American furniture only became acceptable in the early twentieth century. Many museums resisted collecting photography and contemporary art for the first half of the twentieth century. They continued to resist innovations in art or anything that challenged their usual categories. Women artists and artists of color are still fighting for their place.

Photography offers a good example. Alfred Stieglitz liked to tell the story of his attempt, about 1900, to persuade the Metropolitan Museum of Art to include photography. General Cesnola, the first director, was shocked at the idea. "Cesnola gasped. He said, 'Why, Mr. Stieglitz, you won't insist that a photograph can possibly be a work of art?'" Stieglitz replied that, yes, "there were certain photographs that I felt were art." Cesnola replied "You are a fanatic."[18]

Some twenty years later, Stieglitz tried again. The Boston Museum of Fine Arts had accepted some of his photographs, to hang with Goya and Dürer prints—accepted them, but wouldn't pay for them. The Met, wanting to keep up with the Boston museum, decided that it wanted Stieglitz photographs, too—but, like Boston, not enough to pay for them. Stieglitz told the curator that "as the museum bought paintings and sculpture and etchings and other things, I didn't see why, if photographs were deserving they should not be bought with museum funds."[19] But he finally gave in, and found donors who would donate a collection of photographs. Documentary photography took even longer to find a home in museums. It was still controversial in some museums into the 1970s.

Contemporary art has been a challenge for museums throughout the twentieth century. The Met was the scene of protests by artists, critics, and dealers over its refusal to collect contemporary American art in the 1930s. Stuart Davis, an abstract painter, wrote in 1940 that the Met "suppresses modern and abstract art in its policies as effectively as would a totalitarian regime." One trustee advised that the museum should accept a few pieces, to quiet the controversy: "It doesn't matter particularly if we acquire a certain number of mediocre or uninteresting, or more or less radical pictures so long as their purchase can be justified on political grounds. They can hang on our walls for a while and then receive a decent burial in the cellar."[20]

Even museums devoted to contemporary art have found themselves constantly trying to decide whether to accept each new innovation. Each new phase of contemporary art—video, performance art, activist art, institutional critique, and more—has had to fight its way in. Hans Haacke's *Shapolsky et al. Real Estate Holdings, a Real-Time Social System, as of May 1, 1971,* an early work of socially engaged art, provoked a strong reaction from the director of the Guggenheim, who declared it "an alien substance that had entered the art museum organism," and cancelled Haacke's show.[21] The work is now in the collection of the Whitney.

History museums have had their fights, too. In the 1980s, social history expanded what seems collectable, but collecting everyday life remains a challenge: The elegant furniture of a rich family's living room seems more worthy of the permanent collection than the cheap, worn-out furniture of the servants' quarters. Mass-produced artifacts seem less collectable than handmade ones. Items from famous figures are easier to make the case for than those from everyman and everywoman. Some aspects of life, especially traditional women's work like childcare and cleaning, don't create artifacts that seem collectable. Objects reflecting changing ideas about race and religion can seem too controversial. Present attitudes always shape history museum collections; we collect what seems important to us today, to help us make sense of—and use—the past.

Anthropology museums have always collected broadly, both objects and documentation, but their focus has changed over time. They originally tried to collect the "pure" culture of an imagined primitive past, based on theories of social evolution. Ethnographic collecting went out of favor at the end of the twentieth century, and archaeological collecting became more difficult, restricted both by new ethical guidelines and by political realities. Some anthropology museums have begun to focus their collecting on the complexities and cross-connecting cultures of the contemporary world. The Smithsonian's National Museum of Natural History now has a "curator of globalization." Chicago's Field Museum has begun to build contemporary urban collections.

Natural history museums today try to document the whole world. Their precursors, the cabinets of curiosities, wanted exotics, curiosities, exceptions; the modern natural history museum wants the representative, the vast range of the ordinary. They have also expanded their collecting in a different way: they now collect not just skin and skeleton but also tissue from which DNA can be easily extracted, and information on behavior and environment, to go with the physical specimen.

Every museum has a mission statement, and many have collecting mission statements, but those are rarely specific enough to address particular objects. They often use fuzzy language. Some defer to the museum mission more generally, asking that collections be consistent with the purposes of the museum. Or they put it as a negative. The U.S. Army museum system wants to "prevent nonmission objects from being acquired."[22]

Some simply suggest excellence, by various measures. The classic statement of this is Alfred Barr's definition of the work of the museum as "the conscientious, continuous, resolute distinction of quality from mediocrity."[23] (Barr was the founding director of New York's Museum of Modern Art, in 1929, and eager to claim quality for contemporary art.) Boston's Institute of Contemporary Art wants "the very best art being made today." The J. Paul Getty Museum collects "works of art of outstanding quality and historical importance."[24]

Some collecting visions suggest the larger value of collections. They echo George Brown Goode's 1895 definition of a museum as a place for objects "which best illustrate the phenomena of nature and the works of man."[25] "Best illustrate" is a wonderful phrase, and particularly appropriate to Goode's work as the head of the Smithsonian's United States National Museum. It suggests both judgment in collecting and that collections are there to be used, to be learned from, and to be judged by how well they serve that mission. They're chosen not for intrinsic value but for teaching and research value. The Boston Museum of Fine Arts "seeks to acquire art of the past and present which is visually significant and educationally meaningful." The words "significant" and "meaningful," like Goode's phrase, imply a purpose beyond the object itself. The Chicago History Museum demands even more, asking that every new acquisition "present opportunities to reveal or expand on a compelling and significant story or theme of Chicago."[26]

Whether the standard is quality, significance, story, or usefulness, the key thing is to have a standard. It's tempting to acquire in an ad hoc way, making the case for each object as it comes in. The Smithsonian's National Museum of American History website insists that every one of its more than three million artifacts is a "true national treasure."[27] It's easy to see why the museum might make that claim, but by defining everything as a national treasure it offers no guidance for future acquisitions. It defines "treasure" through the act of collecting, rather than forcing the museum to collect things that meet some definition of "treasure."

Shaping collections is difficult, requiring museums to consider fundamental questions about use and audience. Are the acquisitions to be used for exhibition, research, or education? Do they appeal to the visitors that attend the museum, or to those that the museum would like to have attend, or to some aspect of the community the museum is supposed to serve? Is the museum collecting for contemporary audiences or future audiences? More generally, how might the acquisitive drive of curators be channeled into collections of long-term value not just to their own interests and research but to the needs of the museum and, beyond that, to the benefit of researchers and the general public?

Museum-worthiness sets the stage for a curator's collecting work. What makes a particular object worth collecting? That's the topic of the next chapter.

❧ 2 ❧

COLLECTABLE

IN 1874 PROFESSOR JENKS spent five weeks in Florida's swamps with gun and net, fighting alligators to build his museum's collections: "Hardly afloat and a roseate spoonbill rose from its nest and perched beside it. Fred shot her while I poled the scow in all haste, as, the moment it struck the water, watchful alligators made for it on every side. . . . I could plainly see about six feet deep the pink hues of the spoonbill as it was held down by the alligator. Two or three thrusts of my pole so astonished the brute that he let go the bird, and it now graces the Museum of Brown University."[1]

Curators don't need to wrestle alligators for every object, but collecting isn't easy, and the curator's job is to decide when it's worth fighting to bring an object into the museum. This is a curator's everyday work. Gary Tinterow, a curator at the Metropolitan Museum of Art, put it this way: "Curators and administrators make decisions about the formation of the collection every day. We're the gatekeepers. . . . When something gets here, it's because a curator has made a decision to admit this work."[2]

How does the curator decide that an artifact is worth the considerable trouble it takes to collect it, accession it, make it available for use in exhibition and research, and keep it forever? There are several steps. First, the curator evaluates the existing collection: what's there, what's lacking, what might be needed for future research and exhibitions. In some museums, collecting is driven by the exhibition or education program. In others, research interests drive collecting, either the curator's own work or an institution-wide agenda. Next, the curator evaluates the object: What does its provenance and your connoisseurship and research tell about it? Is it of

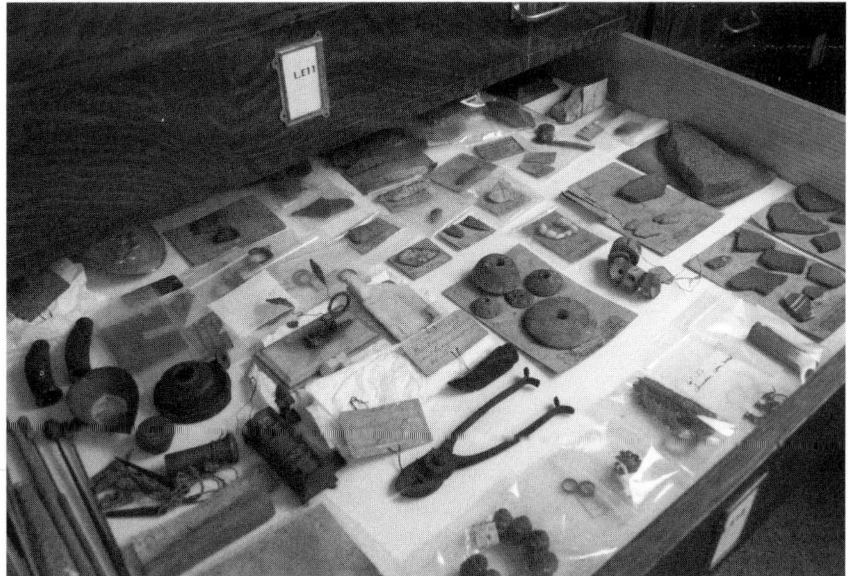

Some of the artifacts Jenks collected for the museum, now at Brown University's
Haffenreffer Museum of Anthropology.

sufficient quality and significance? Is it in good condition? Does it come
with a good story? Is this the right example? How useful will it be? What
does it cost? How much effort will it take to acquire it? And finally, there
is a layer of legal and ethical concerns. Museums must respect the many
laws that control import, acquisition, and display, and pay strict attention
to the ethics of collecting. The curator balances all of this and makes a
case for the object.

Purpose and Planning

Curators working independently, collecting what's useful to their own
work, can build collections that are strong but idiosyncratic. Collections
built this way speak to curatorial and academic interests and aesthetic
sensibilities, but not necessarily to wider concerns. They don't always add
up to more than the sum of their parts and might not support the big stories the museum wants to tell. Collections shaped by donors present similar problems. The building of useful collections requires purpose and

planning. In many museums, curators are required to write a memorandum about an object they would like to acquire, detailing how it fits collecting goals. In the words of the Metropolitan Museum of Art's policy, how will it "significantly further the Museum's stated mission"?[3]

Many museums have a committee that has to approve every acquisition or acquisitions above a certain value or a certain size. These committees, composed of other curators, members of the board, or outside experts and collectors, are designed to make sure collections justify the time and expense of acquisition. The committee weighs the value of the object to the museum against its cost—not just purchase costs, but also shipping, conservation, and long-term storage—and approves or turns down the suggestion. At some museums, the collections committee is responsible for raising the funds for acquisitions they're particularly excited about.

The collections committee uses a collections development plan to make its decision. The American Alliance of Museums (AAM) offers this straightforward—and one might think uncontroversial—advice for collections planning: "The museum strategically plans for the use and development of its collections."[4] Yes, curators might have excellent instincts, but writing them down, thinking them through, and making them explicit is a good idea. It's hard to imagine being against strategy or planning, but this advice actually suggests a radical change from traditional practice. It makes collections less a personal expression of interest, taste, and knowledge, and more a part of official policy. The curator loses some independence and freedom of choice.

Thinking about collections as part of a larger picture can make collecting a part of a strategic plan and collections a more important part of the museum's mission. That is good for collections in the long run. Collections planning puts the users of the collections, museum staff as well as visitors and researchers, at the center of collections work. Different collections are used in different ways, and those uses should shape the collection. In traditional natural history museums, for example, most of the collections are purely of research value and the audience for those collections is other scientists. In history museums, eager to attract an audience beyond collectors and buffs, collections planning has focused on building diverse collections that mean something to the community. Art museums have a range of audiences, from artists to collectors to the general public.

The first step in deciding what to collect is to determine who you're collecting for.

The AAM urges that a museum create an intellectual framework to shape its collecting. This framework is "the compelling vision that defines the unique role your museum plays and provides the context for making decisions about the future of the collections."[5] It should be built on the needs of the users of the collections, both visitors and researchers. It provides the goals that the collection should meet. It allows the museum to say how much is enough and where more is needed. An intellectual framework is the first step in building a collecting plan. One way to think about an intellectual framework is that if the museum were starting to build a collection from scratch, what would it need?

The range of possibilities for collecting can be seen in a delightful thought exercise created by museum writer Stephen Weil in 1981, "Twenty-One Ways to Buy Art." Here's the setup: "You are the director of a municipally funded museum of twentieth century art in a small mid-western city. Your board of trustees has asked that you draft a brief policy statement providing guidance as to how the museum's acquisition funds might best be spent over the next several years." Weil gives some wonderful possibilities, putting them in the voice of the various stakeholders in the museum: We need to fill gaps! We should build on our strengths! We should support young artists and buy new work! We should support local artists! Patronize local galleries! We should support women and minority artists, who are underrepresented in our collection! We should buy only the finest art, one piece every ten years if need be! We should collect that which is useful for our docents to teach with! We should ask the public what it thinks![6]

These are all perfectly fine ways of thinking about what to collect. They represent different ideas about the work of the museum and the way that the collection might support it. General Pitt-Rivers, founder of the museum at Oxford, England, stated this clearly: "Museums may be established for different purposes, and the objects in them should be collected and arranged to further the particular purposes for which they are intended."[7]

To make the case for collecting, curators need to consider the proposed new object in the context of the existing collections. To do that, they need a deep knowledge of what the museum already owns. Often, that's based on a physical sense of what's where in the collections storage areas. Curators spend a lot of time wandering through collections and exhibition areas, building a mental map. George Storey, newly hired as a curator at the Met in 1889, wrote in his first monthly report to the director, "I carried out your instructions and examined with care every picture in the museum [building] very carefully noting the condition of each and every one."[8]

Next the curator looks at how a potential acquisition fits into the carefully considered goals of the museum's collections development plan. A curator probably shouldn't accept a gift, no matter how tempting, if, when the museum last thought through the big picture of the collections that the museum needs, the object wasn't on the list. Perhaps it's something so wonderful no one could have imagined it . . . but probably not.

How might it be useful? Many collections are built based on exhibition needs. The curator must consider what objects tell important stories in convincing ways. If the museum is collecting for research, what might scholars and scientists learn from this object?

Should the curator build on the collection's strengths or fill in gaps? This is an old question, with answers driven by both philosophy and personality. There's an easy tendency to simply acquire more of what the museum already has, to build the collection deeper. That might be right, if the audience is specialists, and especially if the museum's goal is to encourage research. Scientists and scholars will take the time to visit a deep collection. But a broader approach might be better if the museum has a general audience and an exhibition program. For many museums, the best choice is a middle ground of carefully defined areas of strength, along with broad general coverage.

The curator must also consider whether there is room to store the acquisition when it comes in. Can the museum provide appropriate, secure, and environmentally stable storage at reasonable cost? (One of the first pieces of advice I was given as a young curator: never accept a boat.)

But most important is the question of whether this is a useful object. An object that sits in a storage room, unexamined, unloved, is of no value— until it is. The curator must balance usefulness now, and in the future. Lonnie Bunch, director of the National Museum of African American History and Culture, writes that "part of the job of any museum is to anticipate what historians will want to know 50 years from now."[9]

Usefulness means many things. Does it tell a good story? An ideal object comes with a long provenance, a history, attached. There's a slippery slope of "objects like these"—you want to tell a specific story, but all that's available is an object that's similar to the one you really need. How similar is similar enough? Local history museums want to collect something actually used in the area; sometimes they have to settle for an artifact "like those" used in the area, or something used nearby. For mass-produced goods, that can be fine: no local angle, but it can fill the role, tell a general story.

Does it have exhibition possibilities? Does it work for visitors, helping to explain art or history or the natural world, better than some other object that might be used? Can it be put on exhibit? Will visitors notice it on exhibit? Will it fit through the door of the exhibit hall; will it fall through the floor?

Does it have research potential? Might researchers visit the museum, or the museum's website, to learn from it? Does it prove or contradict a theory? Does it add to an artifactual series in a way that enhances or contradicts it? Can you think of ways that it might serve as evidence in a scholarly paper? Herbariums and natural history collections have their own special desiderata: they are most interested in type specimens—the examples that define a species—and in specimens that show a wider geographic range than previously known for a species.

Does it have educational value? Many museums keep a separate collection for hands-on educational use, and objects that aren't appropriate for the permanent collection are sometimes useful for this purpose. Educational collections don't require the same long-term commitment—the museum doesn't promise to keep them forever—and so the hurdle for entry into this collection should be lower.

Finally, what are the opportunity costs of collecting this object, that is, what is the museum giving up? What won't it be able to afford, or have room for, or have time to deal with, because it said yes to this object?

Quality, Significance, and Usefulness

A museum should collect the best objects and those most useful to its mission. That requires a balancing act between what's useful and what's of appropriate quality and significance. Standards of quality and significance are complex. They have politics. They change over time. Many museum collections reflect a long history of changing ideas, each generation of curators building on previous collections with new approaches.

That happened at the Jenks Museum. In his later years, Jenks's ideas about what to collect had come to seem decidedly old-fashioned. Younger professors had new ideas. Taxidermy was out of fashion. Jenks complained to his friend Henry Ward, a dealer in natural history specimens, that "these modern Professors run entirely to skeletons."[10]

Every collection goes through changes like this. American art museums collected the work of living artists, then casts and copies, then old masters, then impressionists, then contemporary art. Industrial museums went from collecting contemporary examples of design to documenting the technical evolution of machines to considering the social history of industrial work. The timekeeping collection at the National Museum of American History, established to document the technical history of clockwork, is expanding with objects that tell stories about "the changing ways Americans have measured, used and thought about time."[11] Natural history museums were once interested in collecting trophy animals, the bigger and more impressive the better; now representative samples are preferred. Anthropology museums, once focused on artifacts that showed precontact native cultures, are now more interested in the interactions between cultures than an imagined "pure" culture.

Nancy Bercaw, a curator at the Smithsonian's National Museum of African American History and Culture, argues one of the reasons for the success of the new museum was that, ironically, it had to build its collections from scratch. While the Smithsonian had been collecting for more than one hundred and fifty years, its collections were embedded in a system of classification based on that long history. The two great eras of collecting were during the western expansion in the nineteenth and early twentieth centuries, and during the Cold War. Collections from the first era were based on colonial and imperial ideas, including an anthropology of racism. The second era, a period of enthusiasm for science, technology, and the middle class, aimed to show American exceptionalism. Neither had much room for black lives. By collecting from black families and organizations, the new museum was able to acquire artifacts that told stories that would not have been of interest to the curators of those earlier eras.[12]

The desire to tell new stories means both new collections, and new ways of using old collections. It might also mean deaccessioning—getting rid of collections. More on that in Chapter 8.

Quality, significance, and usefulness mean different things to different people, at different times. Museums are a battlefield for those debates. When an art museum collects artifacts from a non-Western tradition, whose ideas of quality and significance should be used? The National Museum of the American Indian's "Infinity of Nations" exhibition combines Western aesthetic quality and native tribal perspectives of significance—and some reviewers and visitors find it confusing. Does it show great art or explain native culture? Similar questions arise when museums collect religious objects. Is religious significance important to their selection, or aesthetics? Even within the Western art tradition, how should a museum balance the quality of the art, its significance to an art-historical understanding of art, and—to raise the crowd-sourcing question—its popularity? Who gets to decide?

Art museums have developed a language of quality—or rather, two ways of speaking about quality. There's intrinsic quality: a great object, and a connoisseur-collector knows it when he sees it. Roger Fry, curator of paintings at the Met in the early twentieth century and perhaps the most important curator and critic of his day, wrote that "the aim of the Museum ought to be to set a standard of what may be called serious as opposed to merely occasional and frivolous art." A curator's business, he continued, is "a just appreciation of what is really serious in intention and fine in execution," as opposed to "a kind of painting which answers to ephemeral fiction in literature."[13] No thrillers or romance novels in the museum! Fry's attitudes held sway for a long time. Theodore E. Stebbins Jr., a curator at the Yale University Art Gallery in the 1960s and 1970s, was an unabashed proponent of this notion. "What we must collect," he wrote, "are the jewels, those things of great quality and sensitivity which truly reveal the soul of the artist and the culture."[14]

Edmund P. Pillsbury, director of the Kimbell Art Museum, represented the other side of the debate. He was less certain of a curator's ability to look at an object and see the artist's soul. He had a more complicated notion of quality. "Quality," he wrote, "rests on various criteria, subjective as well as objective; and the only reliable yardstick remains the consensus of informed opinion over time." His was a more pragmatic set of concerns. First, he warned the curator not to be "seduced by a work that is an apparent bargain, purports to fill a gap, demonstrates technical proficiency, or represents a fashionable artist, period, or genre." Rather, a curator must weigh all evidence, taking into account "price, condition, rarity, importance, suitability, and visual appeal." This moves the discussion beyond the level of connoisseurship: "No more is it a question of how much this or that individual likes or dislikes an object, but rather how successfully the work meets the various criteria that have been established to assess the quality of the object." Art museums work within a system of galleries, auction houses, and art-historical writings that define what's appropriate to collect. For Pillsbury, what was most important was that collections reflect a canon of art that has been deemed, by curators and art historians, to be "museum worthy."[15] That's one reason museums can be so conservative in their collecting.

Some connoisseurs have tried to make a method of their work, to make it more than just an intuitive judgment, to establish standards for quality. Giovanni Morelli, the inventor of modern connoisseurship in the 1870s, argued for concentrating on minor details of art—fingers, ears, toes. Look for the tiny clues, he wrote, to know that you're collecting the work of a great artist.[16] Charles Montgomery, director of the Winterthur Museum in the 1960s, outlined "14 points of connoisseurship," including overall appearance, form, ornament, color, materials, techniques, trade practices, function, and style; put those together, along with the attributed date, the history of the object, and the condition, and you could determine authenticity and value—and, presumably, whether or not you should acquire the object.[17]

The language that decorative arts curators use to describe their acquisitions tells us a good deal about how a connoisseur thinks about what makes an object suitable for a museum. Consider the Met's *Notable Acquisitions* report of 1975: A Newport table "in perfection of form and precision of execution . . . stands out from its peers." An easy chair's "faultless proportions and masterly execution make this one of the finest examples of the form." A pianoforte "reflects all of the varieties of cabinet ornament that were typical of the finest New York furniture of the time."[18] These descriptions reflect deep curatorial knowledge of regional forms and wide familiarity with the range of objects in museums and private collections, building on that base to make an aesthetic judgment about proportion and form and what it means to be "the finest."

Curators of contemporary art don't have the luxury of "the consensus of informed opinion over time," the canon of museum-worthy art. Their notions of quality overlap with the way that the word "quality" is used in the art market and with the way artists use the word. But there's more to it than either of those. Clara Lieu, a professor at the Rhode Island School of Design, suggests this as a rule of thumb: "Museum quality work is work that talks about contemporary issues, yet is timeless. . . . Museum works go beyond simply having a visual experience, the works should stimulate thinking in the viewer, deal with tough topics, and address issues of contemporary concern."[19] Museum-quality art, perhaps, is art that seems worth saving for an imagined future audience. It might be the equivalent of Ezra Pound's definition of literature as "news that STAYS news."[20]

Quality in art is difficult to define. It includes social, cultural, and personal dimensions. Brooklyn Museum director Arnold Lehman didn't use the phrase "museum quality" to describe his purchases, but he had a clear sense—an instinctual judgment—of what was good. He bought what he liked—but what he liked was informed by his long exposure to art, strong relationships with artists and gallerists, and a thorough understanding of the existing collections. On a 2015 tour of an exhibition of art bought during his seventeen years as director, he described some of his purchases. A Mickalene Thomas work at the Art Basel Miami Beach he called "a wonderful artist and we had nothing in the collection." A Mark Bradford: "I fell head-over-heels in love with that painting." A Kehinde Wiley: "I thought those paintings were fabulous. . . . I walked up to the dealer and I said, 'We want that one and that one.'" In each case, he knew the artists, knew their work: his collecting instincts, his ideals, his notion of quality, were based on very deep knowledge.[21]

A connoisseur focuses on the object itself, mostly on its appearance. But that's only one dimension of its value for museums. Museums want not only quality, but also significance. They want objects of the highest quality as well as those that let them tell a story. For the largest art museums, this means telling the story, primarily, of the history of Western art and filling in the chronological gaps in the collection. This was true of the Met in 1906, when newly hired painting curator Roger Fry told the board in a "Strictly Confidential and Private" report that the museum should "aim at getting a fairly representative gallery of such paintings of all periods and all schools as will enable the student to understand and appreciate the great creative minds of the past, as well as to acquire some notion of historical perspective in art; to understand something of its evolution."[22] And this remains the model for many museums today. When the Met spent $50 million in 2004 for a Duccio *Madonna and Child*—its most expensive purchase ever—Phillip de Montebello, the director, noted that the painting "will enable visitors to follow the entire trajectory of European painting from its beginnings to the present. . . . The first slide in an art history 101 course is a Duccio."[23]

Over the past few decades, many museum curators have begun to think of new stories, beyond art history 101. More and more, they are interested in objects that tell a range of stories. Indeed, they're not just collecting objects; they're collecting stories, collecting meaning. Sometimes, the story is in the object itself: its creation, its use. Sometimes, it's the way it relates to other objects, the role it played in society. Collecting objects for the stories they tell allows museums to make an argument for the value of collections, and to focus collecting. It lets them make a case for an object being significant: a part of history, useful for teaching and research.

History museums tend to be more concerned with significance than quality. At the National Museum of American History in the 1970s, curators needed to prepare a memorandum to a committee of curators and administrators stating an object's value for research and exhibition. They needed to offer "other historical context, as appropriate, that makes clear why this object is deserving of inclusion in the collections." An object was only worthy of accessioning if it possessed, as the Smithsonian put it, "potential for research and scholarship" or if it would be "useful for exhibition purposes, now or in the future," or if it were "significant in itself so that it merits inclusion." Significance was interpreted widely and essentially left up to the curators, based on their areas of collecting: "Technological, social and historical factors should be weighed. Association, aesthetic merit, rarity, and status in its own particular category should be considered."[24]

Many Australian museums today use significance in assessing collecting possibilities, as a way both of getting beyond the object itself and of connecting with the public. "Significance," according to the Collections Council of Australia, includes "historic, artistic, scientific and social or spiritual values that items and collections have for past, present and future generations." It incorporates "context, environment, history, provenance, uses, function, social values and intangible associations" through consultation with communities with which the object has meaning. Defining significance this way acknowledges change over time. It also allows museums to make a case for the importance of their collections to a wide audience. Significance, the council argues, "is a proven persuader."[25]

One might add further categories in determining historical significance in specific fields. Science and technology curators, for example, consider not just the object's role in the development of the field, but also its impact on society and the economy. Cultural history curators might look to an object's meaning to particular groups or its iconic value, that is, its role in shaping culture. Social history museums often look for objects that change people's lives.

In the 1990s, the Smithsonian considered a new approach to deciding what makes objects worth acquiring, defining significance as an object's value to science or place in culture as well as its fit with the museum's collection. The question became, what would happen if the museum *didn't* collect the thing? Natural history specimens, a committee suggested, should be collected if they "a) are in danger of being irreparably lost; b) fill in evolutionary lineages not yet represented in the collections; and c) generate material for new kinds of analyses and understandings." Cultural artifacts had a different set of desiderata. They were worth collecting, the committee argued, if they: "a) are in danger of being irreparably lost; b) represent and record important historical events; c) have multiple meanings for different segments of U.S. society; d) are judged to have unusually high quality; e) fill important gaps in existing collections; and f) illustrate important expressions of human creativity."[26]

The interest in objects that offered multiple meanings marked a change in museum collecting toward the end of the twentieth century. The 1987 Common Agenda for History Museums called this "interpretive collecting," and many history museum curators noted that they were interested in objects with stories. Technology curators at the National Museum of American History defined as interesting an object that "allows us to tell a good story." A medical curator at the museum wrote that she was looking for artifacts that told patients' stories. At the Henry Ford, curators looked for objects that told of change over time, a way of combining the story of technological progress, a longtime strength of the museum, with a bigger story of social, cultural, and political change.[27]

Some art museums, especially in the opening decades of the twenty-first century, have realized the value of collecting stories beyond art history. Director Lehman of the Brooklyn Museum described the museum's collecting this way: "We don't seek out objects to fill holes, we seek objects that tell a story." He had particular stories in mind. For example: "We are always thinking about taking a more global view, a way to let our visitors know that there are influences from all over the world."[28] The Brooklyn Museum's quest for global stories fit well with both the history of the museum and its desire to reach new audiences. Many museums have followed this path, thinking about how to put a traditional Western narrative into a larger global context. The Peabody Essex Museum also went global, "embracing artistic and cultural achievements worldwide . . . to explore a multilayered and interconnected world of creative expression."[29]

The curator makes the case for collecting using the language of usefulness, significance, and quality appropriate to the museum. But that's just the first step. It's also essential to consider the thing itself, the particular object available. Is this object the best one of its type to acquire?

Ideally, the object is a good example—both typical and with a good individual story. It's really what it claims to be, that is, it's authentic, not a forgery or a fake. It's legal to acquire it, and the museum can obtain clear title to it. It's in good condition. (Good condition is a complicated notion; it might mean as new, or it might mean that it shows appropriate signs of use, signs of its history, depending on the object and the museum.) A conservator has given it a clean bill of health—it's no good acquiring something that will self-destruct, or cost more than the museum can afford to conserve. If it needs conservation work, is it worth the expense? Or would it be better to wait and find one in better shape?

The price is right. Ideally, the object is a gift. But if not, it should be reasonably priced and affordable given the museum's acquisition budget or the budgets of friends of the museum with deeper pockets. If not, it might be better to wait for another opportunity.

And—once acquired—the museum can use the object as it likes. There are no strings attached—the donor doesn't demand that it be displayed, or displayed in a certain way, or with certain other objects; that it be kept forever; that it be used to tell a particular story. The acquisition won't be construed as a commercial or political endorsement—beware of companies looking for some good publicity when they're about to experience some bad publicity. (Museums make exceptions to these rules, but—one hopes—only after careful thought about the long-term consequences.)

Finally, given the other possibilities, is this the best example, in the best condition, at the best price? If the museum collects this object, will it experience buyer's remorse when a better example comes along and it can't collect that one? Perhaps it is best for the museum to consider that now and make some inquiries.

The curator, in short, combines a knowledge of the subject, an understanding of the strengths and weaknesses of the collection, a commitment to the museum's mission, a connoisseur's knowledge of how to interrogate an object to figure out whether it is what it seems to be, a dealer's sense of the market, and a scholar's sense of what stories it can tell. If a potential acquisition passes all of those tests—and if the price is right, and there's room and a use—then the curator can make the case to the museum that the object is worth acquiring, worth moving from the world of the everyday to the immortality of the museum.

Thirty-Seven Laws

No matter how useful an object might be to the museum, no matter how good a story it might let you tell or what valuable research it might allow, you can't accept it if doing so breaks the law. The American Alliance of Museums counts thirty-seven laws "affecting collections and collections management," from the Abandoned Shipwrecks Act to the Wild Exotic

Bird Conservation Act.[30] These laws restrict collecting and even ownership of some property.

Laws on cultural property, endangered species, and provenance shape the curator's collecting. The curator at a natural history museum must pay attention to laws protecting African elephants, migratory birds, bald eagles, and marine mammals, among others. The Convention on International Trade in Endangered Species (CITES) restricts what plants and animals can be brought into the country from abroad. Registered research institutions can apply for exemptions under some laws, and in some cases materials owned by the federal government are released on long-term loan to museums.

These rules can affect other kinds of museums, too. Robert Rauschenberg's 1959 *Canyon* contains parts of a stuffed bald eagle. It's a violation of the Bald and Golden Eagle Protection Act to sell a stuffed eagle. Indeed, it is illegal to possess one—unless you can prove that it was killed prior to 1940, which this one was. *Canyon* prompted an interesting legal battle: What's an unsalable artwork worth? Zero, as the owners suggested for inheritance tax purposes, or $65 million, as the IRS countered? The IRS dropped the tax claim when the work was donated to the Museum of Modern Art in 2012.[31]

The acquisition and ownership of art, archaeological, and ethnographic materials fall under other legal restrictions. Art museums must follow laws governing the importation of art and need to be certain they are not purchasing or accepting as gifts material covered by the Convention on Cultural Property Act, the National Stolen Property Act, and the State Department's sanctions on certain countries. Museums must avoid acquiring illegally excavated or illegally trafficked artifacts. They also need to be certain that the material may legally be exported according to whatever country's laws may apply.

A separate set of rules applies to artwork that changed hands in Continental Europe between 1933 and 1945. It needs careful research on provenance, to make sure that it had not been unlawfully appropriated by Nazi governments or sympathizers. The American Alliance of Museums maintains an online searchable inventory of art in U.S. museum collections—currently about 30,000 objects in 179 museums—to help families recover their property.

The Native American Graves Protection and Repatriation Act (NAGPRA), a 1990 law, concerns Native American human remains and certain cultural items—funerary objects, sacred objects needed for present-day Native American traditional practices, and objects of significant cultural value. Museums that receive federal funding must inventory these collections and—unless they demonstrate that they were obtained with the voluntary consent of an individual who had the right to transfer them—notify the appropriate tribes or native Hawaiian organizations and, if the tribe wants them returned, repatriate them. In NAGPRA's first twenty-four years, museums returned the remains of 50,518 individuals (it has been estimated that there are some half-million Native American remains in American museums); over 1.3 million associated funerary objects; almost 5,000 sacred objects; and over 8,000 objects of cultural patrimony.[32]

Most museums consider the spirit of these laws, rather than simply following them to the letter. Many museums will simply not accept some materials—for example, pre-Columbian art or artifacts—unless the owner can prove, with receipts and pictures, that the object was in the United States before 1970, when the UNESCO Convention on the Means of Prohibiting and Preventing the Illicit Import, Export and Transfer of Ownership of Cultural Property was passed. Many will not acquire or display any materials brought up by underwater treasure hunters who don't document the ships they explore to good archaeological standards.

The Museum of Fine Arts, Boston, established the position of curator of provenance in 2010 to research works of questionable provenance. Victoria Reed, the first incumbent, described her work this way: "We have a two-page, 21-question form that curators are required to fill out for every acquisition. It comes down to risk assessment. . . . The most important thing in making an acquisition is to do so being fully aware of what has been asked, what issues exist, where the object has been and to be fully transparent about the object's collecting history."[33]

There is a lively debate about the merits of controlling the international trade in cultural goods. No one wants to encourage looting. But shouldn't we encourage the widespread appreciation of diverse cultures around the world by sharing artifacts in museums? Striking that balance is challenging. A 1954 UNESCO convention considers art "the cultural heritage of all mankind, since each people makes its contribution to the culture of the world."[34] A second 1970 UNESCO convention focuses more on the cultural heritage of the nation state and urges that works "created by the individual or collective genius of nationals of the State" and "cultural property found within the national territory" should be protected, and that nations should prohibit the "illicit import, export and transfer of ownership of cultural property."[35] Philosopher Kwame Anthony Appiah calls attention to the different points of view in these two conventions. He raises questions about the meaning of the term "cultural patrimony," asking questions about what "national territory" means for art created before there was a nation. He argues for a cosmopolitan, transnational understanding of culture, asking, "If it is good to share art . . . why should the sharing cease at national borders?"[36]

The philosophical discussion is fascinating, but from a real-world point of view it doesn't matter: you can't collect something that the law forbids you to own. Museums have learned this lesson the hard way, with bad publicity, losses in court cases, and objects returned to rightful owners.

There's another set of legal concerns that is more practical. Does the seller or donor actually own the object? That might be easy to establish, or it might be complicated, even impossible. Is he or she the sole owner? Can a valid title be passed to the museum? Is the provenance known, or are there questions?

Ownership of cultural objects can be complicated. Just because you own a physical object doesn't mean you own the rights to reproduce it, for example. You can put a record that you own on display but you can't necessarily play it. You can put a photograph on display, but not necessarily reproduce it without owning the copyright, too. It's important to know what rights are transferred to the museum with the object itself. Is a photograph worth acquiring if someone else owns the copyright and won't let you use the image online or in a publication? The Visual Artists Rights Act of 1990 (VARA) gives artists moral rights to their works, even after they're sold, including proper attribution, and the right to prevent distortion or modification. Some museums ask artists to sign a waiver of their VARA rights.

This is a long checklist, but an essential part of collecting. After the curator has been through it, checked with lawyers and registrars, and is convinced that there are no legal reasons to avoid the object, once the curator has made the case that it's the right object for the museum, that it's useful, significant, and of high quality—and defined each of those terms in a way that makes sense for the museum, it's time to make that purchase or get back in touch with that donor.

ACQUISITIONS

A FEW MONTHS after Professor Jenks's death, faculty and students gathered in Brown University's largest auditorium for a memorial service. Reuben Aldridge Guild, librarian emeritus, eulogized his former colleague. He told stories of Jenks's boyhood: Jenks grew up poor but precocious in central Massachusetts. Visits to Brown and to Amherst, where he saw his first museum, "determined young Jenks upon a college life." He taught school in Virginia, and then joined the Brown class of 1838, its youngest student. After graduation he taught for eighteen months in Americus, Georgia, and then served as a pastor in another town. Returning to Massachusetts, he taught at the Peirce Academy in Middleboro for thirty-three years before returning to Brown to create his museum.

Guild told his audience many stories of collecting, tales that he must have heard from Jenks over the years. Noted ornithologist John Cassin died in Philadelphia, and his collection was for sale; Jenks was on the next train there, beating a letter from Harvard by a day. Reverend Frederic Denison, a Brown alumnus, had a collection of Indian relics; "the Professor was in the pastor's study, his face all radiant with joy as he gazed upon the six hundred relics illustrating the history, manners and customs of the aborigines," and soon the relics were in the museum. Yale's curator arrived a day too late. The Blanding collection, thousands of minerals, shells, coins, and more, was secured through Jenks's "perseverance and zeal."[1]

Jenks worked hard at collecting. He purchased entire collections and taxidermied specimens, spending his own money when he couldn't raise

MUSEUM OF NATURAL HISTORY.

Very valuable additions have been recently made to the Museum, by the purchase of the large collection of birds belonging to the late John Cassin, Esq., and also by the donation of the varied and extensive collection of the late William Blanding, M. D., of Philadelphia. Other accessions have been made to the Museum through the liberality of gentlemen interested in the growth of this department of the University. The Museum is under the care of Professor John W. P. Jenks, who has been appointed by the Corporation to be its Curator. The collections it now contains, as classified and arranged by Professor Jenks, are as follows :

1. In Zoology, including skeletons for comparative Anatomy,
    - Birds, (500 mounted,) - - - - - 4,800
    - Quadrupeds, - - - - - 50
    - Fishes and Reptiles, - - - - - 300
    - Insects, - - - - - - 1,500
2. In Mineralogy, specimens, - - - - 8,600
3. In Geology and Palæontology, specimens, - - 3,900
4. In Conchology, - - - - - - 12,200
5. Indian Implements and Relics, chiefly from New England, specimens, - - - - - 800
6. Implements and curiosities of other uncivilized peoples, 300
7. Coins and Medals, - - - - - 1,700

LIBRARY.

The University Library is in the lower part of Manning Hall. It contains *forty-five thousand* well-bound and carefully selected volumes. In its early history it received additions from donations and legacies made by friends of the College, both in this

The Jenks Museum's gifts and purchases, noted in the *Catalogue of the Officers and Students of Brown University, 1870–1871.*

the funds from the university or donors. He was part of a network of natural history museums and enthusiasts that traded material. He went on expeditions. Jenks both acquired on his own and built networks that brought his museum gifts, trades, and funds for purchase.

Curators today still acquire in all these ways. In this chapter, we watch Professor Jenks and other curators persuade donors to give artifacts to the museum, purchase those they can't get as gifts from individuals and dealers, at auction, at flea markets, or on eBay, and broker transfers with other museums. Chapter 4 considers field collecting, joining Professor Jenks in the Florida swamps and other curators on expeditions around the world, visits to artist's studios, and on the front lines of political rallies.

Gifts Given

The Jenks Museum's annual reports include long lists of thank-yous. In 1876, Jenks thanked Mrs. S. R. Ward, of Assam, for "a large collection of Plants, Insects, Native Weapons, etc." In 1879, he thanked the Reverend David Weston, of Salem, Massachusetts, for his kind gift of "a collection of the Insects of Massachusetts, carefully mounted, numbering more than a thousand." These donors were typical of the donors to the museum. Amateur scientists presented their collections as teaching tools. Brown was a Baptist institution and many Baptist missionaries sent "curiosities" to teach students at their alma mater about the world, as well as, no doubt, to shine a light on their own good work. Some donors saw the museum as a secure place to preserve things. Others saw it as an opportunity to share a story they thought important. Donors of local artifacts found the museum as an appropriate place for treasures they thought should be shared with the community. Some gifts came from the families of collectors. Physician and naturalist William Blanding's family gave the museum his collection after his death, both to honor William's work, and, no doubt, make some room in their home. Some donors enjoyed collecting more than they enjoyed taking care of collections.

Gifts are the basis of most museum collections, and many museums depend almost completely on gifts. Some, like the nineteenth-century Smithsonian, had a "no purchase" policy. Few museums have the funds to buy all they want and certainly not enough to compete in the market for art and antiques.

Though good historical statistics are hard to come by, a few examples show the range of acquisition sources. At the Fine Arts Museums of San Francisco, donations accounted for 82 percent of the museum's collection in 1999—79 percent gifts and 3 percent bequests. Only 7 percent were purchases. (Long-term loans accounted for 7 percent, and 4 percent came from other sources.) The Museum of Modern Art in New York has accessioned and cataloged about 129,000 objects since its founding in 1929. Gifts comprise about 67 percent of the total. Gifts directly from the artist (or photographer, or designer, or architect, or manufacturer) account for about 12 percent of the collection. About 1 percent were anonymous gifts. Purchases account for 27 percent of acquisitions. Bequests and exchanges are each a bit more than 1 percent. (Complex donations like partial gifts make it difficult to be precise about some acquisitions.[2])

Overall, it's estimated that more than 90 percent of the collections of American art museums have been donated. In 2013, the 220 largest art museums in the United States reported about 73,813 gifts and bequests and only about 12,197 purchases—that is, about 83 percent of acquisitions were gifts.[3] Smaller art museums are even more dependent on gifts. So too are history museums, which often get offers based not on what the museum is looking for but on what a potential donor is eager to find a new home for. Natural history museums often depend on the field work of scientists for new acquisitions.

The range of reasons to donate an object to a museum is even wider today than in Jenks's time. Individuals continue to give in order to support the institution or to see something they believe important preserved and presented to the public. But sometimes the reason is less noble: for personal or family glory, or simply to clear out the attic without feeling guilty about throwing things away. And tax lawyers and public relations firms (neither were around in Jenks's time) now present practical reasons to give. Jimmy Durante's publicist made a splash by trying to give a plaster cast of his client's famous nose to the Smithsonian in 1959![4]

Letters to the Smithsonian from potential donors show the wide range of reasons for donating objects to a history museum. Some donors want to see their own work in a museum. Dr. Claude S. Beck, inventor of the defibrillator, wrote to the Smithsonian in 1970 offering "the defibrillator that first reversed death." He thought it belonged in the national museum. (Offers like this can be tricky: getting an invention into a museum collection can be a way to claim precedence or importance.) Others think that their objects should be in a museum because they seem typical, or symbolic, or because they tell what seems to the donor an important story. A schoolgirl proposed giving her favorite blue jeans to the Smithsonian in 1973: "This may be an unusual request. . . . My favorite jeans will be three years old this spring . . . they're so thin, my mother wants to throw them away. . . . Is it possible that you may be able to use them in a display? . . . These jeans aren't art, but they're a sample of costume in America. . . . Best of all, they're authentic."[5]

Making the case that a certain kind of object is museum-worthy often has political implications, and donors can make that case implicitly or explicitly. Collectors of early American furniture fought their way into art museum collections in the first decades of the twentieth century. At the opening of the Metropolitan Museum of Art's American Wing in 1924 curator William Ivins hailed "not only the coming of age of a particular kind of collecting but a new departure in this country's museum practice."[6] The Met both adopted notions of quality from the antiques market and blessed that market with the museum's aesthetic imprimatur, moving antiques almost to the realm of fine art. This served both the market and the museum. It also served a political purpose, ennobling the American colonial past at a time of increased immigration. A visit to the American Wing, its curator wrote, would be "invaluable to the Americanization of so many of our people to whom much of our history has been hidden in a fog of unenlightenment."[7] Historian Jeffrey Trask notes that the collections and exhibitions there "focused the story of American domesticity on lives of the elite, and simplified American history by presenting a conflict-free narrative of progress."[8] A few decades later, the Smithsonian focused on collecting a different kind of American furniture, from the homes of rural folk. Malcolm Watkins, the curator there, called the earlier model "an elitist put down" and offered what historian Briann Greenfield called "a powerful ideological challenge to the dominant collecting model."[9] Even furniture collecting has politics.

The National American Woman Suffrage Association worked hard to be sure that their movement was well represented in Smithsonian collections. It donated mementos of Susan B. Anthony and Elizabeth Cady Stanton in 1920, just after the passage of the Nineteenth Amendment, to "show the origin and development of the greatest bloodless revolution ever known." Edith Fleetwood, the daughter of Sergeant Major Christian Abraham Fleetwood of the 4th U.S. Colored Troops, wanted the Smithsonian to have her father's Congressional Medal of Honor "to serve my race rather than to glorify my father." Friends of disability rights activist Ed Roberts left his wheelchair on the steps of the Smithsonian following his funeral with a note saying it was a gift to pay tribute to his "amazing life"—and to push the Smithsonian to acknowledge their movement as an important part of American history. The National Museum of American History was pleased to accept it.[10]

Some donations come from businesses that see a gift to a museum as good advertising. In the early twentieth century, curator Chester Gilbert encouraged donation as a public relations ploy. He wrote to the Johns-Manville Company, "When you pause to consider the conditions involved in such an exhibit as that proposed, I think you cannot fail to recognize its decided advantages to you, viewed simply in the light of advertisement. The series would show in full, the industrial uses to which your Company has adapted asbestos, and the system of labeling would set forth your claims as to the merits in each instance, and your name would be conspicuously present."[11]

It's common today for a company to suggest a donation to the Smithsonian as a way of making a claim for a product's importance. The proposed press release will often read "inducted into the Smithsonian Hall of Fame." The actual press release will never say that; the museum's press office informs the donor that the museum takes things for many purposes—because they are interesting, or have useful documentation, or tell a good story—not necessarily because they are "first" or "best," as a PR firm might want to claim.

There are times when giving an object to a museum serves exactly the opposite function, not glorifying or exalting but rather signifying closure, the end of an era. "It's history" is a way of saying something is no longer important. A company might make a donation of a product that is "being retired," honoring it but sending a message to employees and the public that the firm is moving on.

This last—declaring something no longer part of contemporary culture by "museumifying" it—can serve political purposes. As Confederate flags and memorials come to seem inappropriate, the cry goes out: They belong in a museum. The *New York Times* called for the Confederate flag that flew on the South Carolina statehouse grounds to be moved to the state's Confederate Relic Room and Military Museum. The editorial noted, "That is an apt name for what should be the battle flag's resting place. It belongs there." Being in a museum can mean something is no longer part of current culture but part of history.[12]

The reasons donors give art and artifacts to museums are not necessarily the same as the reasons the museum accepts them. A donor's interests might be shaped by personal concerns or family dynamics or economics—a whole range of emotional and financial issues that are generally not important to the museum's interest in the object. The museum's interest, on the other hand, is in using the object to tell a story—which may or may not be the story the donor wants to tell. The trick is to find the overlap where the gift makes sense to both the donor and the curator.

Museums' dependence on gifts means that collections are often built based on what's given, not what's best. There's some evidence that purchases—which represent the curator's choice in a way that gifts may not—see more use in museum exhibitions. Donated artworks are more likely to stay in storage. At the Fine Arts Museums of San Francisco in 1999, for example, 6.6 percent of purchased objects were on display. Only 2.4 percent of donated objects were. Excluding the museum's study collections (prints, drawings, and textiles) makes the comparison even more dramatic: 26.5 percent vs. 7.8 percent. Ann Stone, who calculated these statistics, draws this conclusion: "Historically, the Museum has applied looser standards in accepting donated objects than it has in making purchases."[13]

Stone found other differences between donated and purchased collections, too. Most donations were of relatively low value. In the 1990s, curators purchased less than 5 percent of acquisitions valued at less than

$10,000. They purchased more than 25 percent of the acquisitions valued at more than $25,000. Even when accounting for the wide range of values in different departments of the museum, the same general rule held: the least expensive acquisitions were donations. (The most expensive objects were an exception; the museum couldn't afford to purchase million-dollar paintings.)[14]

How to explain this? Stone suggests that the museum was accepting low-value art that it would probably never display, to please donors, in exchange for the gift of the high-value art it really wanted. This benefitted the donors, who got tax breaks and the pleasure of seeing the art they had collected deemed "museum quality." It benefited the curators, too. Museum curators were given annual goals for total monetary value of gifts and purchases to their department. The incentives were clear; curators gained by accepting art of lesser value than the museum would likely display, in exchange for also accepting the valuable, high-quality art they wanted.[15] This is not a new story: in 1930, the director of the Met, Robert de Forest, noted that a good reason to accept a gift is "the cultivation of good will from sources likely to aid the Museum in other directions. Acceptance of a gift not otherwise desirable may be very helpful in cultivating this good will."[16]

Museums depend on collectors, but there are major differences between the ways that collectors and museum curators think about what to collect, how to collect it, and how to treat it. The relationship of curator and donor presents great possibilities for creative expansion of collections, but it also presents challenges. Who really decides what the museum needs for its collection? What does the donor get in return for his or her gift?

There's an enormous literature on why collectors collect. That's because there are so many reasons individuals collect—from objects as souvenirs of family or travel, to objects as a way of controlling and making sense of a confusing world—and so many types of explanations, from the economic to the psychoanalytical. But the reasons museums collect have very little overlap with those personal reasons. And once collected, an object plays a different role in a public collection than it does in a private one. Many collectors are interested in completeness of a collection, "one of each"; that's less important to most museums. Collectors and museums often treat objects differently: antique car collectors want to drive their cars; most museums won't allow their cars out on the road. Collectors buy and sell and trade; museums almost never do. Many collectors obsess over fine details of "grading" to determine value; that's not something that museums worry much about. Some collectors are concerned with making money by buying low and selling high; that's not the way curators think about their collections.

The differences between the way that museums collect and collectors collect can lead to friction. But it can also be a source of strength. The fastest way to build a collection is collect it in one fell swoop. That's not uncommon, both in Jenks's time and today. At the Museum of Modern Art, the catalog lists more than two hundred collections acquired by the museum, totaling some 27,000 objects or 22 percent of the collection. Fourteen large collections (those with more than 100 objects) account for about 20 percent of the collection. Some were purchased, some donated; in either case, they represent a collector, not a curator, shaping the collection.[17]

When museums collect collections, rather than individual pieces, they can acquire not just the art but the collector's in-depth understanding of it. Collectors often lead taste; museums tend to follow. Collectors can take greater risks. They can be monomaniacal, collecting a single artist, or school, in great depth, and focusing their expertise narrowly; museums tend to cover the bases. Museums tend to move slowly; collectors can buy without waiting for the bureaucracy. Designer Glenn Gissler, a donor to the RISD Museum of Art, explains the advantages: "As a donor, you can operate more quickly and stealthily than a curator, who has to follow a whole procedure," says Gissler. "And curators are not paying attention to twelve different auction houses and what's online. Because I am, I also know what things should cost."[18]

And so some of the greatest museum acquisitions come complete, the lifetime's work of a collector. Leonard A. Lauder built perhaps the world's greatest collection of Cubist paintings, drawings, and sculptures. His gift to the Met was, director Thomas P. Campbell said, transformative. "In one fell swoop this puts the Met at the forefront of early-20th-century art. . . . It is an unreproducible collection, something museum directors only dream about."[19]

When museums acquire these large collections they need to consider the biases that shaped them. The missionaries who sent exotic artifacts to the Jenks Museum chose things that spoke to their interests and concerns, often artifacts that showed the success of their proselytizing. Colonial officials, another major source of donations to anthropology museums, likewise often collected to show the success of their work. Anthropologists often wish they had collected differently.

But for all of these challenges, curators are good at finding ways to work with collectors. Curators need to convince an object's owner that the museum's the right place for it. Curators often form long-term relationships with collectors, sharing information, even sometimes serving as informal advisors to shape personal collections in the hope or promise that someday they'll come to the museum.

Thomas Hoving, director of the Met in the 1960s and 1970s, wrote that "the tracking down and winning of a single work of art offers the highest excitement of all—the excitement of a love affair." In his kiss-and-tell memoir, *Making the Mummies Dance,* Hoving tells many stories of convincing donors that the Met was the place for their treasures. Seduction is a good word for these conquests, examples of curatorial acquisitiveness at the highest level. *People* magazine described him as having "the art-politik zest for intrigue of a Kissinger and a high roller's savvy and nerve."[20]

In 1969 Hoving decided that he needed Nelson Rockefeller's collection of primitive art, in part because he thought it would "round out the encyclopedia," but also because he wanted to distract his board from a disastrous exhibition. He came up with a scheme. First, he acquired some Eskimo art, to show Rockefeller that the museum was serious about the topic. He installed an exhibition of Mayan art, organized in the way that Rockefeller liked. He had an architect draw up some "quick, sexy sketches" of a new wing for the collection. Finally, he organized a temporary exhibition of Rockefeller's collection—and was rewarded, at the opening, with Rockefeller's announcement that the entire collection would come to the Met. High-wire collecting at its finest![21]

There are many constraints on what the curator can (or should) promise. In general, the curator should never promise that the art or artifact will always be on display, or that the collection will be kept together on display, or that the donor will have the right to control the display. (An exception to these rules often requires the approval of the museum's board of trustees.) There's a long history to making these promises anyway. Hoving certainly did. Get the stuff, the curator thinks, and worry about the promises later. Or, more commonly, that someone else will worry about the promises later. Generations of museum curators have cursed their predecessors for collecting with too much enthusiasm. Henry Brown, director of the Detroit Historical Museum, lectured his fellow history museum directors in 1956: "Have you promised a donor that you will keep his items always on display? If so, you are in the warehouse business. You have lost the flexibility for change. You are a frozen museum."[22]

The final step in a museum donation, after all the courting, the dancing, the deciding, the back-and-forth, is the signing of a deed of gift. This is the formal legal agreement that transfers ownership of the materials to the museum. It's signed by both the donor and the curator or director of the museum. Deeds of gift are usually fairly straightforward documents. Most are short. In a typical deed of gift, the donor acknowledges that he or she owns the objects being donated and agrees to "unconditionally relinquish ownership to donate, bestow and set over all rights, titles and copyrights held by me, my assigns and my heirs of said items." The deed of gift reminds the donor that the museum can do with the objects what it wishes: they "may be retained, exhibited, loaned or disposed of." Finally, the gift is described, the donor signs, the curator or director signs, and the legal part of the transfer is complete. (Packing and shipping is a separate, equally complicated part of the process.)

Of course, it's not always that easy. Donors may try to control the donation, even after it belongs to the museum. Donors may not own all of the rights to a gift, or be able to prove that they do. Particularly challenging is the question of intellectual property rights to an object. The owner of a photograph, or an archive, for example, may not own the right to reproduce it. Some deed of gift forms have a checkbox for the donor to indicate ownership, or not, of copyright.

There's no valuation attached to the deed of gift. The donor may want an appraisal, for tax purposes, but that's not the museum's responsibility. Museums don't provide appraisals, nor do they select, hire, or pay for appraisals. That would be a conflict of interest, and one could imagine museums competing for donations with generous appraisals. The donor pays an appraiser to determine fair market value, which can be complicated. Fair market value is defined by the Internal Revenue Service as "the price that would be agreed on between a willing buyer and a willing seller, with neither being required to act, and both having reasonable knowledge of the relevant facts."[23] That's a nice bit of legal poetry. But what if, as is often the case with museum artifacts, there is no market other than museums that don't have budgets to purchase collections? Appraisers consider previous sales of similar objects and make their best guess. And for many objects donated to museums, it doesn't matter: they have no monetary value. That's not why the museum thinks they're worth collecting.

Internal Revenue Service rules can shape the interaction of donors and museums in complex ways, for good and bad. On the one hand, the income tax charitable donation rule encourages gifts. It keeps museums focused on their mission; in order to deduct the fair market value of a gift, not its purchase value, the donation needs to be used for a purpose related to the museum's tax-exempt purpose. On the other hand, IRS rules can encourage curators to take donations that they don't really want, to get a donor to donate what they do want. The rules can also shape collections in odd ways: until a recent tax law change, the donor of a taxidermied zebra could include the cost of the safari as part of the value of the gift.

Income and estate tax laws can complicate a museum's work. The tax deduction comes with reporting requirements both for the individual and the museum. And a good attorney can come up with no end of complications: fractional gifts, gifts over a period of time, lifetime gifts, combination gift/purchase/bequest agreements, bargain sales—all have their place, but, as with other complications and restrictions, the museum should be careful. Museum legal expert Marie Malaro sums up the challenge: "The issue of restricted gifts is one of conflicting interests, the interests of the donor versus the interests of public." She urges museums to take a strong stand in favor of the public's interests.[24]

Worth the Price?

The Jenks Museum was built mostly on gifts, but Jenks occasionally found funds to buy things he really wanted. The story of the museum's first major acquisition—an "extremely valuable collection of 4300 birds, (with a few exceptions,) made by an eminent ornithologist, the late John Cassin, Esq., of Philadelphia"—was in some ways typical of the museum's purchases. Jenks bought it impulsively, desperate to beat out other museums. He spent his own money and then raised funds for the university to buy it from him.[25]

Jenks also purchased individual specimens for particular needs, for particular courses, or, more often, just because he wanted them so badly. He bought from individuals, when he couldn't persuade them to make a donation, and worked within a market that supplied museums and individual collectors. Jenks's main commercial supplier was Ward's Natural Science Establishment, of Rochester, New York, the preeminent dealer in natural history specimens of the day. Henry Augustus Ward, its proprietor, was an adventurer and collector, taxidermist, and businessman. (He achieved his greatest fame when he refashioned Jumbo, P. T. Barnum's elephant, into a skeleton and mounted specimen, so that it could continue to tour after the animal's untimely death.) His catalogs listed minerals, rocks, fossils, taxidermied animals, skeletons, invertebrates, and anatomical models.

Jenks and Ward hit it off, and maintained a cordial correspondence for many years. In addition to details of travels and health, they discussed a long list of mounted and skeleton animals Jenks wanted for his museum. Jenks sent animals to be taxidermied and purchased materials from Ward. Jenks's letters to Ward are full of the hard choices he's trying to make. A *Dinornis,* the giant moa? A buffalo bull? A buffalo cow? Why not both? A "Skeleton of the Horse"? "You know my will is a great deal better than my purse," he wrote, and almost every letter bogs down in the challenges of raising funds to buy more collections. "Don't tempt me this year with any specimen from Central Africa for I had to borrow the money I have already expended since January over & above $200 & must first pay that up."[26]

Jenks's dependence on the market for some of the objects he wanted for the museum, and his frustration with that market, is typical of curators. Museum collecting overlaps with the marketplace for art and antiques. Sometimes both collectors and curators are after the same objects—and the collectors usually have more money. Art makes the news when a multimillion-dollar sale breaks records at auction houses, but museums rarely pay those prices. Most museums can only hope that a wealthy collector will someday donate the art to the museum, trading some of his or her fortune for the fame and satisfaction that comes with a "gift of" or "collection of" label on the wall in a museum.

For the most part, museums don't think about the monetary value of their collections. Many curators don't pay attention to market values and most of them like it that way. Art and artifact, they'll tell you, is in the museum because it holds other kinds of value: scientific, aesthetic, historical, art-historical. It has, curators like to think, transcended vulgar notions of "what's it worth?" The price is never included on the exhibition label (though it's often what the public wants to know). In general, once art is in the museum, it can't be sold. It is priceless—and price-less.

Appraisal for tax purposes is one of the points where museum value and price intersect, but there are other times when curators need to think about monetary values. Loans between museums require a valuation, for insurance. Value sometimes comes into play with deaccessions; is it worth the very considerable effort it takes to remove something from the collection? The security office needs to worry about what might be stolen. Some museums make lists of "most valuable" items that should be moved off site in case of disaster or in times of war, defining "most valuable" in a variety of ways. And of course, value comes into play when the museum decides to expand collections by buying something or to reshape a collection by deaccessioning.

While most museums buy very little, if anything, for their collections, there are exceptions. Wealthy founders allowed the J. Paul Getty Museum and the Crystal Bridges Museum of American Art to spend hundreds of millions of dollars to build their collections. *The New York Times* reported in 2013 that the Qatar Museums Authority had a $1 billion annual acquisitions budget. The Museum of Modern Art spent $32 million to acquire art that year; the Met, $39 million. But at most museums acquisitions budgets—often the result of much-debated deaccessions—are small and hoarded. Board members are asked for contributions when something that seems essential comes on the market.

Very occasionally, museums splurge at an auction or in private for objects that can be transformative—and for which there is PR value as well as collections value. A half dozen museums raised more than a million dollars to bid on Sue the *Tyrannosaurus rex* when it went up for auction at Sotheby's in 1997. The Field Museum won, at $8.4 million, with funds it had raised from McDonald's, Walt Disney, and others. The Henry Ford Museum bought the Rosa Parks bus at auction for $492,000, outbidding the Smithsonian and other museums to acquire an object that would not only be a centerpiece of an exhibition but would also help announce to the Detroit community its new, more inclusive, mission. (Even bolder: when museums deaccession collections to fund transformative acquisitions. More on that in Chapter 8.)

But whether or not museums make purchases, they need to be aware of areas where they overlap with the market. History museums lose material to private collectors, theme restaurants, and flea markets. Collectors have created markets for everything from meteorites to microscopes, and curators need to either outbid collectors or make the case that the museum is the proper home for the material. They might argue civic pride, scholarly accessibility, and educational value, on the one hand, or fame, legacy, good care for a treasured heirloom, friendship, or even tax breaks, on the other; but it is necessary to make an argument.

Art museum curators live in a complicated ecosystem of collectors, dealers, auction houses, galleries, *kunsthalles* (museums without permanent collections), art fairs, artist studios, and, increasingly, Instagram and Tumblr. The art world, writes ethnographer Sarah Thornton, is much broader than the art market, a symbolic economy that "blurs the lines between work and play, local and international, the cultural and the economic."[27] Museum curators occasionally buy, but also advise, cajole, and connect, balancing collecting with gathering knowledge and making connections.

The recent rise of online markets has changed the relationship between museums and the markets. As with almost every other area of commerce, the internet has sometimes made it possible to cut out the middleman. This opens new possibilities for direct contact between curators and the creators of art. Instagram and other social media can make the art market more transparent, less dependent on galleries, allowing curators (and other buyers) more direct contact with more artists. It also reminds us of the value that middlemen and arbiters of value bring to the art market. Brad Phillips, a photographer who now sells most of his art on Instagram, writes that "increasingly for young artists, galleries are becoming obsolete." That has its good side: artists "can now be themselves"; they don't need "to play the boring and expensive game the art world requires," or lose the 50 percent of the price that galleries take as their commission. It has its bad side, too: "The art world right now is a youth-fetishizing cannibalistic death cult of speculation and interior design masked as progressive painting." That's what can happen when there are no middlemen.[28]

Just as Instagram offers artists direct access to the buyers of art, eBay, where everyone can be their own dealer, gives history curators the opportunity to connect directly with sellers who might have stories to tell about the objects they've put up for sale. That information, which can be lost as objects move into shops, is often exactly what makes an object useful to a history museum. On the other hand, it can cut out the expertise a picker or a dealer brings to the transaction.

Museum curators' interactions with markets are always fraught. The curator's job is to move objects out of the real world, where things have utility and market value, and into the timeless, price-less world of the museum, where they have only meaning. Oftentimes, though, the way they make that transition is through the marketplace: the gallery or the auction house, the eBay store or the artist's Instagram account. Museum objects often get a price just before they become priceless.

Sharing

Professor Jenks never had much money to purchase collections, but he had a different kind of currency: his contacts, and trades. Some of the first and most important collections at the Jenks Museum were transfers from other museums. Jenks noted in his 1872 report the receipt of "several boxes from the Smithsonian Institution, comprising casts of Indian Relics, Esquimaux curiosities, specimens of building stones, implements and curiosities from the Fegee Islands." Many of the Smithsonian materials sent to the Jenks Museum survive today at the university's Haffenreffer Museum of Anthropology, identifiable by the yellow wax marker the Smithsonian used to number its collections.[29]

The Jenks Museum was one of many museums that received material like this from the Smithsonian. The Smithsonian played an important role in the distribution of museum specimens around the country and around the world, sometimes trading for material it wanted. "No greater aid can be given to the cause of scientific education," wrote the institution's secretary in 1873. (It was also useful politically, a way of showing the Smithsonian's congressional overseers the local value of their support for federal science.) The numbers were remarkable: By the end of 1871, the Smithsonian had distributed 308,080 duplicate specimens (mostly shells, birds, plants, and fossils) from its anthropology and natural history collections. The Smithsonian continued to distribute anthropological and archaeological specimens to schools, public libraries, asylums, societies, and private collectors into the 1960s.[30]

How did the Smithsonian decide what to keep, and what to give away or exchange? In natural history, research came first. Examples of identified species were distributed to other museums so taxonomists could use them in their work. (Chapter 17 follows some of the mice that Jenks sent the Smithsonian to museums around the country.) In exchange, the museums often sent type specimens of species they had identified and representative collections from their region. What counted as a duplicate in ethnographic collections was more complicated. Early ethnographic curators applied a logic like that of natural history curators: When curators believed that indigenous peoples were simply passing through stages of progress, objects represented types, not individual work. The Smithsonian put together museum starter kits for new American museums and traded duplicates for material the Smithsonian wanted—for example, exchanging Native American objects for African objects from major European museums.[31]

Transfers between museums remained common in the first half of the twentieth century. A proposed 1925 code of ethics for museum workers urged that "museums should cooperate by exchange, sale or otherwise so that a very rare object or specimen may be placed where it can best be studied and kept in association with closely related objects."[32] As museums became more specialized and the number of museums increased, some shifted collections. The Fine Arts Museums of San Francisco's Asian art collections made more sense at the nearby Asian Art Museum, its natural history collections at the Natural History Museum.

Large-scale transfers still happen when there's good will and good reason; it's a sign of an institution's focus on strengths and future needs, rather than holding to past ways. One of the largest recent transfers came in 2008, when the Brooklyn Museum transferred its costume collection—some 23,500 objects, one of the best in the world—to the Met. Money drove the deal; costume collections are very labor intensive, very expensive to store and exhibit. They're fragile and can only be on display for very short times, weeks or months. The Met, supported by the fashion industry, had the funding. Brooklyn didn't. But more important: it was a good match. The Met's Costume Institute focused on haute couture, particularly European fashions, Brooklyn on American fashions from before the mid-twentieth century. Brooklyn will be able to use the collections in its exhibitions, but won't need to care for them.[33]

More common are small-scale transfers, as part of deaccessioning. Pruning collections is an essential part of building stronger collections (see Chapter 8), and transfer to another institution is the preferred method of disposing of deaccessioned artifacts. Some museums websites list material available for transfer. In 2016 the National Air and Space Museum had several dozen items slated for deaccession and ready to be "permanently transferred to organizations who will maintain them in the public domain," from an entire Douglas VB-26B-61-DL Invader to rocket engines. The National Park Service also maintains a list of material for transfer.

When curators acquire as gifts or in the marketplace or through transfer, they're bringing in art and artifacts that are already deemed collectable. There's another way to acquire. Sometimes the curator sidesteps the middlemen of the market and other museums and heads out into the field. There's a long history of the curatorial expedition.

❧ 4 ❧

IN THE FIELD

PROFESSOR JENKS WAS NEVER happier than when he was out in the field. His *Hunting in Florida* is an account of "the adventures of a Naturalist Collector in the Everglades." It describes with enormous pleasure his 1874 trip as "a camp life of fifty consecutive days in the miasmatic swamps and everglades around Lake Okechobee in southern Florida." Jenks hunted animals, explored Seminole villages—and even fought off alligators to obtain that roseate spoonbill for the museum.[1] Nine years later, at age sixty-six, Jenks spent a year traveling the United States, visiting every state and territory, including Alaska and Hawaii. This too was an exploring and collecting trip, if not quite as dangerous. Jenks shipped Native American artifacts, geology specimens, and birds and animal skins back to the museum. One of his students described Jenks as "the old-fashioned sort of naturalist . . . who went up and down the world with his traps and gun, collecting and skinning and mounting his specimens . . . quite as much of an explorer and discoverer as he was a collector."[2]

In combining curating with exploration and collecting, Jenks was typical of his day. Natural history museum curators were expected to hunt and trap and prepare their specimens. Anthropologists collected as part of their studies. Museums mounted expeditions around the world to explore archaeological sites, indigenous cultures, and the natural world.

The tradition continues today, though in a more restrained way. Legal and ethical concerns restrict some of the more free-wheeling acquisitions practices curators once delighted in. In their place are more sophisticated collecting procedures that see both the natural world and human commu-

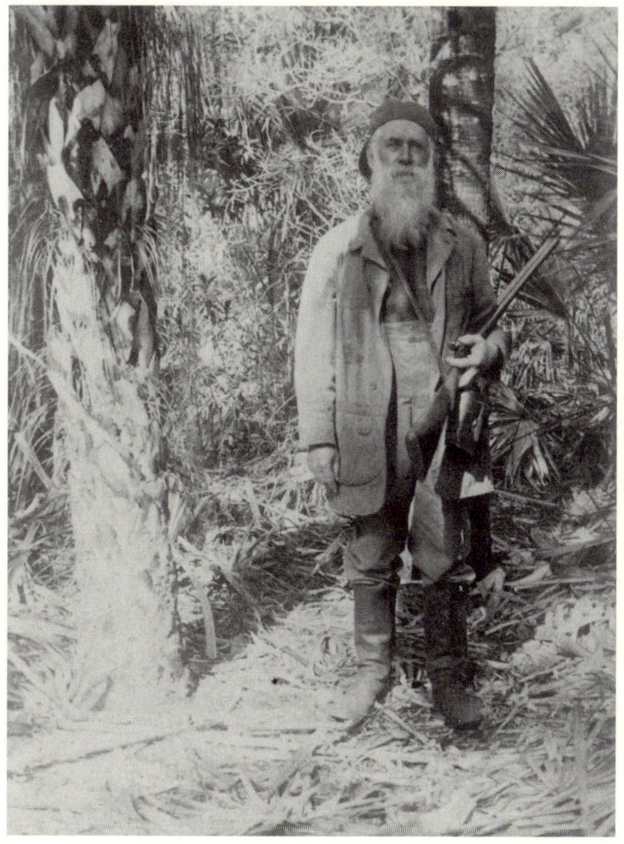

Jenks collecting in Florida, 1874.

nities as complex places. But the spirit of the curator as explorer survives. Field collecting allows a museum to decide what's important, to document collections well, and to keep up with changes in society. It can make the museum relevant to new audiences by building collections that are meaningful not only to future audiences but also to contemporary ones. To collect well, sometimes the curator needs to get out of the museum and into the field—or into the artist's studio, onto the factory floor, and to the scene of historic events.

Hunting Objects

In the nineteenth century, collecting was part of exploration. Museums and cabinets of curiosities in Europe were filled with the products of nature and natives from exotic places. Governments and scientific societies gathered these objects as oddities, as evidence for scientific work, and to support the economic demands of empire. Collecting artifacts from around the world allowed scientists to make sense of larger patterns. Ralph Waldo Emerson, visiting the Paris Jardin des Plantes in 1833, wrote that the "mountain and the morass and prairie and jungle, the ocean, and rivers, the mines and the atmosphere have been ransacked [for] the keen insatiable eye of French Science."[3] Science theorist Bruno Latour explains the power of the collection: in a museum, a curator can "travel through all the continents, climates and periods. . . . common features that could not be visible . . . far away in space and time can easily appear between one case and the next."[4]

The U.S. government saw the value of collecting as part of exploration and expansion. The Lewis and Clark expedition overland to the Pacific in 1804–1806 brought back plant, animal, and mineral specimens as well as cultural items from the native peoples they met. The United States Exploring Expedition circumnavigated the globe in 1838–1842 to explore, map, establish American presence, encourage trade—and to collect. The naturalists on board the ships of the Ex. Ex., as it was called, returned with some 4,000 cultural objects, 50,000 plant specimens, more than one thousand living plants, and thousands of bird, mammal, and fish specimens. Many of these would end up at the new Smithsonian Institution. The Smithsonian shared Ex. Ex. artifacts with other museums, including the Jenks Museum, and so they became the basis of many other collections as well.

Field collecting was an essential part of most natural history museums, as well as some museums where expeditions may seem less likely. The Army Medical Museum, established during the Civil War, had a difficult time persuading army doctors to send specimens, so its first curator, Major John H. Brinton, visited battlefields to collect. He vividly described his work:

> Many and many a putrid heap have I had dug out of trenches where
> they had been buried, in the supposition of an everlasting rest, and
> ghoul-like work have I done, amid surrounding gatherings of won-
> dering surgeons, and scarcely less wondering doctors. But all saw

that I was in earnest and my example was infectious. By going thus from corps hospital to corps hospital, a real interest was excited as to the Museum work, and an active co-operation was eventually established.[5]

Later in the century, large museums sent expeditions around the world to build their collections of the natural world and of the artifacts (and, all too often, the skulls and skeletons) of "primitive" peoples, to collect data and photographic archives, and to do new science based on those collections. Curators from Harvard's Museum of Comparative Zoology and other museums cruised the Atlantic and the Caribbean gathering specimens, first on government ships, and then on the yachts of wealthy supporters. Yale's Peabody Museum sent students to dig for dinosaurs in the West. University museums and the new art museums in major American cities were desperate for collections that showed the roots of civilization in the ancient Near East, especially those that cast light on biblical times.

The museum expeditions of the turn of the twentieth century were great multidisciplinary adventures aimed at comprehensive, systematic, and well-documented collecting. They aimed to find pristine environments and "vanishing" peoples, and to document them scientifically. The American Museum of Natural History (AMNH), for example, launched major research expeditions to northeastern Congo and southern Sudan, the north Pacific, China, and Oaxaca. They returned with specimens, artifacts, photographs, notes, and drawings, building well-documented collections and producing detailed research reports that covered almost every discipline of museum interest.

Expeditions meant museums didn't need to depend on the middlemen—missionaries, traders, and explorers—who had previously made the choices about what to collect. Scientists could observe and collect cultural and natural artifacts in situ. Artifacts collected on expeditions were usually better documented than those purchased from dealers or donated by travelers. The scientists and curators who carried out these expeditions wrote comprehensive reports on their work, careful documentation of the data and artifacts collected and the science done in the field. Recent scholarship has examined more closely the details of the fieldwork and collections strategies of these scientific expeditions, revealing the complexities of their work. They were not, as the published reports suggest, the pure science of researchers and curators meeting unexplored peoples and landscapes. Rather, like all collecting, expeditions were mediated by indigenous collaborators, government bureaucrats, wealthy supporters, and others. Like the earlier expeditions, they sometimes overlapped with political, imperial, and commercial interests. Museum collecting expeditions, like the earlier national ones, had an ideological edge.[6]

But they were successful. Hundreds of expeditions brought back thousands of objects. These, added to gifts and purchases from collectors who explored on their own, made for enormous collections. By the 1960s, one Smithsonian curator estimated, there were some 1.5 million ethnographic objects in American museums.[7] Archaeological collections brought in many more objects (in part because archaeologists count every potsherd as an object), as did expeditions that collected for natural history museums.

American museums continue to support collecting expeditions today, but these tend to be smaller and more focused than the large multidisciplinary adventures of the nineteenth and twentieth centuries and driven by research problems, not mass collecting. Many are undertaken in collaboration with local scientific institutions. The AMNH, for example, works with the Mongolian Academy of Sciences on joint paleontological expeditions to the Gobi. Many museums have reconnected with communities that were the subject of earlier expeditions, sharing information and finding new perspectives on earlier collections. (See Chapter 19.)

Another kind of expedition—the salvage archaeology required by federal law for projects on federal lands or that disturb historic sites—has brought even larger numbers of artifacts to museums. The Antiquities Act of 1906 required permits for excavations on federal and tribal lands and insisted that "the gatherings [of objects of antiquity] shall be made for permanent preservation in public museums." The Archaeological and Historic Preservation Act of 1974 is much broader, calling for the "preservation of historical and archeological data (including relics and specimens) which might otherwise be irreparably lost or destroyed as the result of . . . any federal construction project or federally licensed activity or program."[8] These laws are responsible for many of the over 200 million individually cataloged archaeological items and 2.6 million cubic feet of bulk cataloged archaeological items preserved in museums and other repositories today.

Collecting the Contemporary

While museums rarely send out expeditions for historical artifacts, other than to antique stores and flea markets, there's a long history and a recent revival of collecting contemporary history, of curators going out not into nature but into the everyday world to collect artifacts to document the moment.

Contemporary collecting was central to technology museums in the nineteenth century. Technology and industry curators at the Smithsonian Institution in the nineteenth century believed that their displays should come up to the present day. J. Elfreth Watkins, the Smithsonian's first curator of railroads, corresponded with engineers around the country to create exhibits like his "Development of American Rail and Track" (1889). The engineers advised him on the most important current developments, and sent him plans and samples. Watkins and the other curators of technology at the National Museum saw their work as providing an archive of contemporary practice both for practitioners of their day and for the future.

Many early twentieth-century art museums saw industrial work as a key part of their job. John Cotton Dana collected contemporary industrial production at the Newark Museum, producing exhibitions like "New Jersey Clay Products," in 1915. At the Metropolitan Museum of Art, Richard Bach, director of the Department of Industrial Relations, produced exhibits on design that depended on his persuading contemporary industrial producers to donate their latest projects, focusing on high-style machine-made goods as representative of the age.

State and local historical societies were eager to collect material to capture the historical moment. At the turn of the twentieth century, the Washington State Historical Society noted the need "to preserve historical papers, letters, and other documents that, while of no value now, would be of great value to posterity." Director Edward Fuller took this to heart, building a collection of ephemera by accepting flyers on the street and pulling posters and broadsides from walls and telephone poles.[9]

Curators of the mid-twentieth century reacted against this kind of collecting. They shied away from the commercial and even from contemporary art. Laurence Coleman, director of the American Association of Museums, noted the challenges in his 1939 survey of museums. "While material is still easy to get there is too much of it for convenience; and selection involves problems one may not care to approach." Coleman had mixed feelings about contemporary collecting. He understood the challenges of trying to choose what to collect before the market or history had sorted it out. "Sampling the contemporary scene . . . is much more trouble and much less fun than digging up the lost scene of yesterday." Still, he mused, thinking of the 1929 sociological study of Muncie, Indiana, "it seems odd that no one has set out to make a Middletown collection—a cross-section of one community's life in the present."[10]

In the 1980s museums did begin to think about a "Middletown" collection. Swedish museums created a model for contemporary collecting with the SAMDOK program, founded in 1977. SAMDOK is an association of Swedish museums trying to document everyday life, a philosophy of collecting, and a portmanteau word with several meanings: SAM is for *samordning* (coordination), *samarbete* (collaboration), and *samtid* (the present or the contemporary). DOK is for *dokumentation*. The SAMDOK strategy considers several criteria in collecting. It looks at an object and asks: How common is it? Does it reveal something of culture, and how obvious is that information? Does it show change over time, or is it responsible for change? SAMDOK museums try to collect groups of objects, objects with meanings, and objects that tell a bigger story, supplemented by interviews and photography. To deal with the vast range of contemporary artifacts, SAMDOK organized museums into groups, dividing responsibility for different kinds of artifacts between them. SAMDOK's collections have changed over time. Early on, the effort focused on rescue collecting, such as documenting factories as they closed. More recently, it has focused on emerging trends like globalization and migration.[11]

The U.S. military has also rethought contemporary collecting. The army's *Military History Operations Field Manual* outlines procedures for field collecting, indicating both what should be collected (objects that constitute "evidence of battle experience or other military activity of significance to the United States") and those that must not be (private property from civilians, religious or cultural material). Army curators collect both American objects and those of allies and enemies, with the goal of providing "a balanced and documented historical artifact collection for long-term preservation and for use in research and analysis." Essential to this work is collecting not just things but information about them: dates, location, and descriptions. Military field collecting has unique challenges; curators must be certain their artifacts don't have immediate intelligence value and that they're not booby-trapped.[12]

It might seem that contemporary collecting, at least off the battlefield, should be easy: the curator simply asks the people involved what's worth saving. But most people don't know how to think in terms of objects; they don't know which object will be useful to capture a historical event. That takes a specific curatorial view of the world, gained through experience with historical collections. And even when an object is identified, it's not always easy to persuade the owner to let go of a still useful object or one that holds important memories. That takes a special kind of curatorial persuasion.

Art museum curators don't describe their work as expeditions, but the contemporary curator's visits to artists' studios is in some ways equivalent. Collecting from artists requires making aesthetic judgments without the benefit of historical perspective and without a market setting values. Historian Bruce Altshuler notes that "putting a price on a recent work of art in part is to place a bet on how important this work will be in the art historical future."[13] The curator must know the history and the market, and extrapolate from it to make a best guess about the future. Collecting contemporary art helps shape that art historical future, just as the history curator collecting contemporary events makes strong suggestions about what will be worth remembering as part of history.

Ripped from the Headlines

Curators at the Smithsonian's National Museum of American History (NMAH) developed another approach to contemporary collecting: documenting stories in the news. Starting in 1960s, curators visited political conventions to collect signs and funny costumes to document the presidential nomination process. Curator Lisa Graddy describes the work:

"You're playing hunches about what seems to relate to what has come before and what seems really new and different."[14]

Curators took this strategy beyond party politics to document protest movements as well. This was a controversial move, shaped by progressive politics. In 1968 curators at NMAH collected materials from the Poor People's Campaign's Resurrection City—even though they had to sneak the material into the museum and hid it for years, until museum administrators would consider it worthy of collecting. Some of this material appeared in "We the People," a 1976 exhibition that combined history and contemporary politics and social activism. Its curators broke new ground by collecting contemporary protest material for the exhibition. Edith Mayo, one of the exhibit's curators, remembers that when she started to collect contemporary artifacts, she was accused by her older colleagues of bringing "junk" and "trash" into the collection. But she urges museums to take on the challenge. Active collecting, she writes, is the way to shape the primary sources used by future historians. Contemporary collecting is hard work, Mayo writes, but "one thing is certain: curators must not leave the selection of current historical materials for the future solely to collectors, antique dealers, elites, or to chance." Since then, curators at the museum have collected from many Washington protests, taking advantage of the museum's front row seat on the National Mall to document history in the making.[15]

In the past few decades, many museums have documented current affairs by collecting artifacts that capture stories of political protest, changes in work and home life, cultural events, and even disasters like the September 11, 2001, terrorist attacks or Hurricane Katrina. This kind of contemporary collecting requires expeditions to the places where history is being made. Peter Liebhold, one of the curators responsible for 9/11 collecting at the Smithsonian, writes that "this isn't a matter of waiting for phone calls; to tell these stories, we have to identify what we need and try to find it. Some things will disappear forever if we don't grab them; others will become expensive and hard to get." It can be difficult work, at times dangerous and emotionally wrenching.[16]

Contemporary history collections have three audiences. There's the usual audience: future historians and museum curators and visitors, who will use these objects to understand and explain the past and make meaning from it. What will be useful to them? How can curators use their knowledge of the past to make informed guesses about what the future will want from the present? There's also a contemporary audience. The museum can help society make sense of traumatic events, authenticate them, and provide perspective. And, more so than with most collections, the donors of artifacts are a distinct audience. Curators need to take into account the effects—often positive, sometimes negative—of trauma becoming a part of the historical record. Does collecting their story make those involved feel better, or might it retraumatize them? Contemporary collecting requires balancing all three of these audiences, all three of these uses.

Collecting history-in-the-making requires that a curator extrapolate from past events to develop a sense of which present events are worth documenting. When a 1991 fire at a chicken processing plant in North Carolina killed twenty-five workers trapped behind a locked door, curators at the NMAH immediately thought of the 1911 Triangle Shirtwaist fire, where locked doors caused 146 deaths. They collected the door and other materials that showed working conditions. A 1995 raid at a sweatshop in El Monte, California, reminded curators of the long history of American sweatshops, so they collected sewing machines and the barbed wire that surrounded the factory to make that history part of contemporary debate.

William Pretzer, a curator at the National Museum of African American History and Culture (NMAAHC), describes his work collecting after the riots that followed a 2015 police shooting in Baltimore: "We look for public expression . . . artifacts that are evocative of events—so something that has emotional power, something that may have been attacked or destroyed, something that was damaged in the process." The NMAAHC's collecting reinforced the importance of the riots and the events that caused them; it made them part of national history.[17]

David Shayt, a curator at the NMAH, was a virtuoso of this kind of field collecting. In a 2006 article, he described collecting after Hurricane Katrina hit New Orleans. First, he and his team made of list of artifacts to search for, "things we thought belonged in an ideal material record of the storm." These included an axe used to chop through an attic roof, handmade "Help!" signs, and flotation devices. This list provided initial direction and helped explain their work to officials. Then they set out on a long trip, keeping these artifacts in mind while remaining open to serendipity. They returned with a remarkable collection, including curtains showing the high water mark within a home; the bulletproof vest a SWAT team member wore when he dove into floodwaters to rescue a baby at a housing project; and a storm window spray-painted with the large orange X used to mark searched houses. They collected fifty-eight artifacts, each thoroughly documented, each carefully chosen to tell a part of the story, to capture "a set of meanings." The museum made these artifacts, photographs, and stories available online.[18]

This kind of collecting of artifacts from traumatic events is often supplemented by a less-mediated, more populist style of collecting. Instead of the expert-driven method Shayt outlined, anyone can contribute and everything is accepted. When the Vietnam Veterans Memorial opened on the Washington Mall in 1982, many visitors left tributes, from dog tags and combat boots to teddy bears and liquor bottles. In the 1980s and 1990s, the National Park Service saved almost all of it, with very little curatorial decision-making. Today, the collecting process is much more selective. Objects "with a direct association to the site and the names etched on the [memorial] are more desirable for inclusion in the collection than objects without a direct association." The goal of the collection, some 200,000 items as of 2016, is to "increase knowledge of the Vietnam experience and inspire present and future generations," so the Park Service established these criteria: Items must have a connection to service in the Vietnam War, have been donated by a family member or personal friend of a person listed on the Wall, or have been left in honor of Vietnam veterans and provide context for a better understanding of the War.[19]

After Oklahoma City's Alfred P. Murrah Federal Building was bombed in 1995, authorities decided to follow the Vietnam Veterans Memorial model. The collection of objects left at the Oklahoma City memorial—"wonderfully eclectic material, a cacophony of clashing tastes, from the most traditionally religious to the most embarrassingly bizarre and superficial"—documents individual reaction to the site. Visitors, the site's curator observed, "want to leave a piece of themselves, perhaps to show respect, often with a message."[20]

The outpouring of artifacts left as memorials after the terrorist attacks of September 11, 2001, raised new questions for museums. A group of museum curators met soon after the event to determine what they would collect, how they would discharge what one participant called "their dual obligation to bring history to bear on the tragic present and to help to usher that still-raw present sensitively into 'the past,' where it could be collected, preserved, and interpreted." They worried about collecting personal memorials: might that declare emotional closure too soon? On the other hand, it seemed wrong to abandon the memorials to the elements or to suggest they weren't important. Would their meaning change in the confines of a museum? How much should be collected? Who would choose? A group of curators drew up a list of material they hoped to find, but the actual process of collecting turned out to be "far more improvisational and haphazard" than they had hoped.[21]

The NMAH built a collection of several hundred objects to document the attacks themselves, the experience of first responders, and the recovery effort. The criteria for selection, wrote curator David Shayt, were "stories within stories, lasting physical value and authenticity."[22] The New-York Historical Society photo-documented shrines and memorial tributes, and collected material that documented life before, during, and after the attacks. Eventually the National September 11 Memorial & Museum would build a collection of some 11,000 artifacts and more than 40,000 photographs documenting the history of the terrorist attacks and their immediate aftermath, as well as the rebuilding of the site; the "evolving national significance" of the attacks as they shape foreign and domestic policy; and art, oral history, and ongoing memorial offerings at the site. It envisions its collection "as a deep reservoir of historical facts, trustworthy content, and cumulative insight."[23]

Objects tell only part of the story, and museums have also turned to websites and social media to collect personal stories. One of the first of these was the September 11 Digital Archive, organized by the American Social History Project and the Roy Rosenzweig Center for History and New Media at George Mason University. The public were encouraged to upload images, videos, and memories, eventually amounting to over 150,000 virtual items. The creators of the Digital Archive struggled with the question of accuracy. "Every submission," they wrote, "even those that are erroneous, misleading, or dubious—contributes in some way to the historical record. A misleading individual account, for example, could reveal certain personal and emotional aspects of the event that would otherwise be lost in a strict authentication and appraisal process." Not everyone was willing to let go of curatorial authority. American studies scholar Erica Doss called this curatorial approach both "admirably inclusive" and "critically vacuous."[24]

Collecting contemporary trauma is among the most challenging curatorial work. It's emotionally difficult. It has complicated politics. It's often done with the public looking over your shoulder. As with contemporary art collecting, curators may have a hard time guessing what history will find important. But it's important work, offering museums a way to participate in and be useful in times of national and community trauma. As Doss notes, it also raises difficult questions. What is it that curators bring to the process of collecting that might allow them to be both "admirably inclusive" yet also answer her "critically vacuous" charge? The question of what curators need to know is the subject of Chapter 5.

~ 5 ~

WHO COLLECTS?

JENKS'S WORK AS A COLLECTOR of natural history specimens was acknowledged in the lengthy preface to the first volume of Louis Agassiz's 1857 *Contributions to the Natural History of the United States of America.* Agassiz was the most important natural scientist of his day, founder of the great museum at Harvard, a man who deeply believed in collecting and collections. Jenks was the headmaster of the Pierce Academy and an avid student of nature.

Agassiz's book, despite its grand title, is mostly about turtles. Agassiz took his turtles seriously, filling his backyard with live turtles and his shelves with dead ones. Agassiz extended thanks to those who helped with his collecting. Museums around the country sent him turtles. The Smithsonian not only offered him all of the "specimens of *Testudinata*" it had, but also, Agassiz wrote, "caused special collections of Turtles to be made for me." He thanks "a number of intelligent boys of the vicinity of Cambridge" and Henry David Thoreau. But he singles out "Mr. Jenks, of Middleboro," for the thousands of turtles he sent and for his "notices respecting the mode of life and the distribution of our Turtles" as "among the most valuable of the kind I have received." And further, he writes, "To Mr. Jenks I am indebted for most of the eggs the development of which I have been able to trace. For a number of years he has provided me annually with many hundreds of eggs."[1]

Collecting those turtle eggs would bring Jenks posthumous fame, too. His student Dallas Lore Sharp, a writer of popular natural history, wrote "Turtle Eggs for Agassiz," published in the *Atlantic Monthly* in 1910. The

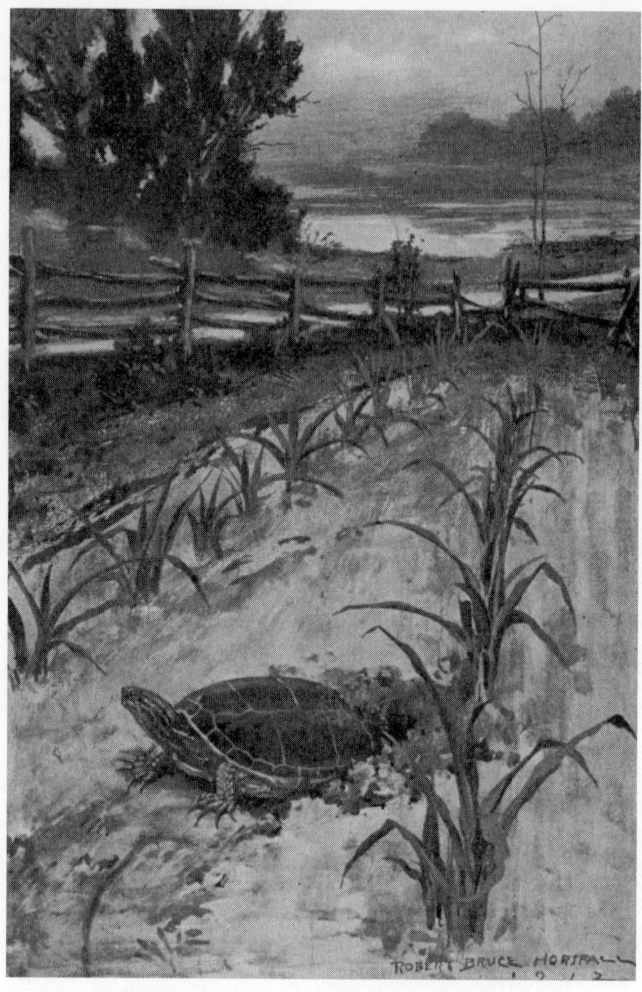

An illustration from Dallas Lore Sharp's "Turtle Eggs for Agassiz": "There in the sand, were the eggs! and Agassiz! and the great book!" From *The Spring of the Year,* 1912.

story captured Jenks's fierce collecting style. Sharp lets him tell his story: "Would I get him some turtle eggs? . . . Yes, I would. And would I get them to Cambridge within three hours from the time they were laid? Yes, I would. And I did. And it was worth the doing."

It took Jenks a month, visiting a local pond each morning. Finally, a turtle left the pond to lay eggs. He followed her: "Then up the narrow cow-path on all fours, just like another turtle, I paddled, and into the high wet grass along the fence." Finally, after hours of waiting: "there in the sand, were the eggs! And Agassiz! and the great book!" He raced his dogcart to the train tracks, forced a train to stop, and jumped on board. " 'Throw her wide open,' I commanded. 'Wide open! These are fresh turtle eggs for Professor Agassiz of Cambridge. He must have them before breakfast.' " The tale continued with a mad dash through the streets of Cambridge, and Agassiz roused from bed. But the turtle eggs made it![2]

Jenks was just as voracious in all of his collecting. He enjoyed hunting and trapping—his collecting—more than any other part of his museum work. His memoirist wrote that "his life work as a scientist was that of a collector, and this work he loved." When Jenks studied robins to see if they were eating farm crops, he dissected hundreds of birds to prove them innocent. When Spencer Baird of the Smithsonian needed specimens for his book on North American mammals, Jenks sent hundreds of mice and shrews and voles. At Pierce Academy he built "a Museum superior to that of any academy in New England, and which attracted the attention of men of science," and which "cost him an outlay of thousands of dollars, besides an infinite amount of time and labor." When Jenks explored Lake Okeechobee in Florida in 1874, he "collected at his own expense, and presented to the Museum, one hundred rare birds, two hundred rare eggs, a miscellaneous variety of fishes, reptiles, animals, insects, and relics of the Seminole Indians."[3]

Curators: Born or Made?

Jenks was a self-taught curator, and that was typical for his day. But as museums grew and took on new missions, directors reconsidered the skills and qualities necessary for curatorial work. They asked: Are curators born or made? Is a natural history curator simply a person who can hunt and trap and do taxidermy, or a trained scientist? Is all that's required of an art museum curator good taste, or does the job require training in art history or museum work, or as an artist? Should history museums hire connoisseurs or historians? In the early twentieth century many fields were professionalizing, requiring academic training. Should curatorship join them?

The first description of curators at a public museum dates from 1761. Curators at the British Museum, a guide said, "are remarked for being a sensible and learned Set of Men, all equal to the Employment, being well versed in the Business of their several Departments, and at all Times willing to gratify the Curiosity of the Inquisitive, with any Information that can be required of them."[4] Learned, knowledgeable about collections, and willing to work with the public: striking the right balance of these three would be a continual challenge for museums.

Curators at the nineteenth-century Smithsonian came from a wide range of backgrounds. Spencer Baird taught himself natural history and worked as a collector for the institution before becoming its first curator. Early curators of photography, transportation, and electricity were trained in those fields, and taught themselves history and curatorship. (Their work included both practical and curatorial duties.) A few had academic training or acquired it. Cyrus Adler, curator of historic religious ceremonial objects, started at the Smithsonian a week after he received his PhD in Semitics at Johns Hopkins University in 1887. George Brown Goode, director of the Smithsonian's National Museum, was a natural history enthusiast who spent a year working with Agassiz at Harvard. He would receive a PhD for his work in ichthyology after a decade of curatorial and administrative work at the Smithsonian. Goode wrote that successful curators needed both "years of study and experience in a well-organized museum."[5]

But the Smithsonian curators' technical expertise was unusual. In the early days of the Art Institute of Chicago, for example, the trustees served as curators. They were wealthy collectors, proud of their good taste; who would know better what to collect? In 1900, the director was also head of the department of art and sculpture. Another trustee was curator of prints and sculpture. A group of trustees ran the oriental art department, and a women's group took responsibility for the "purchase of articles pertaining to the Industrial Arts, such as pottery, china, embroideries, laces, etc." They bought what they liked. A later professional curator dismissed this early collecting as "mostly doggies and flowers" by "forgotten mid-western artists and lady amateurs which are never displayed, and are disposed of whenever possible."[6] (Of course, one might wonder whether this comment is simply a reflection of the taste of the 1970s, when he was interviewed; perhaps the art created by those "lady amateurs" was simply not part of the canon he had been taught.)

The trustee-curators of the Art Institute would have called themselves connoisseurs. The connoisseur's knowledge comes from looking at many things; he or she has a good eye, a gut feeling of right and wrong. Perhaps most important, a connoisseur has "good taste." That's problematic because this kind of knowledge was often tied, especially in the early days of museums, more to class preferences than any more general understanding of what made art museum-worthy.

Bernard Berenson, one of the most important art historians of the early twentieth century, was the very model of a connoisseur and the founder of modern connoisseurship. He knew a painting was good, authentic, and worthy of purchase, because . . . he knew. Toward the end of his career, when he was challenged over his attribution of a painting to Titian, he explained how he knew. "My proofs? They are in my own head [and] cannot easily go into words, for they come from that sixth sense, the result of fifty years' experience whose promptings are incommunicable."[7] (The painting is now widely considered a Giorgione.)

Caspar Purdon Clarke, the second director of the Metropolitan Museum of Art, when asked what made a good museum curator, replied, "A guess, a bluff, and no one to contradict you." Of course, there was more to connoisseurship than that. The curators of the Met in the early twentieth century were connoisseurs, but also trained art historians. Still, they were clearly hired help, there to assist the trustees, who made the real decisions about acquisitions. Calvin Tomkins, in his history of the Met, provides two compelling anecdotes. When curators appeared before the trustees to make a case for a purchase, they were not allowed to sit down. They would make their case and leave. And once a year, the trustees' wives would donate clothes they didn't need to the curators' wives.[8]

Job Descriptions

At the Met, curators and the director spent a great deal of effort debating just who was responsible for what—and it sometimes got pretty nasty. The first director, Luigi Palma di Cesnola, had served as a general in the Civil War and American consul to Cyprus, where he had collected antiquities. He sold them to the Met and in 1879 became the museum's director. In 1882 he hired William Henry Goodyear as the museum's first curator. Goodyear had been trained as an art historian in Germany and had taught at The Cooper Union.

Cesnola and Goodyear did not get along. In 1885 Goodyear wrote a long memorandum listing about fifty duties he was responsible for, clearly looking to lose some of them. Many were prosaic: making lists, filing, writing thank-you letters. The list reflected his annoyance at parts of his job he felt more social than academic, among them "showing courtesy to people who are not able to see the Director" and "answering enquiries and giving advice to people whom it is unwise to treat rudely." It also reflected the challenges of the job: "The Museum is besieged by people with objects to sell."[9] In 1886 Goodyear and curator of sculpture Isaac H. Hall agreed to a list of sixteen tasks that curators were responsible for, mostly the more professional ones, from "safekeeping and preservation of all art objects" to preparing "handbooks and catalogues."[10]

But was collecting the job of the curator, the director, or the trustees? Curators were advisors. An 1893 exchange with a donor is typical. First, a letter from the owner to the director: "We have a fine portrait in Oil, of Danl Webster which we would be pleased to donate to you, if agreeable. If you might send some one to look at it, and if acceptable, to take same." George Storey, the curator, was sent to take took a look. He reported back: "It does not possess sufficient merits as a work of art to be entitled its place upon the wall of the museum." The director sent a note: "The curator reported this morning" that "its artistic merits are not such as to entitle it to a permanent place upon the walls of the Museum."[11]

Many of the curator's monthly reports switch to the passive voice when dealing with collecting: "There was received at the museum," "There has also been eighteen oil paintings received as a loan." One gets the sense that the museum mostly took what it was offered and that curators spent a great deal of time dealing with paintings wealthy patrons left at the museum as loans.[12]

The Museum of Fine Arts, Boston (MFA), presents an excellent example of the next stage of the debate over curatorial prerogative. The museum, founded in the 1870s, determined to buy copies, not original works of art, a fairly common approach of American art museums in that period. There were good reasons. It could never compete with European museums. There were no great collectors to donate art. And besides, weren't copies really better? You could get good photographs of the best old masters. A plaster cast didn't show the discoloration an original sculpture did. By acquiring casts of art already in museums, the managers of the museum would be "prevented from squandering their funds upon the private fancies of would-be connoisseurs."[13]

An 1883 "Special Report on the Increase of Collections" laid out the museum's choices. The museum had been chartered to do three things: collect art, exhibit art, and afford instruction in the fine arts. (The corporate seal showed three intertwined rings representing Art, Industry, and Education.) Reproductions served those goals best, and that was where 70 percent of the acquisitions budget went, mostly for casts of sculpture. The goal of the museum was to "collect and exhibit the best obtainable works of genius and skill," and while the report acknowledged "individual opinions may differ . . . but the general consent of cultivated men furnishes us with a safe and sufficiently definite standard."[14]

But what cultivated men? And was there really a consensus? The board distrusted experts. Martin Brimmer, a wealthy civic leader who became the museum's first director, worried that curators would collect what was fashionable, not what was truly great: "Judgments of intrinsic merit . . . are nowhere infallible. They vary somewhat with individual tastes; they vary more with the shifting tendencies of the time. . . . What we need is a collection of permanent value, and in forming it, it will be well to avoid too strict an adherence to the theories . . . of the day." Curators would collect what they thought was good, not what the public needed: "The function of the managers of a museum is not criticism, but the collection of materials for the criticism of others. While they should carefully guard against the inroad of pictorial rubbish, they should bring to the increase of their stores a broad and tolerant spirit."[15]

This decision made good sense, considering the goals of the museum. It was interested in "collecting materials for the education of a nation in art not at making collections of objects of art. That must be done at a later stage of national development when we are willing to pay for them."[16] And indeed, when the museum had more money—and as museum work became more professional—its collecting began to change. In 1910 the museum moved to its current building, on the Fenway, and instituted a new acquisitions and exhibition policy. It began to focus on originals.

Benjamin Ives Gilman, curator and then secretary of the MFA, was the theoretician behind these changes. Gilman, who had studied psychology at Harvard with William James, taught aesthetics at Clark University, and published on a remarkable range of topics from Native American music to Italian Renaissance sculpture, was the MFA's in-house philosopher. His 1904 article "On the Distinctive Purpose of Museums of Art" insists that the purpose of art museums is aesthetic, not didactic. Art museums should acquire great art and show it without distractions, and that required connoisseurs.

Gilman's influence revealed itself as the debate over the purpose of the museum came to a head again in 1921, when the museum's board considered the purchase of a collection of engraved gemstones—art that was mostly of interest to specialists and hard to display. Some members of the board, including Morris Gray, the president, had proposed a new rule for acquisitions: "No very costly objects should be acquired unless they appeal to the public as well as to connoisseurs." Gilman objected strongly: "I really fear for our standards of excellence," he wrote in a letter to Gray, "should we attempt to follow it."

Gilman explained his reasoning, an unabashed defense of curatorial expertise. First, who can know what the public really thinks? And what about the future public? And more important: the rule would make it impossible to collect the finest art. "One may say broadly that it is the public that wastes the artistic inheritance of the world and the connoisseurs who rescue it." To adopt this rule would mean that "the Museum abdicates its function as an educator of public taste." The museum, Gilman insisted, must depend on "those best fitted to read and value an artistic message and determine whether it has been well delivered. These are the connoisseurs."

It's easy to read this as elitism, but Gilman is more interesting than that. He goes on to define a connoisseur: "The connoisseur is not the pedant." Gilman didn't want art historians picking his art. Curators need something deeper, a true appreciation. "The little girl at your side at the Metropolitan concert," he wrote to Gray, "was a connoisseur of that piece of music that so moved her." Gilman trusted aesthetic responses above all, and even a little girl might have that. But he had second thoughts, before he sent the letter, and took pen in hand to add a correction, a qualification: the little girl "was *within her limits* a connoisseur." For Gilman, curatorship—the selection of art—required both taste and training.[17]

The board of trustees met to resolve the issue. They decided it all in two votes:

> VOTED: To reaffirm the declaration of the Trustees in 1883 that the primary intention of this museum is to collect and exhibit the best obtainable works of genius and skill; and further,

> VOTED: That by the best obtainable works of genius and skill the Trustees understand the obtainable works most approved by persons exceptionally fitted to apprehend at once the intentions of genius and the achievements of skill.[18]

This marked a turning point for the museum, one reflected in American art museums generally. Curators were professionals. Museums were places where the public went to see what experts said was best. Gilman kept open the possibility that a "little girl" might be an expert, in her way, but she would learn those skills of aesthetic discernment from the art chosen by trained curators.

The role, place, and training of the curator began to change at natural history museums, too. In 1891, Sir William Henry Flower, director of the Natural History Museum in London, argued that a curator was a hard-working generalist:

> Now a curator of a museum, if he is fit for his duties, must be a man of very considerable education as well as natural ability. . . . His education, in fact, must be not dissimilar to that required for most of the learned professions. Skill, manual dexterity, and good taste are also most valuable. He must . . . possess various moral qualifications not found in every professional man—punctuality, habits of business, conciliatory manners, and, above all, indomitable and conscientious industry in the discharge of the small and somewhat monotonous routine duties which constitute so large a part of a curator's life.[19]

Francis Arthur Bather, "keeper" (curator) at the Natural History Museum a decade later, tried to combine the specialist and generalist in one person: "a man of enthusiasm, of ideas, of strictest honor, of sincerity, with the grip and devotion of a specialist, yet with the wisdom born of wide experience, with an eye for the most meticulous detail, but with a heart and mind responsive to all things of life, art, and nature."[20]

The 1910 American Association of Museums meeting considered the question. The speaker was Dr. Alja Crook, a curator at the Illinois State Museum of Natural History. Crook, who had a PhD in paleontology from the University of Munich and teaching experience at Wheaton College and Northwestern University, would become director of the museum a few years later. Crook was a progressive reformer, a believer in the museum as an instrument of civic pride and "enlightened entertainment" as well as research and learning.

Crook's scientific background and ambition, as well as his progressive belief in expertise, is evident in his talk to his colleagues in 1910. He was concerned that governors of states, presidents of universities, even museum trustees, lacked a "just appreciation of the preparation requisite for the highest type of museum man." If only they understood how much training a curator required, they would select better staff and support them better.

"In no profession," Crook declared, "is a greater variety of accomplishments and capabilities required." He proposed a set of thirty-five questions to ask potential curators looking to be hired, in three areas: their educational attainments, the nature of museum work, and the big picture of science. Crook wanted curators with a very broad range of skills. They needed to specialize in one field of science but have a "working knowledge" of others. (One of the questions asks for an outline of suitable labels "for *Amphelis cedorum,* Cedar Waxwing; for an army field writing desk used by General Grant during the Civil War; for a fossil plant; for a mineral.") Skill in "mechanical work, photography, taxidermy, and field work" is useful. Suitable candidates should know the history of museums and their present state. ("Name ten of the leading natural history museums of the world and state the essential character of each.") Crook also included practical questions (What should be considered in case construction? In the color scheme of rooms?) as well as administrative (Should a museum accept gifts subject to restrictions by donors?) and scientific ones. (Would you arrange a collection of fossils stratigraphically or zoologically?)

After the candidate had answered these questions correctly—and it seems clear that Crook thought there were correct answers—and wrote the 3,000-word summary of his views "as to the proper organization of a natural history museum as regards (a) personnel (b) care of collections and (c) exhibits," it was time for the hard questions. Could he handle a horse and canoe? Was he an expert proofreader, a good letter writer, something of a cataloger? Did he have artistic taste and sound judgment? A genial personality? Unquestioned character? Did he have good family connections, "able to meet in a proper way, to interest and to please persons of wealth and importance who may visit the museum or be inclined to lend their aid to its development"? How about executive ability, and a certain amount of diplomacy?[21]

Crook was making a point with his list of questions. He believed that curatorial work was a profession, not the work of dilettantes, collectors, or taxidermists who wandered into it. Collecting wasn't about taste or enthusiasm but about expertise. Even someone as knowledgeable and energetic as Jenks would not have made the grade: he had no advanced degree, no scientific specialty. A modern curator collected based on science. Yes, family connections and the right upbringing and a good personality were important, to talk to donors and patrons, but only after you had proved you had the knowledge and technical skills.

The audience at the meeting did not agree.

The minutes provide only a polite summary, but reading between the lines, it seems that Crook's comments caused an uproar: "The general opinion seemed to be that, while varied training and broad experience are essential in the administration of a museum, a successful curator must have pronounced natural gifts for this kind of work." Most important: "experience and personality." Many museum workers still agreed with Flowers's description, from twenty years earlier, of the curator as a generalist.

Frederic Lucas, curator-in-chief of the Brooklyn Museum, concurred with this majority opinion. He had come into the profession the old-fashioned way: he had no formal education but was a fine naturalist, collector, and taxidermist. He had been promoted from preparator to curator at the Smithsonian, and had trained himself in scientific research and writing.

"A curator is born and not made," he lectured Crook. "I do not believe you can train a man to be a curator. He is the result of the combination of natural ability and circumstances. He must be a man . . . who must know something of everything and everything of something. . . . It is not so much what a man knows, where he has been graduated, as what he can do; that is, what he can do to make the knowledge of others available and understandable by the public and his confreres."[22]

The Professional

Lucas's hands-on, up-through-the-ranks, learn-on-the-job training was not to be the wave of the future. Professional education was. In the 1920s, New York University and Harvard offered training for art museum curators. Harvard's museum studies program provides a good outline of the new ideal of museum curator. Paul Sachs, assistant director of Harvard's Fogg Art Museum, agreed with the MFA's Gilman that the museum world needed experts. "In every prosperous municipality in the land, in the next ten years," he wrote, "the call is likely to come for thoroughly equipped curators and directors. Harvard must maintain its leadership in this new profession, the dignity of which is as yet imperfectly understood."[23] Between 1922 and 1948 the Harvard Museum Course would train many of the leading American art museum curators and directors.

Sachs and Gilman were frankly elitist. Museums "must direct public taste . . . and not be dictated [to] by it," Sachs wrote. He wanted museums to remain "firmly in the control of a trained elite." This elite was the guardian of the canon, the art works of "the highest aesthetic power."[24] Sachs and Gilman disagreed about what and how the public should be taught—Sachs was interested in art history, Gilman in aesthetics—but both agreed that visitors came to the museum to learn from art chosen and arranged by experts.

Sachs taught connoisseurship—he would hand around art from his own collection for students to examine closely—but also much more. Future curators should know historical techniques, and so they participated in hands-on workshops on fresco, silverpoint, and the like. They used the Fogg as a laboratory to study the mechanical systems of museums. They gave talks to museum visitors. Sachs summed up his ideal of the curator: he should have "the passion of a collector, the preparation of an art historian, and the public service values and management practices of a modern manager." He emphasized both scholarly expertise and connoisseurship. A curator should be conversant with the literature of the field and publish both learned articles and catalogs. But he was also to take care of objects and to know them: a strong visual memory was essential. A curator must, "each day, study originals," as well as travel, collect, compare, and read.[25]

Sociologist Vera Zolberg, who studied the professionalization of curators, notes that most of the students in these new programs didn't look too different from the curators of what she called the "quasi-aristocratic, patrimonial ethos characteristic of the preprofessional phase" of museums. But this new breed of museum professionals had good training as well as a good family background. A 1956 report on the proper training of curators for the Harvard art museum reflected this balance: "The best curators began training for their profession by unconscious exposure in early childhood, yet it was generally the university which gave them intellectual depth."[26]

Curators also gained new clout through changes outside the walls of the museum. Art history became a university subject, making it easier for museums to argue for the importance of specialized scholarship. New tax laws allowed for deductions for charitable donations. That meant that the wealthy donor wasn't merely giving a gift but was also getting something in exchange—giving the curator new leverage.

Curatorship, especially the collecting part, would never become a science, or even one of the academic humanities. Curators are not, sociologist Zolberg argues, " 'pure' professionals." She makes this comparison: "Unlike university theologians, for example, whose authority is legitimated by intellectual competence, curators require more subjectively-based talents. They are more akin to clergymen, whose authority is partly legitimated by charisma."[27] Curators need both competence and charisma. Competence was originally connoisseurship, with research, scholarship and skills in education added on over the course of the twentieth century; charisma is both the ability to talk "to please persons of wealth and importance" that Crook demanded, and, more and more, the ability to work with the public.

To continue Zolberg's analogy, curators needed to work both with the hierarchy of the church and with their congregation, to understand the creed and to explain it.

Professional curators brought in better donations. Zolberg makes that case by looking at the print collections of the Art Institute of Chicago. Curated by trustees for its first fifty years, it built collections that were later described as doubtful, insignificant, and amateurish. Those acquired after the hiring of a professional, academically trained curator in 1938 were much better: the curator didn't merely accept what was offered, but evaluated potential gifts for quality and fit to the collection.[28]

In 1914, Benjamin Ives Gilman, in his presidential address to the American Association of Museums, looked forward to what he called the "Day of the Expert." Curators, he argued, not boards, should be in charge. Something like that had come to pass by the second half of the twentieth century—earlier in art and natural history museums, later in history museums. By 1950, the largest art museums were professional organizations. Francis Henry Taylor, the first director of the Met who had begun his career as a curator—and the first director at the Met to become a member of the board—had this to say about the relationship of trustees and curators: "Trustees are . . . gladly suffered as highly ornamental, and occasionally useful, sacred cows to be milked on sight. But God forbid that they should have ideas in art beyond their station, for if they ever really got the upper hand, then the public, whose cross-section they represent, might question the omniscience of the expert."[29]

Taylor had his tongue in cheek when he said this, but clearly the situation had changed. A few decades earlier Met curators had to stand when they made a presentation at trustee meetings. In Taylor's era they were invited to join the trustees for dinner.

In 1952 Remington Kellogg, director of the Smithsonian's National Museum (and a former curator of mammals) made the case for professional curators. "The essence of professionalism," he wrote, "is to be found in the strong sense of high purpose and personal responsibility and the strict intellectual integrity that motivate the individual and guide him in the use of his specialized knowledge. These qualities . . . mark the museum curator and are the measure of his stature. As a professional he is a stronghold of individual initiative and responsibility in a world threatened by the ant heap of collectivism."[30]

Kellogg's defense of the idealistic notion of curator as independent expert and moral authority was no doubt shadowed by his knowledge that in fact curators were part of a museum bureaucracy. It is not a long step from professional organizations—reliant on responsibility, integrity, and specialized knowledge—to bureaucratic ones. The Smithsonian was both, with rules and guidelines for the training and experience required for hiring curators. The government hired curators, for the Smithsonian and other museums, and so the U.S. Civil Service Commission had to create a position classification standard for the job.

The very model of a modern museum curator, according to the Civil Service Commission in 1962, included collecting, or more specifically, the "planning necessary to establish, develop and / or expand a collection." The collection should be "balanced," that is, "meaningful as a source of information for scholars and laymen." Like Crook in 1910, the commission acknowledged both the personal and the professional parts of the job: "In some cases efforts to add to collections involve public relations duties, including contacts with persons or organizations who can contribute items to the collections or arrange for such donations, gifts and bequests." The job might also include field work, "the actual physical collection of artifacts or specimens from the field." (Crook's expertise with horse and canoe might still be useful.) And finally, "when budgets include funds for the purchase of material for collections the responsibilities of the curator include such duties as locating and appraising the material, negotiating of purchase agreements and related activities." Connoisseurship and expertise remained important.[31]

The 1962 position description still controls the work of federal curators, and the collecting part of it at least seems a good description of curatorial work. In recent years, for many curators, public-facing work has become increasingly important. So too have the public relations duties, the negotiating, and the fundraising.

But the field continues to debate the skills necessary for curating. New professionals have taken on some parts of what was once the curator's job. Many of the items on Met curator William Goodyear's 1886 list of fifty responsibilities are now done by conservators, registrars, development officers, collections managers, educators, and others, and the balance of work is an ongoing topic of discussion. Another debate is over the degree of academic training for curators. Most curators at large art, anthropology, and natural history museums have a PhD in their field, but often no curatorial training. The training at history museums tends to be more diverse, and there's disagreement about the value of academic training. Some argue that the specialization for a narrow audience demanded by academic programs provides just the wrong training for a curator who will work with and for the public. Public history and museum studies programs provide technical skills essential to museum work. And many curators continue to be self-taught in their fields of expertise, as has been the case since the beginning of museums.

Being a good curator requires an unusual combination of training and personality: wide-ranging and often diverse expertise, both specific and general; a fascination with the details inside an understanding of the big picture; curiosity and a willingness to learn new things, in a wide range of ways; and, often, the inclination to work behind the scenes, building on, supporting, and linking the work of donors, experts, and the public.

Wisdom of the Crowd

Gilman, Sachs, and Taylor saw the curator as an expert, pushing back against the trustees, with their upper-class but untrained taste, and the public, with its lack of knowledge. They would have enjoyed Kellogg's line about the "ant heap of collectivism." But the easy elitism of the early twentieth century is no longer acceptable. One place to see the change is with today's fascination with crowdsourcing, with asking the public what they want, what they believe the museum should collect. Crowdsourced collecting opens up new questions about curatorial expertise and raises anew the philosophical questions of curating that Gilman and others wrestled with at the beginning of the twentieth century.

Today it seems that everyone is a curator, so why not just let the public choose what's important to save, to make the choices about what museums should collect? After all, they're the customers! And, in some ways, they're also the experts—either by living the life that the museum wants to document or because they have built their own collections. The internet makes voting easy and collecting of digital things exceedingly cheap. We have come to believe in the wisdom of the crowd. Wikipedia shows us that it's possible to create an encyclopedia by crowdsourcing; why not a museum collection?

Many museums have gone this route for exhibitions, but crowdsourced collecting is much more difficult. Will the public just pick the easy, popular, artifacts? Does the crowd have the wisdom—or, more precisely, the specialized knowledge of how artifacts shape and reflect material culture—to evaluate the needs of future curators, researchers, and museum-goers? Do we want market surveys to replace the curator?

There's a surprisingly long history to crowdsourcing. Most museums have depended on the public for donations and wanted to attract large audiences, so what's collected has depended, to some extent, on what the public wants to see in a museum. But at times museums have reached out to the public to ask for help. In 1848, even before he officially became the assistant secretary "in charge of Museum" at the Smithsonian, Spencer Baird issued a call for specimens: "The SMITHSONIAN INSTITUTION being desirous of procuring objects of Natural History for its museum would respectfully ask the assistance of the friends of science generally, and especially of officers of the Army and Navy. Any specimens of animals, plants, minerals, and fossils remains will be acceptable." Baird offered instructions to prepare specimens and arranged for the army to ship them to Washington. Baird believed in a democratic kind of science, open to all. Jenks was one of many enthusiastic amateur naturalists who responded to Baird's call for collecting help. (See Chapter 17.)[32]

In the twentieth century, as science and museums professionalized, public participation in collecting became less common, but some museums turned to it to encourage interest in art and science. In the 1910s, the Toledo and Milwaukee art museums asked school children to vote on what new pictures the museums should buy, with one picture each year selected by their vote—"the idea being that thus children have a greater interest in the museums and in art matters."[33]

The Metropolitan Museum of Art turned to a kind of crowdsourcing to deal with the contentious issue of contemporary art in the early 1950s.

First, the museum set up a small committee to bypass the board, which had no interest in contemporary art. The committee then decided among themselves not to buy based on their own interests but rather to reflect the art world in its purchases. "If a large segment of the community interested in 'Art' esteemed the works of an artist, and desired to see them . . . [the committee] would buy and exhibit them."[34]

The Smithsonian tried crowdsourced collecting in 1965, when a curator assigned to an exhibition on African American history realized that the collections in this area were almost nonexistent and appealed to the D.C. public to contribute artifacts for the display. This wasn't well received in the black community. An editorial in the *Afro-American* criticized the "off-handed fashion" of the request and urged that the Smithsonian seek the assistance of "outside experts" in the field. Crowdsourcing seemed disrespectful to the community and especially to the African American historians who were specialists in the field.[35]

When the National Museum of American History started to build collections of popular culture and everyday life in the 1970s, it tried again. The curator of electrical collections reported that "hundreds of individuals and companies responded to a plea publicized in the spring of 1972 asking for early electrical appliances. As a result a fine collection ranging from toasters and egg beaters to stoves and washing machines has been added to the National Collections."[36] A similar call for T-shirts resulted in the museum being overwhelmed with them. That may have dampened the enthusiasm for this kind of collecting.

Where crowdsourcing has been most successful has been with the collection of digital objects, rather than physical ones. The Smithsonian, for example, built significant collections of images and stories about the events of 9/11 and Hurricane Katrina by establishing interactive websites. German museums built collections on the fall of the Berlin Wall. The British Library crowdsourced audio collections of British soundscapes. The Victoria and Albert Museum collected images of clothes worn for weddings. The British Museum / BBC project, "The history of the world in 100 objects," included a digital element: individuals could upload to the BBC website images of personal objects they thought important in the history of the world, and they did, thousands of them, from "one of the only original posters" of Jimi Hendrix's first UK music festival to "the suitcase my mother was carrying when she was forced to leave Nazi Germany in 1939."[37]

Crowdsourcing physical collections is more difficult. It can bring in too many things if it's too easy, or too few things, if it's challenging, or the wrong things, if direction is unclear. Some museums have considered how to use crowdsourcing as a first step in a larger process, balancing popularity and expertise. The Smithsonian's National Museum of African American History and Culture's "Save Our African-American Treasures" helps individuals understand and preserve family treasures—and offers the possibility of finding artifacts that curators might like to acquire for the new museum. The National Museum of American History collected stories and artifacts for an agricultural history project by joining crowdsourcing with focused outreach to organizations and individuals, and with no promise of accepting gifts into the collection.

In 2012 the Canadian Museum of Science and Technology began a crowdsourcing campaign not to collect but to determine what to collect. "Memories Are Made in the Kitchen" used social media to complement curatorial expertise. Curators asked the public "to identify and document artefacts that best reflect domestic technologies used by Canadians today" to help them collect food preparation technologies that would reveal changing attitudes toward food, cooking, and technology.[38] Curators asked the public two questions: What is your favorite small electric kitchen appliance? and, What will this appliance tell your great-grand-children about Canadian lifestyle at the beginning of the twenty-first century?

Most of the answers they received answered only the first question—the second is the kind of question that curators wonder about and know how to answer, not the general public. The museum supplemented this with information from wedding registries and the popular press, and used this to determine what to collect. The answers the museum received reflected the upper middle class respondents: "high-end, well-designed and expensive appliances" that "minimized actual handling of food." (Most popular suggestion: Keurig coffeemaker.) The curator took the research and purchased eight brand-new kitchen appliances for the collection—so as "to preserve boxes, advertisements, warranties, purchase receipts, photos of how these appliances are displayed to customers, and on-line and physical appliance sources."[39]

Thomas Söderqvist, director of the Medical Museion at the University of Copenhagen, argues for a different kind of crowdsourcing—to an expert crowd. Noting that museums have neither the expertise nor the space to collect the material culture of modern biology, he proposes opening up curatorial decision-making to take advantage of expertise beyond the ranks of curators and to make the case for preservation more generally. Instead of seeing museums as a "closed repository for exquisite objects guarded by professional curators," he urges that museums be more open, inviting "every single researcher, technician, and student at the university to become adjunct curators of their own heritage." A "distributed curatorial expertise" would allow "crowd acquisitioning" a networked museum. Museum curators would set guidelines for the network of distributed curators.[40]

Crowdsourcing offers a new perspective on who decides, moving beyond taste and expertise to a new discussion of meaningfulness. Museums have been hesitant to go all in on crowdsourced collecting, but raising the question of who should make collecting decisions reminds us of the importance, and difficulty, of those decisions. After all, museums hold on to their collections for a very long time. The next part of this book explains how they do that.

❧ Preserve ❧

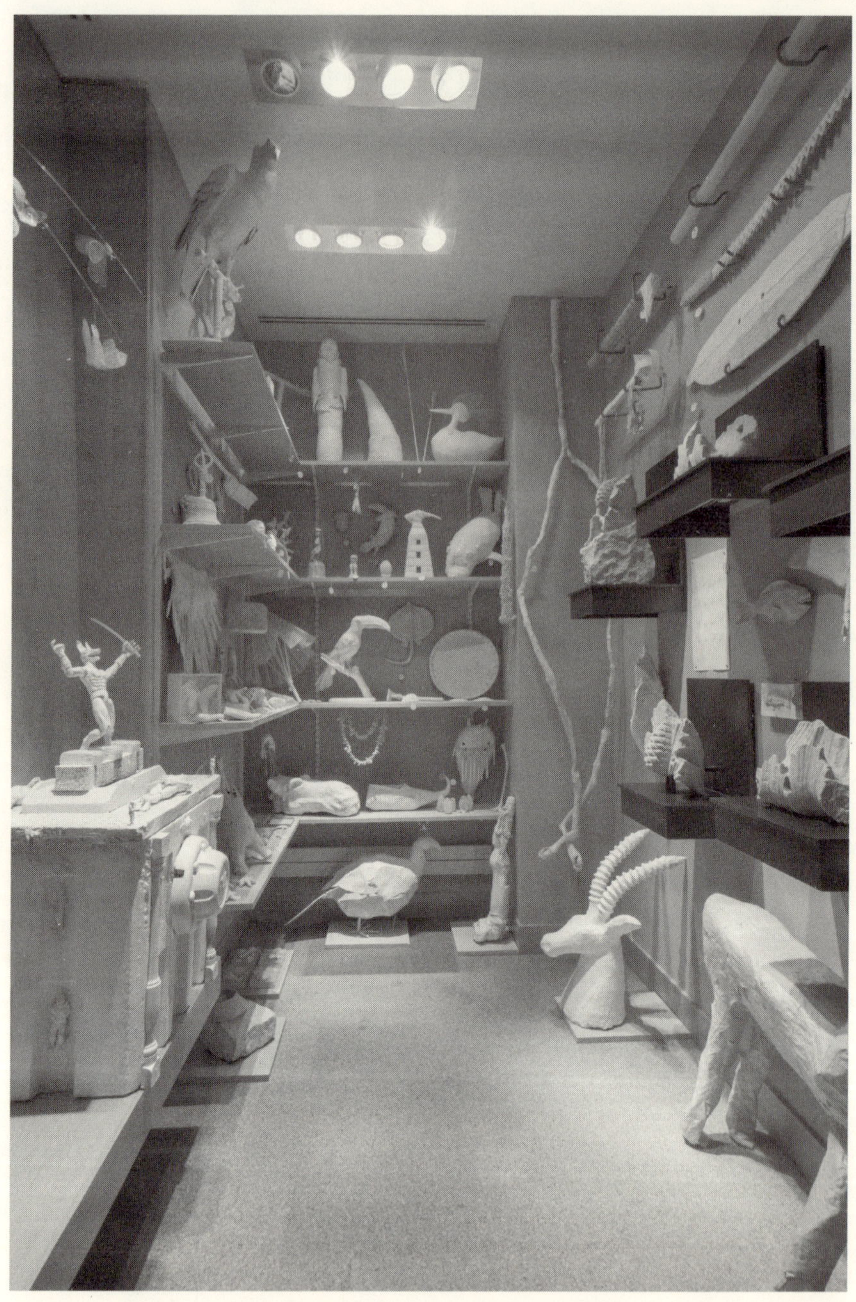

The Jenks Society for Lost Museums gave artists the names of the objects that had been in the Jenks Museum and asked them to reimagine the artifacts.

≽ 6 ≼

INTO THE STOREROOM

JENKS SET OFF ON HIS 1874 expedition to the "miasmatic swamps and everglades around Lake Okechobee in southern Florida," prepared for the collections he would make. "For hunting-dress outfit," he wrote,

> I was provided with a suit of sail-cloth, colored yellowish-brown or butternut, to resemble dead leaves, the sack-coat prepared with ten pockets, besides one, full size of the skirt, for large specimens, the pants with six pockets . . .

> For preserving and transporting specimens, I found a tin knapsack, constructed with various apartments for alcoholic vials, lunches, medicine-box and eggs, very convenient. At least ten gallons of alcohol and twenty pounds of arsenic were provided, besides some hundreds of muslin bags of different sizes, for keeping specimens distinct when thrown into one large jar.[1]

Jenks, with his pockets and boxes and bags for storage and his alcohol and arsenic for preservation, was ready to turn his collecting into a collection. Museums acquire things, and they need to take care of them. Museum collections need to be preserved, cataloged, and made accessible, available for use in the museum, by other curators and scholars, and by the public. They need to be honored for the nature and history and cultures they represent.

At the Jenks Museum, objects were transformed into a collection and the collection into an exhibition. We can sort out his processes of cataloging, storing, and displaying from what survives: a few hundred artifacts from the museum, tags attached; a few dozen more of the tags that have lost their artifacts; a few photographs; and two decades of annual reports.

Jenks, like many museum directors of his day, believed that visitors should be able to see the entire collection. He complained about having unopened crates in the basement. He worried about protecting the collections and must have been troubled by the large objects that he couldn't fit into cases. His museum was packed with display cases, and yet a constant in his annual reports is the need to acquire still more. Jenks's son Elisha must have inherited the concern: he became a museum case builder, selling exhibit cases with his patented locks to the Smithsonian and other museums.

The Jenks Museum had about 50,000 objects. American museums today contain more than one billion objects—some 820 million natural science specimens, 200 million archeological specimens, 48 million historic and ethnographic objects, and 21 million pieces of art. Not included in these numbers are libraries' 1.7 billion books, 1 billion microfilms, more than 700 million photographs, and an untold number of digital files. All in all, according a 2004 report, American cultural heritage institutions have responsibility for almost 5 billion collection items.[2]

Collecting is only the first step toward turning art, artifact, and specimen into museum objects. Before objects can be used, they must be stored, described, and organized so that they can be discovered when needed, and put to use. That's a big job and much of it remains undone. The 2004 report found that two-thirds of museums admitted to substandard storage conditions. Thirty percent of museums, according to the report, had not cataloged any of their artifacts and fewer than 10 percent of museums had cataloged all of them. Many didn't have plans in place for keeping track of what came in or clear guidelines for who's responsible for taking care of it.[3] Many museums have not come to terms with the legal and ethical issues that their collections raise, still working through the legal arrangements for dealing with collections from native peoples, or which might have been seized by Nazi authorities during World War II, or which might have been looted or excavated without proper authority.

But even artifacts without legal issues raise ethical concerns. Museums need to think about the ethics of holding artifacts: Who they are available to? What community can claim them or the stories told about them? Does owning artifacts demand a certain standard of care and accessibility? Organizational consultant Peter Walsh gives this advice to individuals who want to declutter their lives. It's only a collection, he writes, if it's displayed in a way that makes you proud, if it brings you pleasure, if you enjoy showing it to others, and if it's not an obsession that is damaging your relationships or getting in the way of living the life you wish you had. His advice is not perfectly transferable to museums, but the spirit is.[4]

Museums need to bring their collections under control in four ways: physical, intellectual, ethical, and administrative. That is, museums need to take care of what they have; they need to know where it is, what it is, and make it accessible to those who want to use it; they need to do so in a way that honors the objects and the communities they represent; and they need processes in place to make sure they do these things consistently, lawfully, and efficiently. This chapter considers how museums think about the objects they store. Chapter 7 looks at how they keep track of them and Chapter 8 at how they care for them.

Museum Sense and Object-Love

The place to start thinking about museum collections is in a collections storeroom. These are the backrooms of museums, mostly closed to the public, where artifacts sit on shelves, waiting patiently to be put to use. In large museums more than 95 percent of the collections are in storerooms, not on display. Some museum experts suggest that museums should devote at least as much space to storerooms as to exhibition.[5] Storage defines the modern museum.

Mention museum storerooms and scenes from Hollywood movies come to mind. There's mystery there: The Ark of the Covenant, locked away forever in a government warehouse in *Raiders of the Lost Ark*. Museum artifacts coming to life in *Night at the Museum*. Gore Vidal gave us science, seduction, and monsters in the basement storerooms in his novel *The Smithsonian Institution*. Linda Fairstein used the storage areas of the Metropolitan Museum of Art and the American Museum of Natural

History as settings for murder and mayhem in *The Bone Vault*. She offers foul smells and pickled creatures, "prehistoric crawlers . . . unidentifiable wet specimens that glowed against the darkness of the room," and "stacks and stacks of bones."[6]

Look online for stories of museum storerooms, and the words one finds are "secret" and "hidden." It's a visit "behind the scenes," with an implication: *They* don't want you to know about this. This is where the special things are, viewable by invitation only. Museums occasionally play up this aspect. The Boston Museum of Fine Arts (MFA) organized an exhibit in 1990 with a name that suggests mystery: "Unlocking the Hidden Museum: Riches from the Storerooms." The show didn't quite live up to its name, but it explained why some things don't make it out of the storeroom: too many of one kind of thing, poor quality, dirty or damaged, delicate or difficult to display, or just plain out of fashion.[7]

As the MFA exhibition's categories suggest, storerooms tend not to be that exciting—or rather, not exciting in the way that writers of museum fiction imagine. A modern storeroom might be rows of shiny white cabinets, sealed against light and insects. At a history museum, there might be a vista of open industrial shelving, stretching to the ceiling and into the distance, piled high with objects arranged roughly by size, or material, or type: furniture, machines, sealed crates. Textile collections look like closets; open the doors, and they look like extremely neat, well-organized closets. Technology collections resemble old factories, but with machines more tightly packed. Gun collections appear to be armories. Natural history museums have shelves lined with jars of animals swimming in alcohol, closed drawers containing skeletons and skins and dried plants, and freezers full of tissues gathered for DNA analysis. Art museums keep prints and drawings in metal cabinets with many drawers, and paintings packed tight on metal-mesh panels that can be pulled out on tracks. Anthropology collections present the diversity of the world's creativity at a glance, sometimes arranged by source location; you can vicariously visit the Amazon, or New Guinea. Herbaria and natural history museums often shelve by relationship in family-genus-species taxonomic order; you can vicariously recapitulate phylogeny. Many museums arrange collections more practically, by material, or size.

Storerooms are full of treasures, but they are backup treasures, treasures in waiting. Not secrets, as the movies and novels suggest, but rather the raw material for teaching and research, ready for use. More like the stacks of a library or the stockroom of a factory than Fort Knox or (that dreaded comparison) an attic.

Smithsonian photographer Chip Clark reveals the essence of these spaces in his photographs of the National Museum of Natural History's storerooms. A fisheye lens captures an enormous warehouse holding some of the anthropology collections. In the foreground are crates, pallets, an Easter Island stone figure, a totem pole, and dozens of smaller objects on tables. Behind them shelves reach at least thirty feet high, filled with pallets holding large artifacts. His view of the bird storeroom is even more dramatic: a long, long aisle lined on either side with cabinets, drawers pulled open to show a few hundred of the institution's 640,000 bird specimens. Other photographs show similar views: open drawers of butterflies, shells, fossils, fishes. In the spaces shown in these pictures, the museum superimposes a structure—an organizational scheme, a way of making sense—on the organic, disorganized, real world.

Or, more precisely, a way for the museum staff to make sense of it. Clark is careful to show museum staff in these rooms: curators and collections managers overwhelmed by the scale and scope of their collections, perhaps, but doing the essential work of knowing them, cataloging them, and caring for them.

Clark's photographs hint at the emotional connections between curators and collections staff and "their" objects. Curators like to visit their storerooms to show them off. One of the pleasures of curating a collection is taking a visitor behind the scenes to see an object not on public display. When I worked at the Smithsonian, I had the opportunity to show George Armstrong Custer's buckskin jacket to a military history buff and R2D2 to a Star Wars fan. The combination of behind-the-scenes access to an object unmediated by cases or labels allows a powerful emotional connection.

Storerooms are where curators, conservators, and researchers come to know their objects, to bond with them. The curator of fishes at the Field Museum in Chicago told a reporter that he didn't want to use the word "storage"—it sent the wrong signal. "We don't call it storage because we don't collect these things and just store them away. We think of it as 'collections facilities' because we and visiting scholars go into these jars constantly to study the specimens."[8] I changed the name of Brown University's Haffenreffer Museum of Anthropology's off-site storage from "Storage Facility" to "Collections Research Center." It not only sounds better; it's more accurate.

Museum staff connect with their objects in many ways, but in the storage room they connect with their physicality: what they look like, what they weigh, how they're constructed, how to move them, where they'll fit, what conditions will keep them happy. Storerooms highlight the materiality of objects, their heft and presence, a perfect counterbalance to the way that the registrar's files capture their history and exhibitions their meanings.

In the storeroom curators also connect with, and make sense of, their objects as groupings of things. A museum storeroom might be thought of as a kind of memory palace, an extension of the curator's brain. Things are organized, available, visible on shelves or ready to be discovered behind neatly labeled cabinet doors. Knowing the collection often means having a physical sense of the layout of the storeroom—and vice versa. I still remember the locations of many of the things in a storage room I oversaw more than a decade ago, and can remember the collections by walking through that room in my mind's eye.

For natural history museum curators, that physical layout can also be a guide to taxonomy. Louis Agassiz, founder of the Harvard Museum of Comparative Zoology (MCZ), trained his students by assigning them a section of a collections storeroom to study, memorize, make sense of, and, eventually, reorganize. At the MCZ, the storerooms were designed to reflect the order of the natural world. Historian Lukas Rieppel writes, "These cases, as well as the drawers within them, were arranged taxonomically, subdividing the natural world into a nested series of kingdoms, phyla, classes, orders, families, genera, and species. Thus, a naturalist could literally trace the history of life on earth by moving from drawer to drawer and case to case, watching the wisdom of God's divine plan unfold in the material objects before his very eyes."[9]

Curators pondering exhibitions wander their storerooms, absorbing information about the ways the collections look and feel that's impossible to get from a written description or a photograph. Artifacts are more complex, more interesting, than any description or image. Or perhaps more accurately, descriptions never quite capture what you really need to know. Curators need to browse.

[I remember touring storerooms at the National Museum of American History as a young curator working on an exhibition called "A Material World." The exhibit was intended to show something about changing materials in American history, and we were looking for objects that were particularly expressive of their materials. What chair might best show the peculiarities of woodness? What trinket might capture the essence of plastic? What teapot would reveal the purity of porcelainity? No one would ever catalog an object in this way, and it was only by looking at hundreds of artifacts in their storage areas that we could even begin to answer the question. Artist Mark Dion captured the inevitable failings of catalog description and the power of curators' knowledge in his 2005 *Center for the Study of Surrealism* project at the Manchester Museum when he asked curators for objects in their collections that "make you smile, laugh, shake your head in shock and condemnation, or gasp." Search for that in the database![10]] *use anecdotes from the story told by John Shapely*

Otherwise staid and practical curators slip into poetry when they describe what George Brown Goode, nineteenth-century director of the Smithsonian's National Museum, called "that special endowment . . . 'the museum sense.'" Philip S. Doughty, keeper of geology at the Ulster Museum, defined this further: "Hunches, intuition, . . . the apparent mystique is in reality a synthesis of a large mass of detail, the product of generations of talented geological curators who have developed, tested and refined skills and practices." Curators speak of "object-feel," or "a good eye." Nowhere does this better express itself than in the storeroom. [There, curators engage with artifacts directly, thinking through the problems of exhibitions or research in a material way.] The "Material World" exhibition required "the museum sense."[11]

Storerooms are places where curatorship, the "museum sense," exists in its purest form. They are places for what anthropologist Sharon Macdonald calls "object-love": a curator's passion for his or her collection, and for every object in it. Curators love their collections, and storerooms, note museum geographers Hilary Geoghegan and Alison Hess, are ["shaped by the emotion attached to the objects they house."] They allow for the affective, emotional, and sensory relationship between things and the people responsible for them.[12]

Inventing Storage

Most early museums did not have storerooms. Like the Jenks Museum, they displayed all of their collections. John Edward Gray, keeper of zoology at the British Museum, was one of the first to argue for a separate storage collection. His 1864 article, "On Museums, Their Use and Improvement," suggested that most of a museum's collections should not be on public view. A museum had two goals: to teach the public and to provide research materials for scientists. It's hard to do both of those things in one place. Putting everything on display meant overwhelming the public "with a mass of unintelligible objects" so they see "little else than a chaos of specimens." The "man of science" is not well served, either. Not only is it hard to study the mass of specimens on shelves, but it's also hard to carry on your science "uninterrupted by the ignorant curiosity of the ruder class of general visitors."

Gray suggested that most specimens should be stored in drawers, organized by genus and species. They would take up less room that way, be protected from light, and be easier to find. Unmounted skins and bones would be better for scientific work than the "stuffed skins or set-up skeletons" the public preferred. And with the mass of the collection stored away, there would be room to arrange the public material to tell stories. The curator would organize "a series of specimens, selected and arranged so as to present a special object for study; and thus any visitor, looking at a single case only, and taking the trouble to understand it, would carry away a distinct portion of knowledge."[13]

Gray's ideas were controversial. Why would you conceal collections from the public? Were the museum's curators hiding things, denying them to other scientists? Museums were measured by the size of their collections, and public museums felt an obligation to show everything to the public.

Over the next few decades, Gray's ideas won out. Birds, animals, and insects were displayed not as crowded rows of specimens, but to teach about nature, arranged to show the life cycle of the animals, the plants they fed on, their life stages, and more. In 1881, with a new building affording more space, the Smithsonian announced that "the collections in the National Museum are now being assorted and rearranged for the purpose of placing on exhibition a selected series of objects which shall be of interest to visitors, and of making the remainder serviceable for purposes of scientific and technological investigation"—what George Brown Goode, the museum's director, called the People's Museum and the Student's Museum.[14]

Art museums slowly adopted a similar scheme. Roger Fry, the new curator of paintings at the Met, suggested in 1906 (in a report marked "Strictly Confidential and Private") that not everything needed to be displayed. He urged "the temporary withdrawal of a considerable number of pictures." He warned the trustees, "I well know how dangerous and radical this counsel appears at first sight." But perhaps it wasn't necessary to show all of the second-rate paintings, or all of the paintings from second-rate painters.[15]

Some art museums adopted a variation on the idea. By 1912, the Museum of Fine Arts in Boston had separated its exhibition area in two, "one for the exhibition of selected objects in a way to promote their appreciation, the other for the installation of remaining objects in a way to facilitate their investigation." The "more beautiful things, those that have a civilizing value for every man," were placed on the upper floors, nicely displayed, for the general public. The bulk of the collections went into the basement, much more tightly packed, in what the museum called "storage exhibition." Both displays were "open to any one wishing to enter."[16] At the Philadelphia Museum of Art the reserve collection was arranged by material and technique. There would be "no dead storage of works of art," the director assured the public.[17]

By mid-century, the notion of a dual arrangement was standard for art and natural history museums. History museums followed suit: "Historical museums need changing displays and plenty of storage," the director of the Detroit Historical Museum told his colleagues. "The most important space in any museum is storage space. It should be much larger than the exhibit space. . . . Being able to draw on materials in storage makes possible timely exhibits, the covering of special subjects, of telling an interesting story. If your museum has to close some exhibit rooms to make storage rooms[,] do it, you will better serve the public."[18]

But increasingly, the reserve collection was not open to the public but only to "qualified researchers," scholars and others who received permission to visit, in the company of museum staff. It was not a second tier of exhibition space, where second-rate things were hung for all to visit, to see, dense-packed—a different, less narrative, view of art or history or science. At its best it was what the director of the U.S. National Museum called "a great reference library of material objects."[19] At its worst it could be, as the Philadelphia museum director had feared, "dead storage."

Unlocking the Door

Reserve collections were a challenge, and, often, a problem. Once collections were off view and hard to get to, would they be used? Were they really public? The question went to the heart of the way that museums thought about objects.

Let's visit some museum storerooms and take a look.

That's not always easy. There are locked doors, cameras, security guards at night. The Museum of Modern Art, like many museums, allows only "curators, scholars and auction house professionals" into its storeroom: "qualified researchers only." Visitors need to be accompanied by museum staff. Museum storerooms are, after all, full of valuable things unprotected by glass or railings. A curator at San Francisco's Asian Art Museum explains: "There are security issues. We can't have everybody knowing how the locks work and where the security cameras are."[20]

There are other reasons to keep visitors to a minimum. Some museums are embarrassed by their storerooms. Yes, they are full of treasures, but often treasures in the raw. Storage units are expensive, crates and cradles to protect them even more so, and there's always a backlog. Not everything in a storeroom is cataloged as is should be, and some is not cataloged at all: registrars use the phrase "found in collections" to describe art and artifact found in the storeroom but not in the museum's accession records. Most museums storerooms are full. Many are overfull.

Daniel Robbins, director of the Museum of the Rhode Island School of Design, described a visit to his museum's storage area in 1969. He titled his essay "Confessions of a Museum Director" for good reason:

We went to storage. Appalling, stuffed storage. Here paintings are hung floor to ceiling on racks, but no one can see them, can even pull the screens, because everywhere are paintings stacked against the walls, against the screens. Sandbags on the floor prevent the works of art from slipping. No one can be allowed in here, for there is no place to step, one can hardly turn around without endangering a precious object. Here, at the rear of the room, a dark cubicle lined with closets, is the print and drawing cabinet, together with the print library. . . . This treasury is inaccessible to students and the public alike.

Robbins continued in that vein. "From neglect," he lamented, "we have permitted the storage area of our museum to turn into Grandma's attic or, even worse, a junk shop." Why that neglect? In part, it was a problem of funding. As museum collections became less public, they saw less use, and it's easier to raise funds for public uses.[21]

Another reason that kept many out, even qualified researchers, was a question of control. In some museums, collections became the research fiefdom of the curator responsible for them. This is where "object-love" turns to "object-jealousy."

Consider the keeping of the key. In the days before ID cards with magnetic strips, many large museums used a key protocol. Each morning, the curator would visit the guard's office and ask for the keys to his or her storeroom. The guard would check them out, the curator would sign for them. Anyone who wanted to enter the collections storeroom during the day would have to ask the curator for admittance.

Woe be to the curator who didn't return the keys at the end of the day. I remember the first time I took storeroom keys home with me. A guard called, demanding I drive in immediately, so that he could make sure that everything was secure. Not returning the keys—not handing over the collections to the night guards—was a break with the natural order of things. Richard Fortey, for many years curator of trilobites at the Natural History Museum in London, tells a similar story of keys from 1970, a decade earlier, and suggests that key rituals were part of a long tradition. It's a lesson deeply impressed on many curators.[22]

Curatorial keys were not only about security. They showed who "owned" the collections. Collections, for the old-time curator, belonged to the curator, not the museum. The first rules for the Smithsonian, in 1882, spelled that out: "The Curators and Acting Curators will be held responsible for the preservation and proper use of all objects belonging to the departments under their charge." At the Met, curators had "entire charge of their respective Departments and are independent of each other." They were "responsible for the safekeeping and preservation of all art objects," and for keeping a "property book" listing the art in their department.[23]

The curator's ownership of "his" collections was a point of pride. The director of the U.S. National Museum argued the point in his 1952 *Annual Report:* "Each member of the curatorial staff of the National Museum has a professional responsibility to advance through research the knowledge of materials under his charge. . . . As a professional devoted to building up detailed mastery of his chosen field, the curator knows where and how to glean further knowledge from it."[24]

But the curator's ownership of a collection did not always benefit the museum or produce good science and scholarship. It could lead to fights between curators and put collections off-limits to other researchers. (In natural history museums, the desire to be the first to describe new species sometimes exacerbated this problem.) American Association of Museums director Laurence Coleman used a striking metaphor in 1939: in some institutions, each curator's office was "the nest of a sort of learned trapdoor spider." Julian Spalding, director of the Glasgow Museums in the 1990s, captured the symbolic dimensions of this control and its problems: "Currently curators are mini-directors. They sit over 'their' part of the collection like a hen over a brood and dictate what's to be done with it, where and how it's to be stored, how it's to be conserved and displayed, what's to be added to it, where it's to be lent and to whom even if—perhaps especially—they are colleagues in the same museum." It was against this background that Spalding demanded that curators change their ways: they "have to work for the museum as a whole." They must "give up their 'ownership' of their part of the collection. The collection doesn't belong to them; it belongs to the museum as a whole—and to the people who own the museum."[25]

Spalding's demand was part of a more general concern that museums were not making good use of their collections. Collections had grown, but in many museums, their usefulness had not. Storage all too often meant forgotten: out of sight, out of mind; reserved for curatorial, not public projects. As budgets tightened—or as museums began to respond to new demands on their budgets, new missions for which reserve collections weren't useful—museum funders began to ask questions. There was so much in storage that was never put to use. "The magnitude of the task of assimilating the millions of objects already at hand, and not satisfactorily recorded," wrote Brooke Hindle, director of the Smithsonian's National Museum of History and Technology, in 1978, "has no obvious solution."[26]

And so museums continued to obsess over storage: it was necessary, but always a problem. There was never enough space, and it was hard to explain why collections were necessary. Many storage areas did not meet museum environmental standards. And it was expensive, as much as two-thirds of museum operating expenses going to managing and caring for collections. One of the first systematic investigations of the cost of holding collections, in 1983, added up "accessioning, cataloguing, periodic inventory, maintaining accessible records, environment and pest control, storage hardware, security, conservation, insurance, and general overhead including management and building expense" to determine an average cost of $60 a year per object. Douglas Greenberg, the director of the Chicago Historical Society (CHS), offered a rare insight into the costs of storage in the 1990s. Maintaining some 122,500 square feet of storage space, all of it full, took almost two-thirds of the CHS budget. Greenberg reports the advice of a new trustee who worked in retail: "You know what your problem is? You have too much inventory." Greenberg agreed: "And that is precisely correct. In fact, our main business is inventory."[27]

Museums might be in the warehousing business, but the public doesn't know it, or appreciate it, or understand it. Surveys show that most museum visitors don't know that museum storage exists. A 1995 survey of visitors to Chicago's Field Museum of Natural History found that almost none knew that the museum held collections in storage for research work. Most thought that the only reason for storage was to hold things that wouldn't fit on display, and they guessed that there might be thousands of things in storage—not tens of millions. Three-quarters of those polled in the United Kingdom in 2003 didn't know that museums stored collections beyond what was on exhibit.[28]

Museums responded to the storage problem in two ways. On the one hand, they began to pay more attention to the objects in their storerooms. Curators began to explore what might be learned from looking closely at specimens and artifacts, reinvigorating the fields of organismic biology, material culture studies, and technical art history. Registrars took advantage of new information technology to rethink access to collections. Exhibit designers started to explore the possibility of visible storage. Educators rededicated themselves to teaching with objects. The internet offered new kinds of access. Museums also began to think about deaccession and disposal in a new way.

But museums also responded by doubling down on storage. If storage at the museum or nearby was too expensive, perhaps cheaper space could be found farther away. The question of storage went to the heart of the museum mission. Were objects for display, for the education of the general public, or were they for scholarly use? Or was the preservation of objects an end in itself, a form of documentation of the past? The changing answers to these questions shaped storage planning.

If objects were for use by students and scholars, a reserve collection made sense. It was easy to access. Objects might be tightly packed, but they were visible, labeled, and in some order. If the goal of the museum was preservation, pure and simple, it was better served by "deep storage," or that other, scarier, term: "dead storage." Objects might be in crates, stored far off site, organized by size or date shipped, not by subject matter. This was not a place for browsing but for protecting the objects of the past for the future. They could be retrieved, but at considerable expense and inconvenience.[29]

As objects moved away from curatorial control, "ownership" of collections moved from the curator to a collections management bureaucracy and to the museum as a whole. In part, this was a practical requirement. Off-site storage, increasingly common, required that staff other than curators have easy access. Collections managers need to check environmental conditions. Museum educators give tours to the general public.

Some curators mourned their loss of control over "their" collections. The "Campaign for Good Curatorship" in England argued that the specialized knowledge of the curator, combining both deep expertise as well as collections care, is essential to museums and under attack. Curators, they argue, know the collections, their context, and the academic subjects they pertain to. They know "what the strengths and weaknesses of the collections are . . . what is important and what is not so important, what should be kept, what should be disposed of and how the collection needs to be developed." The debate over object-feel, object-love, and object-jealousy continues.[30]

New digital technologies allow collections to find new uses even in storage. As collections move online, with excellent photographs, perhaps three-dimensional scans, and good descriptions available to all, they gain new, virtual lives. Other kinds of technology allow old collections to provide new answers to old questions and suggest new questions. This is most famously the case in biology, where DNA analysis allows collections that had seemed of narrow interest to become essential evidence for answering important questions about climate change and evolution.

New questions and new audiences make old collections valuable, too. Carla Sinopoli, curator of Asian archaeology and ethnology at the University of Michigan's museum, quotes a dean at the university as saying that there was no point in keeping collections after they had been studied—and then rebuts that assertion by giving many examples of the ways in which collections gathered for colonial projects, or in attempt to prove now-discredited theories of human cultural evolution, can serve very different ends. Indigenous peoples use collections to rediscover and revive long-lost traditions. Photographs taken for colonial purposes prove useful for local genealogies. Archaeological materials cast light on climate change.[31] As scholars moved beyond the famous and important and became more interested in the parts of society that didn't leave written records, and as historians and art historians become interested in issues of materiality and affect, collections become increasingly valuable.

But before you can answer those questions, before you can put the collections to use, you first need to find them and know what they are. Professor Jenks simply remembered what was in his collection. Today, computer systems help keep track of things. That's intellectual control, the subject of Chapter 7.

PAPERWORK

WHEN PROFESSOR JENKS RETURNED to the museum with a specimen, he would prepare it, either as a skeleton or as taxidermy. He would write a card for it, checking his field notes and the tag he had attached to the animal when he collected it. The card had "Museum of Brown University" preprinted at the top. Below that, Jenks would write the name of the animal—usually both scientific and common name—and his own name and a number. Cultural objects would be described in a few words on the tag, usually a name, place, and donor: "1537 / Water Bottle / Irebu, Africa / Donor, Rev. C. G. Hartsock, '89." Occasionally, there would be a bit more contextual information: "747 / Lamp for idol worship / discarded in the great / Telegu revival, Rev. D. Downie." Jenks would fasten the tag to the object with glue or string.

Next, Jenks transcribed the information about the object into a catalog book, sometimes adding more description. The catalog books of the museum were lost, most likely in a fire. We can guess, based on what has survived at other museums, what they looked like: a large volume, columns drawn on each page, one line per object, each line starting with a number, and then, in succeeding columns, the name of the object, the donor, the date, and the place collected. There were probably separate catalogs for different types of collections.

Jenks would refer to the catalog books to write his annual report for the university. A typical summary looked like this: "About 150 specimens of Anthropology from Irebu an interior town of Africa collected by Rev. C. G. Hartsock class of '89 were presented by his widow in his name. Accompa-

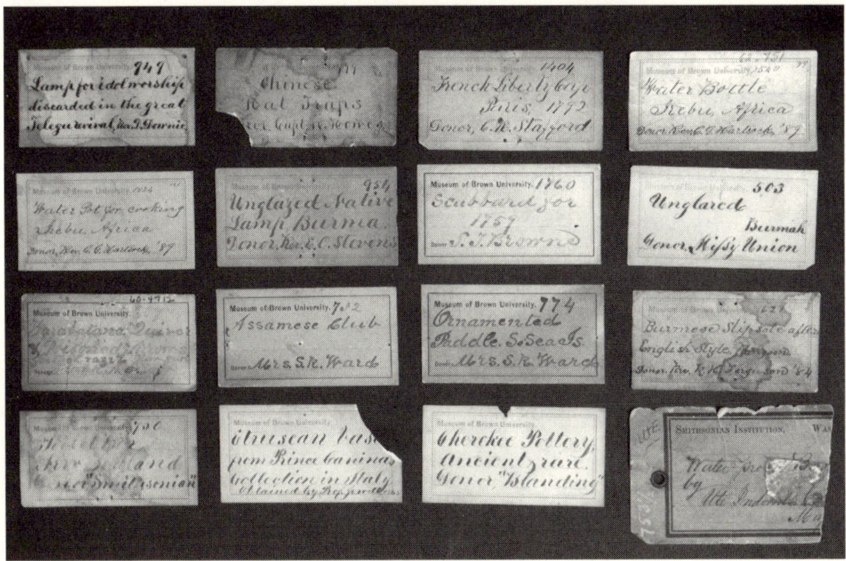

Object labels from the Jenks Museum.

nying this most valuable and unique gift illustrating in its variety the customs of the natives in respect to their every day life is a fine variety of Insects Reptiles and Osteological specimens for the Museum of Zoology."[1]

Jenks kept a lot of the information about his collection in his head and was happy to explain objects to visitors. That's bad policy. It doesn't work for a large museum or for a museum interested on reaching a large public. It didn't work for the Jenks Museum and was one of the reasons that the museum did not long outlive its creator.

Files that connect information about the object, past and present, to the object: that's the essence of the intellectual control necessary for museum collections. In 1895 George Brown Goode, first director of the U.S. National Museum, summed up the importance of good records in two aphorisms:

1. The value of a collection depends in the highest degree upon the accuracy and fullness of the records of the history of the objects which it contains.

2. A museum specimen without a history is practically without value and had much better be destroyed than preserved.[2]

Museums are, in a fundamental way, collections of information as well as collections of artifacts. Just the objects themselves tell us very little. Without information about them, they're interesting, but not fully useful. Some might serve aesthetic, social, or emotional purposes, but that's only part of the work of the museum. With information we can use them for teaching and research. We want to know when and where the object was from. If it is a plant or animal: How did it live? If cultural: Who made it and used it and how? What did it mean to them? We also want to know about its museum life. Who collected it? When, where, why, and how? How has it been cared for? How has it been used? The combination of the two parts of an artifact's life are key to preserving, understanding, and using collections.

Collections information systems record all of this. The earliest ones collected only the basics—name, location, donor, date; a single line in a record book. Modern ones record dozens, sometimes hundreds, of kinds of information; the computer cataloging system offers many screens, each blinking with fields to be filled in. Modern museum objects swim in a sea of metadata, information about the thing itself. They gain meaning from the words that we use to describe them. And the words help us find them, sort them, make sense of them, use them.

Many museum workers admit that their favorite book about museums is E. L. Konigsburg's 1967 *From the Mixed-Up Files of Mrs. Basil E. Frankweiler*. It's a children's book, the story of a brother and sister who run away from their home in Greenwich, Connecticut, to hide in the Metropolitan Museum of Art. Once they're settled in—curled up in a nice bed in the "French and English furniture" exhibit, washing in the fountain—they begin to investigate a newly acquired work of art, an "Angel," a possible Michelangelo. Their research leads them to Mrs. Frankweiler, the donor, and her files. From her they learn that documentation—in this case, "rows and rows of filing cabinets"—are the secret to understanding art, museums, and, perhaps, life. Mrs. Frankweiler's files hold not only the history of the object, but also the story of her collecting it and, by the end of the book, the story of the two children.[3]

For the children, the lesson is the layers of story in the world, the "palimpsests of meaning," as children's literature expert Virginia Zimmerman puts it in her analysis. Coming of age means becoming responsible curators not only for things, but also for one's self. For us, the lesson is the importance of files that tell the stories of objects. Mrs. Frankweiler's files are not really that crazy, or mixed up. Or more accurately, they're mixed up in a very particular, museum-like way. They combine and connect past, present, and future in a way that makes them useful.[4]

Orderliness and Organization

"The good curator," Albert Smith, former director of the National Museum of Natural History, told his colleagues in 1966, "is inevitably a victim of ataxophobia—fear of disorder. He has a mania for orderliness and organization."[5] Orderliness and organization means that both artifacts and information about artifacts can be tracked and found. From the handwritten accession books of Jenks's day to today's online databases, museum information systems have been one of the museum's greatest challenges. A 1762 description of the British Museum acknowledged the difficulty of cataloging. There were, it noted, three departments in the museum, "besides many Articles . . . which are not comprehended in any particular Department."[6]

In the eighteenth- and nineteenth-century museum, the curator was responsible for keeping track of all object information. Today, in all but the smallest museums, different staff members take responsibility for different aspects of that information.

The registrar is responsible for information about the status of the thing in the museum: proof of ownership, its movement from place to place, what's been done to it. The Code of Ethics of Museum Registrars calls these records "the cornerstone of the registrarial function" and requires that they be "meticulously complete, honest, orderly, retrievable, and current."[7]

The curator is responsible for the description of the object before it came to the museum: its history, references in the literature, historical images. He or she compiles this through a combination of connoisseurship, expertise, and new research informed by knowledge of art and artifact and historical, cultural, or scientific context. (This knowledge, anthropologist Hannah Turner points out, is informed by biases about what information is worth recording. She notes the lack of indigenous knowledge in many museum records.[8]) Oral or video history often plays a role for recent objects. The first Smithsonian rules, from 1882, describe the job: "The Curator, after receiving an accession lot [from the Registrar], shall, at his earliest convenience, and as a matter of urgent routine business (if possible the same day), fill up the accession card with the data necessary for the 'Descriptive List of Accessions,' and return it to the Registrar."[9]

Other museum staff add more information to the files. Educators write about ways to teach with the object. Conservators track the condition of objects, including detailed notes about work that they've done. Exhibition staff use all of this in planning exhibits, and add the research they do and the public life of the object to the mix. Photographers add images. In some museums, information about visitors' interactions with exhibits, in person and online, is included in the object records.

Each department of the museum provides a different kind of information, and each thinks about the collections and the information in different ways. In an ideal system the information generated by these diverse ways of viewing and understanding art or artifact gets added back to the data files and made available to all, across the museum, and to the public, either as information on individual objects or as part of a larger picture. Registration makes an object usable in the museum. Cataloging makes it understandable and useful. Objects have many lives and many meanings, and a good collections management system keeps track of them all. It serves as the museum's memory bank, providing advice that can help with making decisions about work across the institution.

Cataloging considers the collection as a whole. It's about systems. David Scott, a museum planning consultant working at the Smithsonian in the 1980s, turned to Arthur Koestler's 1945 essay "The Yogi and the Commissar" for a metaphor to explain the different ways that curators and registrars thought about their collections. The commissar—the registrar—wants system, order, even science. The yogi—the curator—wants to consider each case, each piece of art, on its own. The commissar believes that "all the pests of humanity . . . can and will be cured by Revolution, that is, by a radical reorganization of the system." The yogi wants to focus on the individual—in this case, on the individual object, not on the systems around them. Scott argued that curators and registrars cared equally for their collections but had incompatible value systems.[10]

The museum needs both approaches, and the collections management system connects the two. The registrar gives every object accessioned into the museum's collection a permanent number. In most museums today, that number includes the year, a number for each group of objects given by one donor at one time, assigned chronologically (called an accession number) and then a number for each object in that group. An object's number looks like this: 1994.0388.01. Old museums have other systems, sometimes organized by department, material, type, location, or even schools of art.

That number is written on the object. Tags are sewn onto textiles; numbers are written directly onto ceramics. Where should the object be inscribed so that the number is least visible in an exhibition yet easy to find? Registrars and collections managers know the answers.

The accession number is used to track the object and to connect all of the information about it. There's an accession file for each object. In a perfect world, the file holds the deed of gift for a donation or the receipt for a purchase, information on the way the object got to the museum, and the initial condition report prepared when the object arrived at the museum. Some museums add information discovered later, for example research and conservation reports. (Loans are treated similarly but are given loan numbers, not accession numbers.)

Each object has a name as well. The name should both describe the object and make it easy for others to find it. Starting in the 1970s, just when computers began to play a role in registrarial work, museums began to adopt a standard terminology—a controlled vocabulary to describe their objects—based on published nomenclatures. The Getty Research Institute publishes standard names and nationalities for artists (for example, "Titian" or "Tiziano Vecellio," or one of thirty-three other documented names; and "Italian" is preferred over "Venetian"); terms for art and architecture ("Catherine wheel" or "rose window"); cultural objects ("Mona Lisa" or "La Gioconda"); and geographic names ("Thebes" or "Diospolis" and which of eight Thebes?). The Getty vocabularies not only "provide authoritative information for catalogers, researchers, and data providers" but also, as collections move online and data is linked, offer "a powerful conduit for research and discovery for digital art history."[11]

The American Association for State and Local History supplies standard names and styles of description for historical collections. Its *Nomenclature for Museum Cataloging of Man-Made Objects,* first published in 1978, is built on three fundamental assumptions: catalog records are most useful if objects are named consistently; names are most useful if they're organized into functionally defined groupings in a hierarchical format; and consistently cataloged records facilitate the sharing of data.

Nomenclature furnishes not only standard ways of naming objects but also a hierarchy for those names. There are ten categories, from "Built Environment Objects" and "Furnishings" to "Recreational Objects" as well as, of course, "Unclassifiable Objects." A "Doll" is a "Toy" is a "Recreational Object." A search for "furniture" might bring up chairs, rocking chairs, and sofas; a "wedding dress" might show up not only as "Clothing" but also "Ceremonial Artifact." *Nomenclature* classifies by original function and provides both consistent cataloging and a degree of generalization. The system makes it easy to find things—at least for those familiar with the system.[12]

But an object's name is only the first step toward description. Also needed is information on who made it; where was it made and how, and of what; where was it used, by whom, and for what purpose; and, most generally, what it is about and what stories it tells.

Object cataloging is one of the underappreciated skills of the curator. Rhode Island School of Design Museum director Daniel Robbins described it in 1969: "Each object is obliged to carry its full set of associations, and a weird poetry results; the combination of pedantry and sentiment that can be read in the entries is the serial imagery of history."[13] Micah Walter, at the Cooper Hewitt Museum, gave object descriptions the name "curatorial poetry" and set up a Tumblr blog to post them as found art.

Here's an example of the brief descriptions that capture the essence of an artifact, taken at random from the Cooper Hewitt Tumblr. What is it? "Plate (Netherlands), 19th century. tin-glazed earthenware," puts it into museum categories: what, where, when, materials. A curator described it further: "Circular plate with wide rim; painted in underglaze blue on white, center with chinoiserie scene of pavilion in landscape, cavetto with diaper pattern, outer rim with dot & line border."[14] This is art-historical shorthand that would allow an expert to compare it to other similar plates, to fit it into the history of art, and to make sense of it, even without a picture. Description in art museums uses a technical vocabulary, capturing, in a few words, physical and aesthetic properties like color, size, decoration, and shape, as well as complex histories.

Description in natural history follows carefully prescribed rules: a name, details of collection, and the like. When a specimen comes into the museum a specific set of information is recorded. At Harvard's Museum of Comparative Zoology, the catalog had a column for each of the important facts about a specimen, to make certain that the curator recorded them. The Mammalogy Department started with numbers: the catalog number assigned by the museum and the number given by the collector. Next came information about the specimen: name, sex, locality, whether skin, skeleton, or preserved in alcohol. This was followed by information on the collector (who collected it, where, and who gave it to the museum, and when), and, finally, "Remarks." Most of this was recorded at the time the specimen was received, but the catalog also shows later changes, such as objects transferred to other museums or lost, changes in scientific names, and corrections.

Other kinds of museums collect different information. Geology collections demand information on stratigraphy. Archaeology collections need context, method of recovery, dates and dating method, and inscriptions. History and anthropology collections want detail about the use and users of artifacts, and, increasingly, about the way they were collected. Art museums want provenance: who owned this artwork, all the way back to the artist.

Museum systematization depends not only on descriptions but also systems of classification. Museums, like encyclopedias, were creations of the Enlightenment, and classification was essential to both. Otherwise, wrote an early nineteenth-century scientist, naturalists are "mere collectors of curiosities and superficial trifles . . . objects of ridicule rather than respect."[15] Classification, French philosopher Michel Foucault argues, both creates and limits discourses, and that was the effect of classification in the museum. Classification shapes perception and action; what museums find worth collecting, and what category they put it in, plays an important role in the way that artifacts are used in research and exhibition.

Cultural artifacts present a good example. In the nineteenth century, some museums arranged them according to race or by systems of "the stages of civilization." The connection was explicit in the Field Museum's description of its procedures: "Finally, the collection is entered alphabetically in a large volume, under the names of the collector, the locality, and the race. Thus one can find out at once from what races or parts of the earth the museum has collections; also what collectors, donors, or dealers are represented." Early museum anthropology cataloging was built on a system of categories that explicitly connected race to civilization, a system that shaped not only collections but also exhibitions.[16]

Classification matters. The category an object is assigned can shape the way it's used. Cataloging a machine as "technology" makes it less likely to be considered as a device that shaped labor. Cataloging furniture as "decorative arts" makes it less likely to be thought of as evidence for social history. Deciding whether ethnographic objects are art or ethnographic evidence determines what kind of museum collects them and what stories they tell. Classification can determine what storeroom an object is in, who discovers it in a database, what kinds of questions are asked of it. Museums need to classify, but they also need to constantly push back against too-simple categorization and to consider ways to overcome the hazards of any classification.

Cataloging History

The earliest surviving American museum records are from the Charleston Museum. They include a list of objects, by donor, presumably in the order in which things were donated—a ledger, not too different from what a storekeeper of the time would have kept:

> Presented 16th June 1798 by Dr. Jas Lynah—
> A Quiver with poisoned Arrows from Sierra Leone
> Part of a human thigh bone with oysters growing out of it
> Species of the Moss & King Crabs—from St. Domingo.[17]

Published catalogs followed the same general system. The 1821 *Catalogue of the Articles in the Museum of the East-India Marine Society, in Salem* lists most of the collections of the museum in the order in which they arrived. Shells were listed by genus and species, coins in chronological order—except for coins from exotic places like China, which seemed outside of history.

Systems like these would serve many museums for the next century, but as museums grew, and as curators realized the importance of collecting more information along with their objects, it became essential to build better systems. Lee Barker Walton, of the Kenyon College Museum, summed up the problem in a 1907 talk. He cursed the "systems commonly employed, even in prominent museums," which furnish only "a number, name, and locality of uncertain value . . . more or less heterogeneously arranged in cumbersome and often inaccessible volumes." He could have been talking about the Jenks Museum.[18]

The early twentieth century was a time of system, and museums participated eagerly in building cataloging systems. Museums were becoming more businesslike. "Around 1900," writes historian Lukas Rieppel, museums had evolved "into sprawling, multi-unit organizations that were run by a small army of salaried managers [and] employed sophisticated accounting techniques to keep track of specimens as well as employees." American museums were famous for their cataloging systems: Frances Bather, a curator at London's Natural History Museum, admired the "thoroughness of cataloguing that may be a lesson for most of us."[19]

The reports of the annual meetings of the American Association of Museums are full of presentations on the details of organizational schemes. Members debated catalog books versus index cards, and how big and what color index cards should be, and who should get copies of reports and in what order. George Brown Goode, first director of the U.S. National Museum, spelled out the importance of system in his 1895 *Principles of Museum Administration*. "Catalogues," he wrote, "are the keys to the treasure-vaults of a museum."[20]

Museum boards wanted the institutions to be run efficiently, like businesses, and museum staff responded with increasingly complex systems. The Metropolitan Museum of Art shared its methods with the museum community under a title explicitly connecting museum work to the commercial world: "Some Business Methods in the Metropolitan Museum of

Art." The tone of the piece is slightly defensive: "At just what point legitimate business methods become red tape has never been clearly defined, but doubtless a business house would place it where system ceases to show a profit. Surely neither such red tape nor slipshod unbusinesslike methods of administration should be tolerated in a museum any more than in a factory." The article outlines a system of paperwork that combined those used in libraries and express companies. There were forms designed to make the information easy to gather, and a system of blue-and-white cards that collected information about artwork from the registrar, the curator, and the museum photographer. The registrar, the curator, the printer, and the treasurer received copies of the cards, each filed in a way useful to them.[21]

An admirer of the even more complicated system at the Field Museum gave this explanation: "So complicated a system . . . demands much writing labour but its cost is amply repaid by the ultimate economy of time due to the excellent order." Not everyone felt that way. Bather, the British curator who appreciated American thoroughness, also worried about curators becoming drudges: "The outside world, it is to be feared, usually regards the museum-curator as but one of the strange dry specimens to be found within the walls of the museum. . . . Each of us has to guard against falling into a mechanical method of work, against reckoning results by the number of specimens mounted or catalogued, against sinking into the crevasses of specialization where the sympathies of humanity no longer reach us. The routine work of a museum, especially of a great museum, drives one in this direction."[22]

The "routine work of a museum"—the systems that supported the desire for "business methods," standardization, consistency, the challenges of cataloging complicated material in a way that allowed it to be found and, even more important, shared—seemed an ideal place to apply the power of computer systems. But early attempts to computerize museums were mostly unsuccessful. Dreams of order and access proved hard to capture in computer code. This failure highlighted the need for standard nomenclature and terminologies, though, and it may well be that the most important results of early attempts to move museum collections to the computer was attention to better cataloging and cataloging systems generally, not computerization. The attempts made clear that many cataloging schemes worked only because curators and registrars knew the collections so well.

Computerizing collections turned out to be harder and took longer than anyone thought it would. It has been two hundred years since the first American museums made simple listings of their objects, over a hundred years since business methods were applied to try to make museum collections easy to find, and about fifty years since computers were brought to bear on the problem. Each step has been a challenge, and an opportunity. What changes when we move from thick catalog books, or even catalog cards, to digital systems? What is gained, and what is lost?[23]

Why was computerization so difficult? The diversity of objects (compared to libraries, say) was clearly a problem. Museum objects are complex, varied, and hard to describe. Their individual histories are important. But perhaps more significant was that computerization was aimed at solving an internal problem—and, often, for people who didn't want it solved. The curators knew their collections, and some no doubt feared that computerization would lead to their losing control over them to the director's office or the registrar's office. (Object-jealousy at work.) More people with access to collections information meant less power, and perhaps more work, for curators.

The situation began to change when computerization moved beyond internal use. The internet—along with a great deal of museum politics—made collections information a public good, not simply a tool for internal control. The Library of Congress's American Memory project set an example with its mass digitization of archival materials in the early 1990s, raising private funds with the promise of wide distribution, first via CD-ROM, then on the internet. Museums followed, and by the turn of the century, museums were beginning to offer online access to their collections. The British Museum proudly announced its COMPASS project, with 1,400 objects online, in 2000. But it was slow going. A 2002 survey found that only 62 percent of museums had websites and very few had a searchable database of collections available online.

The potential of public use and a realization of the value of data in every part of museum work bear hope for bringing together data that had traditionally been siloed within the museum. There's much to be gained in a system that integrates information from all a museum's departments and makes it available throughout the museum, and (with some exceptions, like object locations and values) to the public. Seb Chan, former director of digital and emerging media at the Cooper Hewitt Museum, likens the levels of information at a museum to a pyramid. At the bottom are what he calls the two "sources of truth." One is the collection management system: "curatorial knowledge, provenance research, object labels and interpretation, public locations of objects in the galleries, and all the digitized media associated with objects, donors and people associated with the collection." The other is visitor data: names, visit history, and even giving history. Together, these are the data foundations on which the work of the museum stands. On top of them are built the tools that allow access to the museum's data for internal use and for public-facing applications both online and in the galleries.[24]

Finding Things

In his 2012 science fiction adventure story, *Mr. Penumbra's 24-Hour Bookstore,* Robin Sloan has his hero, Clay Jannon, track down the long-lost sixteenth-century Gerritszoon type punches at the center of the mysterious Unbroken Spine cult by bribing his way onto a seat at the mysterious Accession Table. He visits the California Museum of Knitting Arts and Embroidery Sciences and takes a seat before a bright blue computer terminal, the link to "an enormous database that tracks all artifacts in all museums, everywhere." Every museum in the world uses the Accession Table, we're told; it's "the Bloomberg terminal of antiquity."[25]

If only.

Museum collections are there to be used. The first step is cataloging, a kind of putting things away. The second step is finding what you want. One way to consider the challenges of finding things is to think about the ways things can get lost. There are many, many ways that museums can lose things. Looking at what can go wrong tells us what needs to go right.

There are many ways for artifacts to get lost, beginning with being miscataloged. There might be a simple mistake, a slip of the pen or keystroke. A cataloger might misidentify a specimen: a recent study estimated that more than half of the natural history specimens in the world's museums could be incorrectly named.[26] New research rewrites old stories: a curator moves a painting from "by" a famous painter to "school of," or a conservator uncovers a fake. A taxonomist moves a species to a different genus. Cataloging takes deep expertise, and time.

Things might be cataloged correctly, but in a way that's not useful. It was common in the eighteenth century to make lead models of important diamonds to use as placeholders when designing settings. The lead model of the Hope diamond was—not incorrectly, but not usefully—cataloged under "lead specimens" in Paris's Muséum National d'Histoire Naturelle mineral collections.[27]

More prosaically, museums find things in their storerooms that they don't know they have. Anna Ademek, a curator at the Canada Science and Technology Museum tells this story: "We found the oldest existing Ferranti transformer. . . . It was not catalogued at all, so we didn't know we had it." The transformer, the size of a small sofa, came in with a group of 7,000 objects. One day Ademak just noticed it—it "stopped me in my tracks."[28] The Natural History Museum of Denmark found seventy-seven barnacle specimens Charles Darwin had sent a Danish colleague as a gift when he returned a loan—but they had been stored with the loan. Not until a researcher read the correspondence were the barnacles "discovered."[29]

And then things can be just . . . misplaced. When, in the summer of 2015, the Boston Public Library could not find two lost etchings—a Dürer and a Rembrandt—among its 300,000 prints, the director, Amy Ryan, resigned. They were found, a day after her resignation, misfiled—but not until fourteen library workers had spent weeks searching through about 60 percent of the print collection.[30]

There are at least as many ways to find things as to lose them. One of the pleasures of being a curator is being able to wander the storeroom. You might have an idea of what you're looking for, where it is, but you're also open to suggestion. What catches your eye? It's an opportunity to be inspired by the collection, to find things when you don't really know what you're looking for. It's browsing rather than searching; but a browsing inspired by object-feel, that deep knowledge of how things look and how the storeroom is organized. It offers both an overview and the opportunity to explore in detail, distant looking and close looking superimposed. It presents a thorough description in one way—the thing itself—but lacks the detail of the information, the back story, that's on the catalog cards. There's nothing like wandering through the storeroom with an expert who knows it well.

But wandering the storeroom—browsing—only works for some kinds of search. Curators and registrars want to be able to search in other ways, too: by date, material, donor, or date of donation, say. They also need to be able to match up the information in the database with the objects; museums do occasional spot inventories, to see if what's in the database can be found, to check security and system.

The cataloging systems with their white and blue catalog cards, so avidly discussed at early twentieth-century museum meetings, were all about search: an alphabetical card file by donor, card files organized by category, a chronological listing by date of accession.

Online digital systems at first reproduced this way of accessing the collection. But before long, they changed the nature of search and suggested new ways to explore collections. In a modern collections information system, browsing can replace searching, images become more important, and it becomes possible to see connections between groups of artifacts, not artifacts as single, unrelated objects.

Most digital searches of museum collections starts with a search box: type in a word, and, if you're lucky, you'll find something related to it—if you use the same word as the registrar, if you're searching in the right fields. (Indeed, in the *Mr. Penumbra* fantasy of the Accession Table, Jannon is mystified by the possible categories he might explore: "Someone somewhere has categorized everything everywhere." Are the type punches Jannon is searching for filed under tools, or metal, or under political, religious, or ceremonial since they are at the center of a cult?) Mitchell Whitelaw of the University of Canberra describes the problem: "Decades of digitization have made a wealth of digital cultural material available online. Yet search—the dominant interface to these collections—is incapable of representing this abundance. Search is ungenerous: it withholds information, and demands a query." You don't really get a sense of what's there, what universe of data and metadata you're searching in.[31]

Search hides the collection.

One problem is words: the way the public describes art and artifact can be very different from the way curators and registrars describe things.

Some museums let the public add a new layer of description ("tags") for the museum's artifacts. That offers an alternative search, using popular descriptions, built on top of the museum's description. The curator might use an art-historical technical vocabulary to describe a painting's style; the user might be looking for pictures of dogs, or happy people, or yellow paintings, or even something affordable. The word "folksonomy" was coined to describe this vernacular taxonomy.

Computers can describe museum collections, too. (Jannon finds his type punches by cataloging them, and letting the computer find the right categories—and the museum where they are stored. Again: if only.) The most popular way to search the Cooper Hewitt website is by color. The computer analyses each photograph to determine up to five most prevalent colors, and you can search by clicking on color boxes. "Color," writes Aaron Cope, a data wrangler at the Cooper Hewitt, "is an intuitive, comfortable and friendly way to let people warm up to the breadth and depth of our collections."[32]

Expanding the search possibilities can make things better. Some modern collections interfaces will guide your exploration by suggesting categories to consider. (This replicates, in some ways, browsing the storeroom.) Some provide, within those categories or in the collection as a whole, a "faceted" search: you can limit the items the search returns by other categories, like date, material, or geography.

New kinds of online browsing and searching offer new possibilities for research, for understanding not only museum objects but collections as a whole. They can provoke new questions. Online digital systems can move beyond simple search and provide ways of visualizing collections. Using the Cooper Hewitt's color classifications, a programmer produced a chart of the collection's colors by decade. One can imagine more sophisticated connections, such as what objects were on display together at various times, what objects have never been on exhibit, what the museum has from particular countries, or time periods, or schools of art. Or online systems can be playful: The Cooper Hewitt allows users the opportunity to explore "our tallest, shortest, widest or narrowest objects."

Digitized collections also allow searching across museums. (That part of the database fantasy in *Mr. Penumbra's 24-Hour Bookstore* is starting to become true, for some collections.) The Digital Public Library of America serves as a portal to information on more than 13 million items from some 2,000 institutions. The Google Cultural Institute's Art Project lets users explore more than 45,000 works of art from hundreds of museum collections around the world, searchable not only by name and artist, but also by color, material, date, popularity, and category. Biology collections have gone the furthest with this. IDigBio offers a national search for 50 million specimens in hundreds of museums. This works best, of course, for objects that have a structured taxonomy, like plant and animal specimens.

Some museums have begun to make their collections data available not only for search but for deeper exploration. The Museum of Modern Art provides information from their museum collection system publicly as a spreadsheet with 125,000 rows and 15 columns: title, artist, date, dimensions, credit line, date acquired, and so on. Some museums allow access through a public Application Program Interface (API) that lets someone outside the museum explore the museum's collections database. APIs also facilitate inquiries to more than one museum at a time. An accessible API is the first step toward the museum goal of "linked open data"—of opening up museum databases for exploration and cross-connection.

The possibilities of search are exciting, and it's easy to get carried away with thinking of museum collections not as objects but as information about those objects. The twenty-first century is fascinated with metadata, sometimes forgetting the objects themselves. Physical things seem less important than information, physical control less worthy of investment than intellectual control. Digitization is all the rage, which raises the question: why not digitize and discard? The Smithsonian's Wilcomb Washburn proposed this in the 1980s. "When we are collecting objects we are collecting information," he wrote in a controversial article in *Museum News* in 1984. "What I am proposing is that the emphasis be put on the information, not the object."[33] Libraries at the time were microfilming books and discarding the originals, which turned out to be a bad idea. Fortunately, museums didn't follow suit.

Information is important, but it is worth stepping back to consider what is lost when we think of museum artifacts as just information, and a visit to the museum as the equivalent to an interaction with a web page. Objects offer us emotional, affective, physical, possibilities that information about them cannot capture. We react to them more deeply, on different levels, in different ways. We can explore them in unending depth. John Updike extends, in his 1967 short story "Museums and Women," an appreciation of the mystery of the real thing, in a real place, a new world awaiting our discovery. "In museums," he suggests, "we seek the untouched, the never-before-discovered; and it is their final unsearchability that leads us to hope, and return."[34]

It is the varied combinations and overlaps of object and information—the thing itself, curatorial object-love, and registrarial order; online search and wandering through the storeroom and the exhibition hall; finding what we know to look for, browsing to find new things, and enjoying Updike's "unsearchability" in communing with objects—that define the modern museum. Museum collections are marvelous on many levels: as objects and groups of objects, images and scans, information and metadata; and finally, as sources of knowledge and foundations for wisdom. Keeping museum objects safe and making them accessible are essential parts of museum work, and museum ethics. So too, as described in Chapter 8, is using them wisely.

THE ETHICS OF OBJECTS

THE JENKS MUSEUM DID NOT set a good example for the care of collections. Professor Jenks admitted to the "chaotic condition of the overcrowded cases." The collections weren't protected against disaster, and a 1906 fire caused extensive damage. A biology department memorandum supplies the sad details: "A fire, starting in the basement and almost immediately breaking out in the walls at the top of the building, scorched or burned much of the material in the Museum cases and in the attic."[1] The catalog was not very useful—Jenks had kept most of the information in his head—and it too was probably destroyed by the fire. When the museum no longer seemed useful to the university, the remaining objects were unceremoniously dumped.

In 1968 J. Walter Wilson—biology professor, author of the history of the Jenks Museum, and the administrator responsible for the final disposal of the museum's collections—wrote that a natural history museum is "maintained at the price of eternal vigilance."[2] Museums without that "eternal vigilance"—museums that neither care for their collections nor make them available—are not really museums, only temporary rest homes for misplaced objects.

Ethical control is not always on the list of the way that museums think about their collections, along with physical and intellectual control, but it should be. Museums must meet all of the ethical concerns of other nonprofits: transparency, accountability, efficiency. But because of their collections, they must also address additional concerns. The American Alliance of Museums (AAM) notes that the "distinctive character of museum

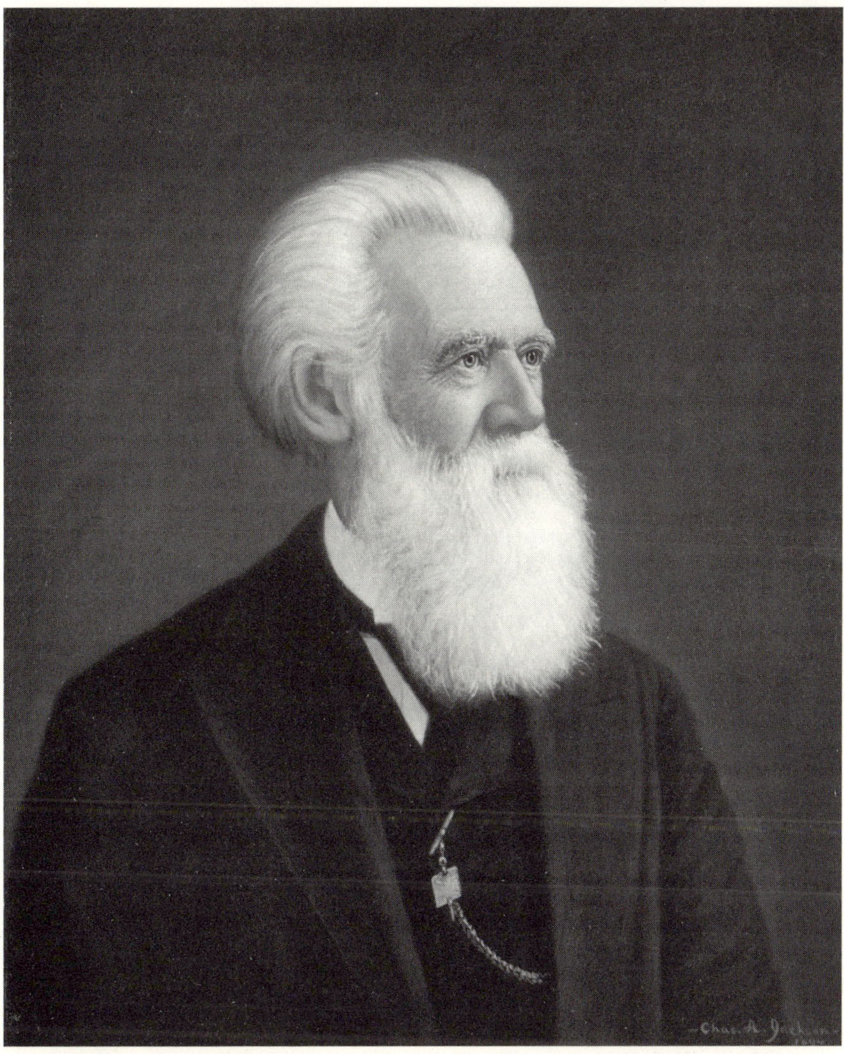

Professor Jenks, by Charles A. Jackson. Painted from a photograph and presented to Brown by Jenks's son, Elisha Tucker Jenks, in 1895. The nameplate reads, "Founder of These Museums."

ethics" is based on the "ownership, care, and use" of collections. When museums take on collections, they take on a weighty responsibility. The museum, the AAM says, must ensure that its collection supports its mission and the public good; that objects are lawfully held, cared for, documented, and accessible; and that human remains and funerary and sacred objects are treated appropriately.[3]

Museums promise to care for their objects and to use them. Both of these promises are more complicated than they sound, and they sometimes push in opposite directions. Using objects can mean using them up. Not using them can mean losing the means to care for them. Museums need to justify their existence, and objects need to earn their keep. Caring for objects appropriately may limit their use or even require giving them up or returning them to their rightful owners. Finding the proper balance isn't easy.

When museums acquire objects, they're making an implicit agreement to keep them. Not forever, quite, though that's the shorthand many museum people use. The decision is a weighty one, and thinking about it as a forever decision is wise. Michael Mares, former director of the Nebraska State Museum, writes, "It is the moral obligation of administrators to maintain their museums at whatever cost. . . . Museums are forever, and no one ever said that the road to 'Forever' would be easy."[4]

To be more precise, museums keep artifacts for as long as they are appropriate and useful to the museum's mission and meet ethical guidelines. When that's no longer the case, they dispose of artifacts in a way that honors the objects and the museum's commitment. While they possess them, museums need to do five things to keep their collections in an ethical way. They need to take good care of them. They need to keep track of them. They need to meet the obligations that artifacts bring with them, both those of the objects themselves and those of the communities from whence they came. They need to put them to use. And they need to have administrative systems in place to support all of these things.

Almost Forever

For an object to stay useful it must be cared for. It's not talked about much in public, but museum collections crises are not uncommon. Some are fast: overnight, an entire collection can be put at risk by a natural disaster like fire, an earthquake, a tornado, or a flood, or by a manmade disaster—a malfunctioning air conditioning system can result in mold that covers every organic object in the collection. A fire can destroy a storage area, and so can the water used to put out a fire. Museums can be casualties of war, or targets of terrorism. Some crises are slow: poor environmental conditions, lax security, insufficient preventive maintenance.

The AAM standard facility report, used when museums loan objects to other museums, includes several pages on storage, and it can be read as a nightmare of things that can go wrong. Is the space locked? alarmed? above ground? climate controlled? Is there a fire detection and suppression system? Is there a backup for the heating and cooling system? And on and on, a terrifying litany of potential disaster.

But it is also a litany, of course, of the ways and means to guard against disaster. Museum collections require constant vigilance, and every museum needs a disaster plan. These plans spell out the procedures to follow before and after a disaster: training, drills, who decides when to implement the plan, who's in charge.

Even without disasters, there are many ways that artifacts can go bad. It is the nature of material things to decay—every object in the collection is decaying, just at different rates—and museums must worry about slow decay as well as fast. Some objects, such as cellulose acetate film or acidic paper, decay by their very nature—inherent vice, it's called, in one of conservation's more poetic phrases. More common is decay that can be prevented or at least long delayed.

Preventive conservation—providing a good environment for museum objects, worrying about their handling, storage, and management—is the most efficient form of conservation, but also the least visible. It has a long but spotty history. Bavaria's 1826 Alte Pinakothek was built outside the city of Munich in the hopes that clean air would better preserve the paintings. In 1850, a select committee that included Michael Faraday, England's most important scientist, was appointed to investigate "the dirty and obscure state" of paintings in the National Gallery. The filthy air of London was bad for them, and the committee proposed that the paintings be enclosed in glass boxes. They also considered whether small children should be barred from the museum, to avoid "all the little accidents that happen with children." Ten years later, almost all of the paintings had been put under glass. (Small children, though, were still allowed.)[5]

Scientists began to explore the way that climate affected art in the late nineteenth century. Arthur Church, a British chemist, explained in 1872 how changes in temperature and humidity hurt art: "The canvases, frames, and panels become altered in shape and size each day," causing "a multitude of minute fissures, and the final flaking off of portions of the paint." Humidity was also fingered as the cause of both accelerated light damage and bronze disease (a form of corrosion of copper alloy objects). By the end of the nineteenth century, it was clear that lower relative humidity and temperature slowed decay, and that climate fluctuations were bad. Scientists recommended metal and glass enclosures.[6]

But it wasn't until after World War II that museums began to take environmental controls seriously. During World War I artifacts from British Museums were stored in railroad tunnels for safety, but were damaged by the high humidity. During World War II, the National Gallery's art was moved to a slate mine in Wales, with excellent conditions—only to suffer when it was moved back to the dry galleries.

The first rules for preventive conservation appear in a surprising place: a 1943 manual issued by the "Monuments Men," the curators responsible for rescuing so much art during World War II. "Notes on Safeguarding and Conserving Cultural Materials in the Field" was "a guide for giving first aid to buildings and monuments and to the contents of libraries, art museums, and scientific collections" jeopardized by "the operations of war." The first step was to take precautions "against immediate damage or disintegration." The next step was to "establish the best environment possible for the physical welfare of the monuments, etc., in question." The guide offered general rules that would become standard for museums.

1. Extremes of heat and cold are to be avoided.
2. Extremes of dryness or dampness in the air are both dangerous.
3. Changes in relative humidity are more dangerous still.
4. Strong light, especially strong sunlight, should always be avoided.

After warning against smoke, chemical fumes, and insects, the manual insisted, "Make frequent, periodical inspection of everything under your care, and check over not only the objects, but everything that may affect them."[7]

William George Constable, one of the authors of the manual and a curator at Boston's Museum of Fine Arts, created a version for civilian curators after the war. Conservation, he asserted, was one of the first responsibilities of the museum. "Unfortunately," he wrote, "that obligation has often been disregarded." "Possession or acquisition of an object," he continued, "carries with it the duty of keeping it in good order." Constable argued for "one of the most important recent advances in conservation . . . analogous to one made earlier in medicine—that it is more important to keep an object in good health than to restore it when it is sick." "Among the first aims of a curator," he wrote, "should be to secure a suitable physical environment for the objects with which he is concerned, not only in exhibition galleries but also in storages."[8]

Museums were slow to act. The 1968 *Belmont Report on America's Museums* worried that most museums were "presiding over the steady deterioration of that which they have been instituted to preserve." Twenty years later, *Museums for a New Century* voiced a similar anxiety: "We are concerned that collections care and maintenance are not a high enough priority within the museum community."[9]

In recent years most museums have taken Constable's message to heart. While there is still much to be done, the principle is accepted: museums must try their best to keep art and artifacts in good shape. There are ongoing debates over the details, such as how to balance the conservation needs of various kinds of materials, how to accommodate the temperature and humidity preferences of museum visitors while doing what's best for museum objects, and how to balance energy and environmental costs against preservation—but preventive conservation is now a central part of collections care. In the late twentieth century, many museums tried to keep a very tight control of temperature and humidity. More recently, standards have been relaxed as conservation science has shown that larger, slower fluctuations are not harmful to most objects. Today, the Smithsonian maintains relative humidity between 37 and 53 percent and temperature between 66° and 74° in exhibit spaces.[10]

Taking good care of a museum artifact is cheaper, simpler, and better than fixing it after it deteriorates. Keep the temperature and humidity within a close range, with no sudden fluctuations, and objects last longer. Paint doesn't flake, metal doesn't rust, wood doesn't crack, mold doesn't grow. Replace old wooden storage cabinets with metal ones and acid fumes don't damage objects. Shield the collection from ultraviolet light and inks don't fade. Filter dust from the air and objects don't need to be cleaned as often. Freeze or fumigate artifacts before they enter the museum and bugs won't enter with them.

Preservation is the least sexy kind of conservation. It's undramatic. It's not hands-on. But—analogous to preventive maintenance, or eating healthy food, or public health—it's the best way to take care of collections.

Even with the healthiest diet—the best environment and good care— some objects need to visit the conservator. Conservators are best known for the technical part of their job, in-painting an old master painting or piecing together a broken statue. As challenging, though, is what happens before that: deciding what should be done.

Barbara Appelbaum, a hands-on conservator and a theoretician of the field, argues that conservation must reflect meaning. That's complicated, because objects mean different things to different people over the course of their histories. They also have a range of meanings to the museum that holds them and to other stakeholders. Conservators must balance those meanings to determine an ideal state for the object. Only then do they choose a treatment. Their goal is to get as close as possible to the ideal state that best captures the history and use of the artifact. Along the way, of course, conservators document all of their work and make all of their actions reversible, so that as the object's meanings change, as new ideal states seem appropriate, and as new treatment technologies become available, future conservators can do their work.[11]

Conservators employ different conservation styles on different types of artifacts. A painting conservator will make only the smallest possible changes, trying hard not to let them show. Antiquities, on the other hand, are pieced back together to show clearly what is original and what isn't. Decorative arts are often more thoroughly restored than other art objects. Curators may want to run historic machinery, raising hard questions about the balance between preservation and research and educational use. There are national differences, too. In the 1970s, for example, conservators in England generally decided to keep darkened varnish on paintings. American conservators wanted to remove it. (Is the ideal state of the painting the way it looked when painted, or the way it looked for most of its history?)

Conservation practice changes over time. It was once common practice for antique dealers to refinish early American furniture. Today, that's anathema. (The *Antiques Roadshow* experts are always telling us that the rare desk would be worth millions—if only it had the original finish.) Historic automobiles, once only showable if restored to pristine condition, can now compete in the "preservation" class. There, as the *New York Times* describes it, they can maintain the "patina that tells a story of decades of service, their faded finishes, worn seats, stone chips and rust specks verifying their biographies" and be valued for their originality and historical significance, not for the quality of a restoration.[12] Museum conservators generally try hard to keep as much as possible of an artifact's patina and signs of use to preserve its history, its passage through time. French preservationists use an evocative phrase for this kind of preservation: *dans son jus,* or "in its juice." Museum conservation tries to make an object as true as it can be to its historical meaning while keeping it stable and usable.

Museums are both places of preservation and places of education and engagement. Sometimes, that puts objects at the center of a tug of war. Are objects happiest, as a conservator might argue, when they are kept in a dark room, at constant temperature and humidity, untouched? Or are they happiest, as a museum educator might argue, when they are being useful, when they're on display, teaching, inspiring, interacting with the public and with other objects? This is a false dichotomy. Objects, like the museum, can do more than one thing at a time. The challenge is to balance the two.

Respecting the Spirit

I had the opportunity to visit the National Museum of the American Indian (NMAI) storage building shortly after the museum became part of the Smithsonian in 1990. Room after dark room held floor-to-ceiling shelving packed with some 600,000 archaeological and 170,000 ethnographic artifacts. George Gustav Heye, who built the museum, had been a voracious collector. He wrote that his museum's "sole aim is to gather and to preserve . . . everything useful in illustrating and elucidating the anthropology of the aborigines of the Western Hemisphere."[13] Like many anthropologists of his day, Heye assumed that native cultures were dying and that the artifacts were all that would remain. The artifacts in the storage building were dusty, were tightly crowded on ancient shelves, and seemed to have been undisturbed for decades. The Bronx warehouse seemed the very definition of "dead storage."

But within that dead storage there was, in fact, new life. Clifford Trafzer, author of a book on the Smithsonian's reanimation of the museum, puts it this way: "The items are alive, just like Indian people are alive."[14] The vast collections of the NMAI are a case study of traditional museum collecting and collections—and of the way objects maintain their meaning even through museum captivity, and how they can be brought back to life.

It would take 177 people five years to inventory, clean, photograph, re-house, and move the collection to Washington, DC. Even while the move was underway and the new museum was being designed, some of the collection was being given new life in a new way. Native peoples were invited to reconnect with these objects, breathing new life into them, even in the museum storage: A native collections manager described praying before objects and fanning smoke from a small pot to carry the prayers to the sky.[15]

The NMAI designed the new museum's Cultural Resources Center as a place where the objects could be both museum objects and part of the outside world. It housed collections in a manner sensitive to both museum and tribal requirements for access and preservation. They have a dual life; preserved, but still available for their original use. W. Richard West Jr., the museum's founding director, wrote that "the Cultural Resources Center strikes a unique balance between museum curatorial practices and traditional Native care," providing "a nurturing environment that properly honors the wishes of Native peoples for the care and protection of the collection."[16] And so native peoples can come and feed corn to a wooden mask or smudge a sacred object with sage. They can use objects in tribal ceremonies.

Native American museums have led the way in allowing artifacts to live a dual life of museum object and living object, but the movement is spreading not only in anthropology museums but also in art and history museums. At the University of British Columbia's Multiversity Gallery, communities are invited to carry out ceremonies they believed important to the vitality of the objects on display. The museum's conservators write of the challenges the museum has in balancing "cultural processes and conservation practice" in this work. A Hindu priest wanted to reactivate a bronze image of the Hindu God Vishnu by conducting an *abishekam*, or ritual renewal. The ceremony involved bathing the piece "in a succession of auspicious liquids including water mixed with turmeric, milk, oil, honey, and finally fruit salad"—"elements that are contrary to the standard care of bronze material," the conservators dryly comment. The museum decided to go ahead anyway: "While it was very possible that the applied liquids would have had a detrimental effect on the metal, it was decided that the spiritual revitalization was an important part of the life of the figure and that it served to further the preservation of the spiritual and cultural integrity of the piece." After the ceremony, attended by members of the Vancouver Hindu community as well as the general public, the image was cleaned, dried, and returned to storage.[17]

Anthropologist James Clifford speaks of museums as "contact zones," spaces of encounter between scholars and communities, mediated by artifacts. "Objects currently in the great museums," he writes, "are travelers, crossers—some strongly 'diasporic' with powerful, still very meaningful, ties elsewhere." He urges us to rethink "collections and displays as unfinished historical processes of travel, of crossing and recrossing." Anthropologist Rosemary Joyce turns this into an ethical argument. Curators, she argues, should "think of objects as having itineraries." The part of their life in the museum, under curatorial control, is just one point in their long journey. That means taking care of them in culturally appropriately ways.[18]

When museums collect things, they take on ethical obligations not only to the communities those objects have come from, or are meaningful to, but also to the objects themselves. On the one hand, we want to honor an object's history. We need to research it well, to understand it. If an object had meaning to those associated with it, we want to remember that meaning. If it was sacred before it became part of the museum collection, we want to respect that sacred value. If the people who used it needed particular skills and knowledge, we should acknowledge that and record them as best we can as part of our acquisition of the artifact. An object's community connections, its context, needs to extend into the museum. On the other hand, we also want to honor the museum's claim to an object. It has a new purpose in its reincarnation in the museum. It represents its past in a material way. Its continued existence becomes its essence. A large part of its new job is to provide an opportunity for historical, cultural, or aesthetic insight.

One might think of this from the object's point of view. The object is torn from its home, from its surroundings. It's immersed in a new world that is physically perfect and environmentally controlled to keep the object from further change. The object is appreciated for its aesthetics or meaning, not, generally, for its usefulness or religious meaning. It is no longer doing what it was designed to do. But its new work does not come easily. The museum needs to help.

Sometimes, a community's claim—the artifact's sacred value—trumps the museum's values. Often the strongest argument is for returning things—either a legal argument, as with Native American sacred objects in the United States, or an ethical one, as with Maori *mokomokai* (preserved heads) or, perhaps, the Parthenon marbles now in the British Museum.

Sometimes, the museum's educational purposes are preeminent. Museums have made a strong case for the value of public ownership by asserting the value of public documentation and display and open scholarship. The best argument for keeping things in museums is this: that they serve the largest community.

Sometimes the community and museum's needs can be combined. Can we overcome the decontextualization of the artifacts by recontextualizing them, or by making new contexts? Can we follow the model of NMAI and share museum collections both with communities and with the general public? Museum scholar Christina Kreps offers a way to think about this. "Today," she writes, "curators and the curatorial process are defined not only by their relationship to visitors but also by their relationships to the people represented in their collections."[19] This might be expanded to respecting those people by respecting their objects and the full history of those objects. If we do that, we might reimagine the notion of object-love for a new era of museums.

Putting Collections to Use

Jenks, like other naturalists of his day, was responsible for the deaths of thousands of animals. Putting nature on display, he believed, by showing visitors the glories of God's creation, more than made up for that: "Whatever induces the young or old to turn their attention to the study of nature, is a gain to society at large, as substituting truth for fiction, and leading the mind to the contemplation of Him whose devising wisdom and sagacity are manifested in all His works."[20]

Today's museum scientists continue to worry about how to make the lives of the creatures killed for museums meaningful. Biologist Karen Haberman writes of the debt scientists owe to the specimens collected over the centuries. In a moving essay, "On the Significance of Small Dead Things," she asks, "So, what happens to all the dead bodies?" The fortunate ones "make their way to neatly organized collection. . . . They are properly arranged, labeled, and catalogued in state-of-the-art, temperature-controlled" museums. That, Haberman suggests, is a fitting memorialization of the lives of the creatures sacrificed in the name of science. They become "treasures for naturalists, systematists, and ecologists, who use them to address fundamental questions in systematics and biodiversity as well as current and crucial issues from public health to climate change."[21]

A similar ethical argument might be made for all of the things in museum collections. What can museums contribute in exchange for moving things from the world of the living to the museum storeroom, and for the significant expense of keeping them there? A 2015 Manifesto for Active History Museum Collections claims that too many artifacts in museum collections aren't pulling their weight—they are taking up room that could be used for better objects, and requiring funds that could be put to other uses—and asserts that "collections must either advance the mission or they must go."[22] Historian Sheila Brennan suggests taking artifacts out of storage, where they're slowly degrading to no end, and using them in programs. Let visitors interact with objects—the real thing, not replicas. Objects are tools for the work museums need to do—not an end in themselves. Get rid of what you don't need, she says, and "make the good stuff sing."[23]

To be worth saving, a museum's things need to be useful. They need to earn their keep. This is true of all the objects preserved in a storeroom, but it is especially true of objects that come with deep ethical quandaries attached. Some art is war booty. Many ethnographic and archaeological items were acquired under circumstances that today would be considered unethical, such as skeletons and burial goods pulled from graves and sacred objects purchased from individuals who did not have permission to sell them.

Perhaps we should think of the objects themselves as having moral rights. One might imagine a bill of rights for a museum object. An object needs to be provided a safe home: good environmental conditions, secure, discoverable. It deserves proper attribution and the promise of integrity. It should be allowed to be itself, to keep its connections with the community that created it. Perhaps some of the protections of the Visual Artists Rights Act should apply to all museum objects: they should be protected against mutilation, misattribution, or destruction.

Objects also need to be available, so that people might love them, or learn from them. Museums need to make sure that their collection is used and used appropriately; that access is allowed for those who need it; that things can be found, and that they are kept in good condition.

Museum artifacts require not just physical care and access, but also knowledgeable caretakers. As budgets shrink and museums focus on areas other than collections, curatorial expertise can be lost. Historian Allison Marsh worries that "shifting priorities, stretched budgets, and debates about the purpose of the museum have resulted in fewer curators and neglected collections." She raises concern about what she calls "orphan collections," those without a curator to care for them, to love them. They may be physically secure, but they are not fully available for use because there is no curator who appreciates, understands, promotes, and speaks for them.[24]

All too many specimens suffer what biologist Haberman calls the "ignominious fate" of not being useful. They end up in museums that don't take care of them, don't keep good records, or don't have staff that know about them. We owe our specimens the obligation of usefulness. Anything less, suggests Haberman, is "curatorial malpractice."[25]

The Rules

Administrative control is the final element of ethical control. Museums need written rules about how collections are handled and who's responsible for them. Staff needs to follow the rules, and management needs to make sure that the staff is following them. Since 1984 the American Association of Museums has required that accredited museums have a formal collections management policy, "a formal and appropriate program of documentation, care, and use of collections."[26]

A collections management policy lays down procedures and standards for collecting and collections. It outlines board and staff responsibilities, and specifies proper physical storage, management, and care for the collections and associated documentation. "Effective collections stewardship," notes the AAM, "ensures that the objects the museum owns, borrows, holds in its custody, and / or uses are available and accessible to present and future generations."[27]

The policy covers both practical matters and more general ones. An important piece is authority—who can make decisions about the collections. Curators recommend accessions to a collections committee composed of board members and others; the collections committee recommends accessions to the board of trustees, which has fiduciary responsibility for the museum as a whole and so must make the final approval. (In many museums, in most cases, this is pro forma, and the curator, or the collections committee, is given freedom for purchases below some value or size.) The board also must make final approval of deaccessions—and make any exceptions to the rules.

The policy also covers loans, incoming and outgoing, covering what kinds of organizations the museum can lend to, what conditions the borrower must guarantee, and what the museum agrees to when it brings in loaned objects. Deaccessions are often covered in detail: who approves them, how are they disposed of, and what restrictions are put on the use of funds raised from the sale of deaccessioned art and artifact.

The collection management policy covers details of collections care. The museum promises to have systems in place for documenting all pertinent information about an object and to maintain the integrity of the objects in the collection, to keep them secure.

Finally, the policy addresses staff policies relating to collections, including issues of conflict of interests. Are museum staff allowed to collect in the same fields as the museum or in their area of specialty? (If so, they often must offer the museum the first opportunity to make a purchase.) Most museums do not allow staff to deal in any collections area.

These administrative actions put ethics into practice. They provide the scaffold whereon the collections work of the museum meshes with the larger issues of governance, the control by the board that has fiduciary responsibility. Just as the storeroom provides a physical place for the collections, the collections management policy supplies an ethical and administrative framework for them.

Pruning

What happens when an item in a museum collection is no longer useful, when it's time to let it go? Museums acquire things aiming to keep them for perpetuity, but as missions change, or as better artifacts come along, it is the ethical duty of the museum to prune its collection. This is called deaccessioning. When a museum deaccessions an artifact, it removes it from the permanent collection and transfers, sells, or (in a few cases, like fakes or frauds) destroys it.

There's a long history of concern that museums waste space and money on collections that are not useful, and that something should be done about it. In 1939 the director of the AAM wrote, "There is hardly a museum without stuff that should be thrown away . . . and certain museums would be improved if half of their accumulations were shoveled onto the dump."[28]

In 1974 museum legal expert Stephen Weil followed up with a more measured suggestion for how the "stuff" might be put to use. He wrote that "no museum can afford today to clog its scarce storage with unconsidered collections that have simply been allowed to accumulate and lie fallow." The concern wasn't just with scarce storage space; it was also a question of money. "Over the years," he wrote, "the wealth of a museum tends to become concentrated in its collection. Unless some of that wealth is allowed to flow back into new programs, the museum may end with a first-class collection ill-cared for in a dilapidated (or even uninsurable) building by a second-class staff. The public interest might be better served by a more balanced approach." Weil argued that museums should use funds from deaccession for operating expenses: curatorial salaries, storage, securing, and building maintenance. Perhaps exhibits and publications, too: "these can give the museum vitality and attract the donors and purchase funds through which the collection will ultimately grow further."[29]

Weil thought better of this argument a few years later, and changed his mind. He wrote that what he had proposed would have been a "recipe for disaster." Museums would lose the confidence of the public if they deaccessioned too readily and put the funds into other museum activities.[30]

Weil's second thoughts—that funds received as a result of deaccessioning should be reinvested in the collection—became the standard policy for museums. Deaccessioning can be controversial, with the potential for bad publicity. Indeed, for many museums, deaccessioning is third rail of museum publicity.

It's not that deaccessioning—pruning and focusing the collection—is bad. In fact, it's good. It's a question of doing it in accord with best practices. To do that takes time and resources. Each artifact must be researched to determine what exactly it is, whether it is worth keeping, whether promises were made to its donor that it would be kept, and who would be upset by its sale. (It's good policy to inform donors, and sometimes the heirs of donors, but bad policy to return an object to them.) Some museum directors argue that deaccessions are not worth the trouble. Curatorial time and attention is scarcer than storage space, so some museums consider deaccessioning only when the financial rewards make it worthwhile, selling a few valuable things to fund new purchases. The National Park Service puts deaccessioning in perspective. Its *Museum Handbook* states that "the best deaccession policy is a good accession policy." Museums that collect wisely will have fewer worries about deaccessioning.[31]

Some examples can illustrate the possibilities of deaccession. When the Henry Ford Museum realized that many of the one million artifacts Henry Ford had collected were essentially duplicates it disposed of over 28,500 items between 2000 and 2002, selling them and using the funds to build collections in areas beyond those Ford had thought interesting. The Walker Art Museum sold off its nineteenth-century collections when it decided to focus on contemporary art. The Guggenheim sold works of art by Kandinsky, Chagall, and Modigliani to raise more than $30 million to acquire some 340 pieces from Giuseppe Panza di Biumo's collection of postwar art in the early 1990s—transforming the museum into a center for the art of the 1960s and 1970s and declaring that it was no longer focused solely on its earlier twentieth-century strengths. In 2005 the Columbus Museum of Art deaccessioned a Thomas Eakins to acquire the Philip and Suzanne Schiller Collection of American Social Commentary Art. It was a bold move, "the second most expensive purchase in its history" for "a group of relatively unknown artists whose compositions had difficult, and to many in the museum community, unappealing content." The museum spent five years building support for the plan, using the opportunity to expand its community support by reaching out to new groups.[32]

There are also examples of museums selling off things they wish they hadn't, or doing it in a way that they shouldn't have. The Boston Museum of Fine Arts dispersed most of its Native American art in the early twentieth century, a decision it would come to regret. (It was able to retrieve objects lent to other museums, almost a century later.) In the 1950s, in what art critic Lee Rosenbaum calls "one of the most infamous art-selling sprees in American museum history," the Minneapolis Institute of Arts sold off some 4,500 objects, including some of its Hudson River School paintings, because the director wanted to focus on what he thought was more important.[33]

A good deaccession frees up space in the storeroom, improves the quality of the collection overall, and refocuses the museum. In 2007 the Indianapolis Museum of Art set a good example when it systematically evaluated its collections, looking for objects that were inappropriate, duplicates, beyond repair, or did not meet "established thresholds for the quality and caliber of the IMA's collections today." The IMA did this publicly, posting on its website very clear guidelines for deaccession and the results of the evaluation.[34]

The IMA's deaccessions offer a fascinating case study of what seems museum-worthy. Curators considered each piece in the collection and decided whether it was appropriate and of sufficient quality, and decided to deaccession some 700 pieces. Many were sold at auction. A few were transferred to other institutions. Several dozen glass pieces were transferred to a university glass collection—they were "of good quality but many of the artists and movements are already represented in the IMA collection." Several Robert Indiana screen prints, labeled "redundancies," were sold to another art museum. A variety of decorative arts objects, "of modest quality and design not appropriate for an arts collection" but "suited to a historical collection," were transferred to historical sites or the state museum. Decorative arts and textiles considered "duplicates," or "secondary examples," or "of modest quality," were sold at auction. A few objects were considered more appropriate for use as exhibit props. A very few pieces of contemporary art were deemed to have been collected in error: "not mission relevant, poor quality, never shown, no historical value."[35]

Where controversy most often occurs is the question of what to do with the proceeds from the sale. The Association of Art Museum Directors sets a high standard for the use of funds raised by selling objects from collections, requiring its member museums to use funds obtained from the sale of artwork to acquire other art. The AAM has a more generous attitude toward the use of the funds: under its rules, proceeds from deaccessioning may be used for either further acquisitions or for direct care of the collection. There's vigorous debate on exactly what "direct care" includes—indeed, the phrase was left undefined by the 1993 Ethic Commission that coined it because it was impossible to reach agreement. A 2015 report acknowledged different rules for different types of museums and defined direct care as "an investment that enhances the life, usefulness or quality of a museum's collection." The AAM sums up its basic principle on deaccessions this way: "the museum is there to save the collection; the collection is not there to save the museum."[36]

Michael O'Hare, a public policy professor at the University of California, Berkeley, makes the opposite case—one strikingly similar to Weil's 1974 argument about deaccessions. A museum's refusal to consider the financial value of its collection, and to sell the collection to fund other parts of its mission, O'Hare points out, is not a legal argument but "an administrative decision elevated by little more than assertion to the level of a professional ethical principle. . . . It cannot be supported from a public policy perspective except as a source of comfort for curators and managers of the largest museums."

gues that the purpose of art museums is not to hold collec-
ther "more, better engagement of more people with art." "Our
istitutions are cheating us of our artistic patrimony every day,"
"and if they wanted to, they could stop." For example, large museum ave so much in storage; why not transfer some of those collections
to smaller museums, where they would be eagerly put on display? Even
more radical, why not sell some and use the proceeds to fund the work
that museums should be doing? If, as O'Hare calculates, the collection of
the Art Institute of Chicago is worth somewhere between $26 and $43
billion, selling off 1 percent would create an endowment for free admis-
sion. Sell off 1 percent more and you could fund 200 new employees to do
the work of the museum.

O'Hare suggests that we should value museum art and artifact in a
purely economic way, and that seems both too mercenary and potentially
fraught. Monetizing the collection makes it not a cultural asset but a
financial one, overturning generations of promises to donors and communities. Would donors support museums, either by donating art and
artifacts or cash, if the collections looked less like a cultural asset and
more like an economic one?[37]

O'Hare probably goes too far. But his proposal, like that of Stephen
Weil in the 1970s and the Manifesto for Active History Museum Collections, reminds us of the importance of balancing use and preservation,
short-term and long-term value. The rest of this book considers the ways
that museums use objects—in exhibitions, in Part 3 and in research,
teaching, community-building and more, in Part 4. The ethical considerations outlined here shape those uses.

III

❧ Display ❧

❧ 9 ❧

OBJECTS, STORIES, AND VISITORS

A HALF-DOZEN PHOTOGRAPHS of the Jenks Museum survive. They show exhibit cases filled to overflowing, chaos barely contained. Sharks, about a dozen of them, are lined up on top of the cases. Birds are grouped together, in cases; a small plaque in front of each specimen provides its scientific name. Ethnographic items are hung on a wall, apparently organized by type: spears together, knives together. Some of these have tags attached: a name, a place, sometimes a donor, rarely much more. A Japanese palanquin hangs from the ceiling, surrounded by large stuffed animals and skeletons. Many things are simply stuck where there's room: posters showing African wildlife hang on a wall behind a moose skull, next to a diagram of a cow. A camel faces a walrus across an aisle, a sea turtle above. Butterflies take flight above bird eggs.

The museum had no storage space, so everything it had was squeezed into the display. "For the last three years," Jenks wrote in his 1881 annual report, "the want of cases wherein to make a proper display of the contributions, has been an increasing trouble, till at last all classification is broken up, and the specimens in most of the cases are simply packed confusedly."[1] Jenks knew order was important, and when he had room, he worked hard to display things in a way that made sense to him. He wrote a few years later, "As Curator I have given full time to the rearrangement made possible by the completion of the new cases for Anthropology, and hope before the end of the year to have the greater part if not all of the tedious but necessary work completed, and the specimens placed in their

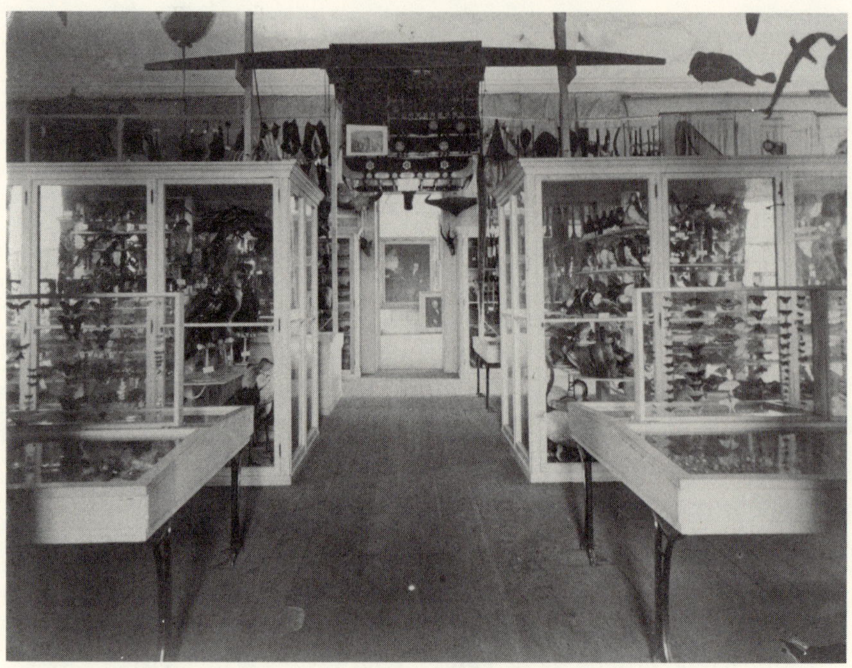

Butterflies, birds, and a Japanese palanquin in the Jenks Museum, about 1890.

respective cabinets according to the latest classification."[2] He died before he finished the job.

If he had to choose between collecting more things and a neat, well-ordered display, Jenks chose collecting. He would worry about orderliness later. He believed in showing off his things. Nature showed the glory of God, and so the more of it, the better. Jenks wanted more room, but he would have packed the new space just as tightly.

The photographs, though, provide only a partial vision of the museum. Consider the Jenks Museum not as seen in a photograph but as a space you might walk into. Start with smell: the museum was also a taxidermy workshop, and students remembered the "strength of the odor" and the "foul atmosphere" when Jenks was working on a specimen. Then there's touch: many of the specimens were in cases or out of reach, but students certainly would have touched what they could, and Jenks opened cases to show off his work. His taxidermy classes gave the students a very hands-on experience. Sounds, too, are missing from the photographs. The museum was a busy place, with students and visitors from off campus going in and out, classes on the first floor, and labs for physics, chemistry, and biology nearby, as well as Jenks's workshop. There would have been conversations, questions to the students, probably laughter, sometimes Jenks's booming voice.

There were few labels in the museum to explain the objects on display but it would be wrong to think of the museum as a place of things without explanations. One of Jenks's students remembered the experience of visiting:

> Professor Jenks could be seen at the museum at any season of the year except while he was abroad collecting specimens for it. Often when visitors entered the room and became interested in some object, suddenly there would arise a loud voice from across the room, and looking around they would see a venerable old man with a long beard as white as snow, having on his head a small cap ornamented with blue trimming. In a pleasant and interesting manner he would explain all about the specimen, where it existed and how it was secured for the museum.[3]

The still quiet of the photograph fools us. The exhibition was a place of sound, touch, and smell. It was, most important, a place where people interacted, as visitors and explainers. The museum was not simply things in cases. It awakened the senses. It was didactic, social, a place for teaching and learning, for connecting to things and people. Jenks knew the artifacts' stories, and shared them. Families visiting the museum talked about what they saw. Students came on class assignments, with questions to ask and answer. Imagine the museum full of people pointing, talking, touching, sniffing.

The Jenks Museum was fairly simple, as museums go, yet like all museums it was a place where visitors engaged with objects and ideas and each other to create something remarkable. Museum exhibitions are not just spaces for looking but also opportunities for engagement. The wonder of museums is how open-ended they are. Start with people in spaces with things, and you can go everywhere. This part of the book explores the ways that museums create exhibitions from artifacts and stories, how visitors use the exhibitions for knowledge and pleasure, and how those two activities intersect in the making of meaning.

More Stories than Specimens

Dallas Lore Sharp, one of Jenks's last students, remembered the old museum in an article in the *Atlantic Monthly* decades after it had closed. Jenks, Sharp declared, brought back from his collecting adventures "more stories than specimens." But he was a "museum-maker," and, Sharp lamented, "No museum has had a section, or even a glass case, for stories."[4]

Today, museums *are* places that tell stories. Exhibitions and programs—what museums call interpretation—are how museums tell stories, and story-telling has become increasingly important in the century since the Jenks Museum closed. The artifacts have stories, and the exhibit curator's job is to put those objects and their stories together in a way that tells a larger story. Or more precisely, in a way that allows visitors to learn, experience, engage with, and explore the museum's art, artifacts, and specimens, to see them, and the rest of the world, in a new way. Objects, stories, and visitors work together in a complex dance of showing, watching, listening, discussing, teaching, and learning. Curators, designers, and educators shape exhibitions, and exhibitions shape the ways that visitors engage with art and artifact.

The photographs of the Jenks Museum don't capture its stories. They don't capture the learning that took place there. They don't show that man with the long white beard and small blue-trimmed cap and "pleasant and interesting manner," the hands-on taxidermy workshops, the sensory understandings that the specimens conveyed, or the stories that visitors told each other as they wandered through the museum. They don't capture the narratives the visitors created and the meanings they made as they wandered through the museum, drawn from stuffed shark to exotic bird. Indeed, like so many photographs of exhibitions, they leave out the visitors altogether, the personal experiences and social interactions where museums do their work.

In his *Poetics of Music in the Form of Six Lessons* Igor Stravinsky argued for imagination and intuition in understanding and using the past. "A living illusion is more valuable . . . than a dead reality," he wrote. "The past slips from our grasp. It leaves us only scattered things. The bond that united them eludes us. Our imagination usually fills in the void."[5]

Stravinsky was writing about music, but his suggestion applies to museums as well. Exhibitions animate museum objects, retrieving them from their hibernation in drawers and cabinets, introducing them to a new audience, telling their stories. The museum assists our imagination by offering intellectual and physical contexts. Intellectual context is the system, order, and overall meaning: the big picture. Physical context is how objects are put on display. The art of exhibition is deciding how to combine these two types of context to create Stravinsky's "living illusions" and thus allow visitors to use their imagination.

A museum's mission, history, collections, expertise, stakeholders, and visitors define the range of stories it tells and how it tells them. What does its mission call for, and what do its collections allow? Who is the audience and what does that audience need or want? Precedent also shapes exhibitions. Curators and critics gasp when a museum tries something new: mixing art and decorative arts! adding videos or interactive touch screens! painting the walls white (before about 1930) or any color other than white (more recently)! Museums change slowly, bound by the weight of tradition as well as the weight of collections.

Within the range set by tradition, though, there's nuanced debate about what to show and how to show it. I have sat at many conference tables where groups of museum staff, outside consultants, community members, and board members add their ideas to a list of subjects and themes for exhibitions.

It's a big, crowded conference table. Curators want to show off their best objects and their most recent research. Educators might call attention to curriculum standards. Scientists and historians concern themselves with the big picture, how this particular story fits with larger themes, what they can say that's new. Designers think about the big picture, too, as well as how it all fits together and the details of display. Audience advocates think about how make the show appeal to visitors, worrying that it will be too dry, too complicated, too academic. Evaluators provide front-end and formative evaluation, reporting the results of audience surveys and tests of prototypes and mockups. Marketers contemplate what might attract an audience; community representatives worry about how to attract new audiences. Public program staff begins thinking about events. The development office wants to be sure that they can pitch the new exhibit to a donor, or that the donors already committed will approve of how their money is spent. Business development thinks about what products might be sold, and what other museums might be interested in a traveling version of the exhibit. Public affairs think about how to get more visitors. Government affairs people and lawyers fret about what might go wrong. Directors consider the museum as a whole and wonder how the exhibition advances the mission. Project managers plot resources and budgets. Collections managers and conservators start to think about preparing the art and artifacts to be displayed. Some museums ask each department to vote on each section of a proposed exhibit, to decide what to include. Some undertake surveys to gauge audience interest.

Who sits at that table is important. Some get a seat because of subject-matter expertise, or because they represent a community that has a relationship with the artifacts or the story being told. Others are there because they speak for or understand the museum's audience, or because they put up money for the project. Anthropologist James Clifford describes museums as "contact zones," places where meaning is cocreated by both curators and communities.[6] There are also many other groups, inside and outside the museum, eager for a seat at the exhibition development table, and their inclusion usually leads to better exhibitions and more audience engagement.

The process of deciding who decides, and how, takes up a lot of time and energy in museums. Museums have learned the hard way what can happen when the right people aren't at the table or don't feel listened to: not only controversy and a backlash that can hurt the museum, but also a story not fully told. For example, when the Metropolitan Museum of Art didn't include Harlem residents in the planning of its 1969 "Harlem on My Mind" exhibition and the exhibit was met with wide-ranging disapproval. "A white man's view of Harlem," a local black newspaper called it.[7] Vietnamese-Americans protested because they didn't feel their story was told in a 2004 exhibition on "California and the Vietnam Era" at the Oakland Museum. "They should have given us a chance to add another dimension to the story," a Vietnamese-American curatorial consultant told a newspaper reporter.[8] It's only very recently that indigenous communities have had a say in the way they were portrayed in museums, correcting an absence that had led many earlier exhibitions to tell misleading, inappropriate, and incomplete stories.

Other examples show the challenges of engagement. In 1994 the National Air and Space Museum found itself in political difficulties because Air Force veterans didn't feel they were listened to in a planned exhibit on the atomic bombing of Hiroshima and Nagasaki. The result was a political crisis for the entire Smithsonian, and a revised exhibition that was more technological than historical. But there were other communities that were as involved in the story—Japanese survivors of the atomic bombing, for example—that found themselves cut out, without the political clout to shape the story.[9] The challenge is not only getting the right people at the table but also taking the time to listen and respond to them. Exhibition development requires conversation and, often, compromise. It's not easy.

Consider a local historical society deciding on its next exhibit. First, they must decide who sits at the table: do they represent themselves, everyone who feels a connection to the historical society, or everyone in town? Where should they draw the lines around "local"? Next, they must consider what will draw a crowd, what historians think is important about the town's history, and what story will be interesting, or useful, to which groups in the town. They need to look at what their museum can do that other organizations can't, what they have funding for, and whether the exhibit might bring new collections to the museum. They should also consider what the existing audience would like to see and how the exhibit might allow the museum to expand its audience.

Then there are the philosophical questions. Limit the stories told to those supported by the collections, or tell an important story in ways that push beyond what the collections allow? Tell a traditional story, or a new one? One that captures what makes the town unique, or one that shows it as part of a larger regional or national story? The answer to all of these questions will depend on the first item: who gets to decide.

For a very different, but overlapping, set of concerns, consider the decisions that the Whitney Museum of American Art made in its 2015 "America is Hard to See." The exhibit, an overview of the museum's collections, inaugurated its new building. Holland Cotter, reviewing the exhibit for the *New York Times*, noted the importance of who made the decisions: "it's the institutional thinking, inherently political, that matters—determining first what and who goes into the collection, then what goes into the public spaces and lastly what new, alternative eyes will be coming on board to oversee these things."[10] The museum needed to weigh curatorial judgments based on aesthetics and art history against educators' expertise about the way that the audience understands and appreciates art. It needed to consider its various audiences: experts, American and international tourists, locals, students, and more. The museum curators, educators, designers, and board balanced these considerations to create a display that has something for almost everyone—except for those who wished the museum had presented a more focused exhibition, rather than an overview.

Once museums determine which stories to tell they consider the best way to present them. The historical society might use period rooms, artifacts in cases, photographs, labels, or hands-on activities. It might focus on a few case studies or try to cover an entire history. It might focus on biography, or on buildings, or on tools and equipment. Perhaps there could be a costumed interpreter, pretending to visit from the past.

The new Whitney faced a different set of challenges, such as how it should split the exhibit space between its deep historical collections and the contemporary work for which it was now best known. Should it present the works in chronological order or thematically and if chronologically, what and how many chapters would be appropriate? The museum might try to show the full diversity of American art or focus on what seemed—either at the time or in retrospect—most important. Do the collections do a good job of representing the history the museum now thinks important, or should the museum acquire more works for a show this ambitious? How much explanation will the art need?

Every museum faces similar issues. The Whitney, the Museum of Modern Art, and the new Met Brauer may show similar art, but each displays it differently to tell different stories. The same object might be shown in an art, history, or anthropology museum to make different points. Objects can be shown to enhance their beauty, to increase their accessibility, or to revel in their mystery. They can be shown with related objects, encouraging visitors to think about a moment in time, or with those that provide a contrast, calling attention to change over time and place. Labels can provide aesthetic, historical, or systematic information. (No labels at all send a different message.) Interactives can provide new perspectives or more information, connect to other objects, or show details. The physical context—a quiet room or an immersive setting, a timeline or a treasure show—might encourage contemplation, active learning, or a call to action. Different audiences appreciate different exhibit techniques.

Museum educator Margaret A. Lindauer suggests that there are four basic types of exhibition. *Laissez-faire* exhibitions invite visitors to explore, attracted to whatever interests them. They may or may not follow the path the curator had in mind, but that's up to them; they're there to enjoy themselves, to learn what they want to learn.[11] The opposite are *narrative* exhibits, which tell a story. Some museums put footprints on the floor, to show visitors the path to follow. A third type of exhibit, which Lindauer calls *authoritative,* is designed to teach. The curator has something he or she thinks is important for the visitor to learn and arranges the artifacts and contexts to make that point. Success is judged by how well the visitor acquires the knowledge. Finally, constructivist exhibits give visitors a chance to solve problems, to learn how to think about issues. Constructivist exhibits are places for dialog, a chance for visitors to develop new ideas or to compare perspectives on a problem.

Each of these exhibition types has its place, and part of the exhibition developer's job is to consider the exhibition's goals and audiences, weigh them against the artifacts, expertise, and funds available, and make choices. Exhibition development is an art, not a science, and no exhibition will work for every visitor.

Visitors Making Meaning

Museums are more than just places with things: they are places with things *and people,* that is, social spaces. Exhibitions are created for an audience, and museums need to know their audience. The American Alliance of Museums claims 850 million visits to American museums in 2015, and is eager to note that that's more than the attendance at all major league sporting events and theme parks combined. Two-thirds of Americans visit a museum each year. It's a diverse group, but not as diverse as it should be: it's older, whiter, richer, and (the best predictor) better-educated than average. The percentage of the non-Hispanic white population who visited an art museum in a given year is roughly twice that of the African

American or Hispanic population. Museums are not places where everyone is welcomed, or feels welcome.[12]

Museum visitors are also diverse in their range of interests and learning styles. There are many ways to categorize these. One useful approach is a Smithsonian study that finds four types of things attract visitor interest in exhibitions: *ideas* (concepts, abstractions, facts, and reasons); *people* (human connection, stories, and social interactions); *objects* (aesthetics, craftsmanship, and visual language); and *physical sensations* (movement, touch, sound, taste, light, and smell). Smithsonian researcher Andrew Pekarik calls this the IPOP theory of experience preference. It helps explain what visitors notice and what they do in exhibits, and also how they judge the quality of their experience in the exhibit. Understanding IPOP preferences and interests reminds exhibition creators of the audience's diversity and that visitors come to museums not as blank slates to be written on by curators. Visitors participate in a range of experiences—intellectual, artifactual, social, and experiential—in their own distinctive ways.[13]

Exhibition makers and visitors cocreate meaning. The exhibition developer chooses the objects, decides the story to be told, and designs the presentation that he or she believes gives visitors the best opportunity for new understanding. Visitors choose how they experience the exhibition, making choices based on what interests them, what attracts their attention. They piece together a story that makes sense to them. An exhibition is best considered not as a place for teaching but as a place that allows for a range of engagement.

How best to do that? Psychologist Mihaly Csikszentmihalyi advances a useful theory of visitor engagement in museums. He uses the word "flow" to describe circumstances where people are engaged. People in flow tend to feel that they are both fully expressing the self and that they are connected to others. The flow experience depends on clear goals, appropriate rules, and immediate and unambiguous feedback; the challenges must match the skills of the individual. Csikszentmihalyi suggests that museum exhibits can create the flow experience by capturing visitors' curiosity, addressing their interests, making links with their lives, and proposing alternative perspectives. "When the visitor is interested in an exhibit and engaged through sensory, intellectual, and emotional faculties," he writes, "he or she should be ready to experience an intrinsically rewarding, optimal experience."[14] Museum experts talk about free-choice learning and curiosity-driven visitors; both terms suggest the importance of letting visitors explore at their own speed, choosing their own path, to make the connections to objects and stories that offer meaning and fit their interests and ways of learning.

Organizing objects in a way that encourages flow is the museum's challenge. Exhibitions have the advantage, over other media, that visitors do literally flow through them. They are free to move through exhibition at a speed, and on a path, that keeps them interested. Exhibitions turn the motion of the visitor into narrative. History museum exhibition developer Richard Rabinowitz offers a view of an exhibition as a stage on which visitors perform, on which they cocreate meaning. "The museum curator," he writes, "becomes a theater director operating in two time frames at once. The contents of an exhibit case are transformed into an animated field of action. To interpret is to imagine one cast of historical actors stepping out of the document, and another set of modern-day visitors coming across it. Historical time and exhibit time flow together."[15]

A good museum exhibit starts where the visitors are. It builds on their interests and knowledge to let them engage with the past, or art, or science—and with each other. The National Park Service's *Interp Guide*, the Bible for park rangers, describes their work this way: they use "the history of an individual visitor's experiences, in conjunction with their contemporary experiences with heritage resources, to provoke visitors to consider a broad range of ideas and perspectives about the resources, and to arrive at their own conclusions about them."[16] This builds on the work of Freeman Tilden, whose 1957 *Interpreting Our Heritage* still shapes modern practice. Tilden recommends this general principle: "Interpretation that does not somehow relate what is being displayed or described to something within the personality or experience of the visitor will be sterile."[17]

Put the right things next to each other, in the right order, offer visitors the right kinds of opportunities to interact with them, provide thought-provoking advice and information and direction, and a rewarding, pleasant place to wander, and you move beyond making an argument. You allow for thoughtful conversation, for significant engagement, for the making of meaning.

This doesn't mean that visitors will follow your arrangement, or that the ideas in your head will be magically transferred to the heads of the visitors. Jay Rounds, a professor of museum studies, defends what he calls the meandering visitor. Visitors, he writes, summarizing many museum surveys, "meander about the museum, sampling randomly here and there, ignoring most of the exhibits, choosing in a seemingly haphazard manner those to which they do attend carefully. . . . They fail to use the exhibitions in the thorough and systematic way that should reward them with the greatest educational benefits." While some museum professionals see this as (in Rounds's words) a "source of despair," he argues that it is in fact a smart way for the visitor to see the museum. Visitors are there to pique and satisfy their curiosity. They seek "wide but shallow learning," not to memorize facts or imbibe a curator's narrative. They pick and choose what they want to see. Museum visitors are free-range learners.[18]

Learning from Visitors

Museum exhibitions entertain, educate, and engage. Visitors seek opportunities for learning, enjoyment, reflection, and inspiration, as well as a pleasant place for a social experience. When both the museum's and the visitors' goals are met, the exhibition is a success.

But exhibition is not an easy medium. Almost every discussion of exhibitions laments the difficulty of teaching through the display of objects.

Museums can be exhausting. Henry James, the American writer who was fascinated by the world of museums, describes a weary museum goer in *The American:* "He had performed great physical feats which left him less jaded than his tranquil stroll through the Louvre. . . . His attention had been strained and his eyes dazzled, and he had sat down with an aesthetic headache."[19] Charlotte Brontë, in *Villette,* laments that a museum visit resulted in "a combined pressure of physical lassitude and entire mental incapacity."[20] Benjamin Ives Gilman of Boston's Museum of Fine Arts coined the term "museum fatigue" in 1916 to describe the result of the "inordinate amount of physical effort" demanded by the way museums display things. Museum fatigue, Gilman postulates, causes the visitor to quickly lose interest and "resign himself to seeing practically everything imperfectly and by a passing glance."[21]

It's not just that museums tire visitors out. They're also difficult places for learning. Consider the challenges: trying to concentrate while standing up; having a transcendent, contemplative experience in a crowd; learning from an experience designed for an average visitor. Nelson Goodman, professor of philosophy at Harvard, gave a damning description of the museum as a place of learning at the American Association of Museums 1983 convention. "The circumstances for viewing in a museum are at best abnormal and adverse," he told the assembled professionals. The museum was a "place hostile to the achievement of its own main purpose."[22]

A Smithsonian overview of audience surveys found that these critics are right. Exhibitions seldom meet their goals: "One of the most striking results of this generation-worth of museum audience studies is that the explicit aims of exhibition planners are rarely achieved to any significant degree. In study after study . . . researchers found that the central goals of the exhibition team (which are usually learning goals) were rarely met for more than half of the visitors."[23]

Visitor studies suggest what works and how museums can do better. Henry Hugh Higgins, honorary curator of invertebrates at the World Museum and the first president of the British Museums Association, was also the first to undertake visitor evaluation. Some two thousand people per day visited the museum, and no one had thought to ask them about their experience. He wrote in 1884 that "a series of observations on the constituents of this irregular procession of visitors, combined with overtures suitable for inducing them to make remarks on the objects exhibited . . . might lead to much valuable information."[24]

And so he watched visitors in exhibits and asked them what they thought of the objects they saw. Only about 1 or 2 percent of the visitors were "students," there with a particular learning goal in mind. Observers, "who fix their attention with more or less intelligence on the objects displayed," made up about 78 percent of the visitors. The other 20 percent, including most children, were "loungers." It was easy to see this last group, especially the children, as "unmitigated plagues," but he noted that they occasionally noticed "a bird or a butterfly or a fossil with desultory but by no means vacant looks." Higgins watched his visitors, talked to them, and came away impressed with the enormous variety of ways that they used the museum.

Since Higgins's time, many researchers have tracked visitors and used that information to divide them into groups based on behavior. There are serious shoppers, window shoppers, and impulse shoppers. There are ants, butterflies, grasshoppers, and fish. Diligent visitors, and drifters. Streakers, strollers, and studiers. Different people behave differently.[25]

Researchers at the Smithsonian's Office of Policy and Analysis use three overlapping sets of rubrics to help understand visitors' exhibition experiences. They consider four general categories of experience: social, object-based, cognitive, and introspective. They look at eleven types of exhibition interaction, including being moved by beauty; connecting with the emotional experiences of others; imagining other times or places; enriching understanding; and seeing rare, valuable, or uncommon things. They ask visitors to rank exhibit features like design, lighting, computer interactives, organization, and amount of text. Together, these three kinds of analysis offer a sense of visitor experience, including what visitors get from the exhibit, whether they get what they wanted, and whether they were pleasantly surprised. Of interest to the exhibit developer, it also suggests what worked, and why.[26]

Museum evaluator Beverly Serrell presents another set of criteria for judging exhibits: balancing museum goals with visitor experiences. First, she considers visitor comfort, both physical and psychological; it's hard to learn if the ergonomics are bad, you're not sure what to do, and you don't feel welcome. Next, she looks at engagement: whether the exhibit invites exploration, social behavior, and a connection through a variety of sensory modalities. Then reinforcement: do visitors have "abundant opportunities to be successful and feel intellectually competent"? And finally, meaning: whether the exhibition is relevant, important, and useful; whether it changes visitors' beliefs or attitudes.[27]

There's another kind of evaluation of the museum experience that takes a longer view. If the museum's goal is to teach or to change lives—not just to give visitors a pleasant, informative, or meaningful experience— then it's important to see what visitors remember weeks or months later. Audrey Hepburn captures the sense of this when she writes that "living is like tearing through a museum. Not until later do you really start absorbing what you saw, thinking about it, looking it up in a book, and remembering—because you can't take it in all at once."[28]

Studies show that museum visits do have long-term effects. Museum evaluation expert John Falk writes that "beyond all reason, people remember their visits to museums." Months and years later, they remember not only that they went but surprising details of the visit. One study found that more than half of the memories were about objects or things, one quarter were related to the visit, and one quarter were associated with feelings and judgments about either the visit or the exhibit. About 10 percent of the memories were "summary memories," reflecting on the visit afterward. Over time, specific memories tended to disappear, replaced by "big picture" memories. Falk writes of the importance of identity in the museum experience, and that's true of museum memories too: we remember what connects with who we are and what we care about.[29]

Feedback from studies like these can be useful for understanding visitors and their experience. But using information about visitor experience to improve that experience is a challenge. In that first survey, in 1884, Higgins assures his readers that of course one would never allow the "interests of ordinary observers . . . to take the lead in determining the construction or arrangement of a museum"—"a groundless apprehension," he calls it—and yet he suggests that it might be wise to, at least, take it into account.[30]

Museums are still trying to figure out how to do that. Should exhibitions give visitors what they want? Some argue against treating visitors as customers and in favor of considering them students, or citizens, or cocreators of meaning. Should the museum offer something for everyone? That attitude can attract a wide audience but leave many unmoved and make its most fervent supporters unhappy. Should it concentrate on what it can do that no other institution can, and focus on art and artifact, and a social space? Or should it provide what a community needs most, even if that is beyond the museum's narrowly defined mission or strength? Should it shape the visitor experience for those who already know and love museums, or attract underserved audiences, a harder audience to reach?

With the goals decided, it's the job of exhibition developers to organize the relationships of objects, stories, and visitors to try to meet them. Chapters 10 through 14 outline the tools they have to do this work.

OBJECTS ON DISPLAY

Professor Jenks called almost everything in his museum that wasn't a natural history specimen a "curiosity." Curiosities, in the nineteenth century, were things from exotic lands, historical objects, artifacts that told stories. In the Jenks Museum, some of these objects had labels that introduced the stories. The objects and labels together hinted at adventure, exoticism, narratives beyond easy imagining: "Point of a Bomb Lance bent by explosion in a Whale"; "Cherokee Pottery, Ancient & rare."

Artifacts in a museum carry with them the power of authenticity. Recent surveys find that people trust museums more than almost any other source of information on history and science. In part, that's because museums have real things. They offer, many visitors believe, direct, unmediated access to truth.[1]

But objects are not simple things nor do they reveal simple truths. They carry many meanings and can tell many stories. Consider the ways we relate to objects: we use them on our own or with others; we buy, sell, and inherit them; we give them as gifts and receive them as presents; we store them, both as custodians of memories and for their economic value; and we throw them away. Some objects serve symbolic roles, sacred or profane. Some are treasure, some trash. An object plays many roles, and different roles for different people. It's the museum's job to understand objects in all their complexity, display and explain them to visitors, and put the object to use in a new way, a museum way.

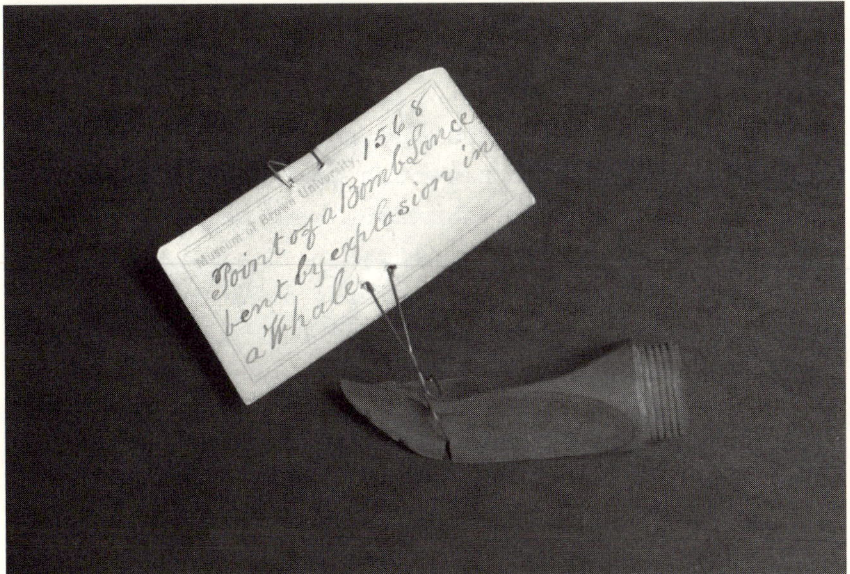

"Point of a Bomb Lance bent by explosion in a Whale," from the Jenks Museum.

Works, Specimens, and Witnesses

In the museum, objects carry not just their real-world meanings but also an additional set of museum uses. German museum scholar Thomas Thiemeyer suggests that artifacts are used in three basic ways in museums. Some are works, that is, works of art. Some are specimens, representatives of their type. And some are there as witnesses, bearing a meaning and telling a story because of where they were in the real world.[2]

Works of art stand on their own. Stephen Greenblatt, a literary historian, proposes the word "wonder" to describe our reaction to this kind of object. "By wonder," he writes, "I mean the power of the displayed object to stop the viewer in his or her tracks, to convey an arresting sense of uniqueness, to evoke an exalted attention."[3] Benjamin Ives Gilman of the Boston Museum of Fine Arts used a religious metaphor in 1918: "A museum of art," he wrote, "is in essence a temple."[4] Art museums display objects as works to be appreciated.

Not every object has the power to stop us in our tracks. Most offer instead what Greenblatt calls "resonance . . . the power of the displayed object to reach out beyond its formal boundaries to a larger world, to evoke in the viewer the complex, dynamic cultural forces from which it has emerged and for which it may be taken by a viewer to stand."[5] Specimens and witness objects possess the power of resonance.

Specimens are objects that stand in for their kind. The animals in the Jenks Museum were not there representing themselves, telling their own story. Rather, they represented their species, an ecological niche, or a particular kind of behavior. (This is what Jenks's student Dallas Lore Sharp meant when he said that the Jenks Museum had no case for stories.) Specimens might provide an illustration of a point made in words or an opportunity for discussion; that's common in history museums. Arranged in timelines, in groups, or in settings, specimens tell stories: this happened, and then that, or this with that. They might make an argument. They might be gathered to create an environment, a space that seems real, like a diorama or a period room.

Witness objects, the third type, have a different type of resonance, telling their own story or offering a connection with a larger story. This is the most complex kind of representation. History and anthropology museums use witness objects to get at larger and deeper stories, to get beyond use and move toward meaning and culture. They ask us to consider where this particular object has been, what it meant to the people who used it, and how it reflected and shaped their lives. Objects as witnesses connect us with people and history.

Relics are the most extreme sort of witness object. They're *only* story. The Jenks Museum had many such artifacts. Surviving labels read "Part of Old Ironsides cut out in 1847. Donor Blanding" and "Block Island Wood. From the House of Simon Ray"—bits of wood that only had meaning and value because of the words attached to them, the story that came with them.

Mark Twain both explained the value of relics and poked fun at them in Huckleberry Finn's explanation of a family treasure:

> Uncle Silas he had a noble brass warming-pan which he thought considerable of, because it belonged to one of his ancestors with a long wooden handle that come over from England with William the Conqueror in the *Mayflower* or one of them early ships and was hid

away up garret with a lot of other old pots and things that was valuable, not on account of being of any account, because they warn't, but
on account of them being relicts, you know.[6]

Museums once "thought considerable of" such things, too, and old museum collections contain many artifacts that are of interest only because
of the words written on their label—or, in many cases, on the object itself.
These objects are often more about the label, about the story, than the thing.

Museums lost interest in relics. They seemed unscientific. In 1917, a
speaker at the annual American Association of Museums meeting lamented the display of relics: "Nothing can be more uninteresting than the
usual case of personal relics . . . thrown together without rhyme or reason
except that all were once connected with some personage of distinction.
Such memorials are pathetic survivals of an unscientific past."[7] A half-century
later, another museum expert echoed the warning: "Relics, curiosities, personal memorabilia, glorification of specific individuals or specific families . . .
do not belong in a public museum."[8]

These writers, in their attempt to be scientific, to teach the material
facts of history, left out emotion and beliefs and, indeed, much of culture.
Relics have made a comeback. They tell us what seemed important,
what seemed worth saving, illuminating a path into how people thought
about the past. Relics, and witness objects more generally, let the museum
tell stories. In some museums, for example the Holocaust Museum, they
simply say: this is real; this happened. They serve as markers of authenticity. Rachel Maines, a persuasive advocate for relics in the museum,
writes:

> [Relics] tell stories; the stories they tell are the stuff of social and in-
> tellectual history—the interaction of emotions, ideas, and beliefs
> with material culture. . . . The emotional force of association objects
> is their validation of memory and physical connection to the past.
> They concretize abstract memories, especially those of intense
> experience not easily captured in words: danger, suffering, birth,
> marriage, and the thrill of achievement. Places and artifacts associ-
> ated with personal and communal marker events provide a means of
> understanding and coming to terms with an emotionally charged past.[9]

Today's museums connect relics with stories in many ways. Oral histories can bring relics to life, making it possible for the visitor to relate not to the thing but to what it stands for. Stories tied to objects can encourage greater inclusiveness, and many museums now make a point of collecting stories when they collect artifacts.

Mix together the many uses and meanings of artifacts—as work, specimen, and witness, as useful thing and symbol, as holder of truth and teller of stories, as wonder and resonance—and you get a sense of their complexity. The way they are displayed needs to capture those meanings. At the same time, the museum needs to protect the object and to capture visitors' interests. Over the past two hundred years, museums have experimented with ways to do all those things.

Look, Don't Touch

Objects are both physical things and carriers of meaning, and those two roles need to be balanced in any exhibition. The first impression many visitors have of museums are of the things that keep us from the objects: cases, "Please don't touch" warnings, and guards. And there's good reason to protect objects. Some things are fragile, and even durable objects can be harmed by the oils on visitors' fingers. Museums need to protect their collections.

But protecting objects from touch also means we interact with them in a limited way. Today, we look at things using what's called the museum gaze: respectful, sequential, critical. We play down the other senses. But that wasn't always the case, and the emergence of the museum gaze offers insight into the nature of the modern museum experience and what's lost when we neglect other ways of interacting with artifacts and exhibitions.

Visitors want to touch things. Julie Stein, director of the Burke Museum of Natural History and Culture, reports that visitors always have two questions when they first visit a natural history museum: is it real, and can I touch it?[10] Touch is more stimulating than sight. We learn different things from touching than from looking. Compare the experience of shopping for clothes (fit, drape, details of construction, the feel of the fabric on the skin) with looking at a display of clothes in a museum. Consider the robust language of touch, the emotional connections that the words like "stroke," "fondle," "caress," and "pet" imply. Contrast the way that the same valuable art work might be handled at a museum and an auction gallery. In a museum: don't touch. In an auction gallery, a potential buyer—or just a curious visitor—can handle the art. "Be aggressive," advises a *New York Times* article on buying at auction. "Have the upholstery torn off furniture so you can check the quality of the wood."[11] Touch lets us connect with things in a way that is often lost in museums. Auction houses and clothing stores know they'll sell more if customers can touch things. Touch allows for deeper engagement.

Museums were once much more open to senses beyond vision. A visitor to a cabinet of curiosities could pick up the objects on display. Visitors toured early museums with their curators, and it seemed only hospitable to share objects physically. In 1760, a keeper at the Ashmolean Museum at Oxford reported that a visitor "desired me to take the Glass from off several of the Drawers, which I was somewhat unwilling to do, lest anything be lost by that means; which she perceiving she told me that I was not quite so civil as might be; that the last time she had seen the Museum . . . she had handled and examin'd the Curiosities in the Cabinet as long as she pleas'd." Visitors used all their senses to try to understand the objects in the museum, connecting viscerally with art and artifact. One visitor to Sloane's museum in 1710 reported tasting a bird's nest.[12] In the mid-eighteenth century, the Library Company in Philadelphia allowed its patrons to check out artifacts, just as they checked out books.[13]

This tradition of touch continued into the early nineteenth century. The British Museum allowed its earliest visitors—mostly members of the upper classes—to touch the objects on display. "Seeing by the sense of touch" was commonplace, according to one 1808 essayist.[14] An 1845 drawing shows the curator of the National Museum, in Copenhagen, in a top hat, giving a tour to a family; he's about to hand a Bronze Age neck ring to a young girl.[15]

But by the mid-nineteenth-century, touching was considered lower class. "We can all remember the loiterers and loungers [who] strutted about . . . talking, flirting, peeping, and prying; lifting up the covers of chairs to examine the furniture; touching the ornaments—and even the pictures!" a writer on the arts reminisced in 1844.[16] By 1880, there were

"Do not touch" signs in museums. "Touch what you like with the eyes, but do not see with the fingers" read one at the Picture Gallery of Oxford's Bodleian Library.[17]

At the Jenks Museum, Jenks handled his objects for teaching. He wrote to the taxidermist mounting skeletons that "I want them mounted on stands with set-screws &c. so that the skull and limbs could be handled separately for class illustration."[18] Still, he put as much as he could behind glass. (He had good reason to worry about students, who didn't always pay attention to the "do not touch" signs. A stuffed giraffe, reported the local newspaper, "became a pet of the student body, who sometimes would smuggle it out of the museum and take it for a walk through the campus, its funny little head scraping the branches of the elms."[19]) Brown's Museum of Classical Antiquity had many large signs reading "The Casts Must Not be Touched."

"Do not touch" reflected a change in the way museums thought about their artifacts. In early museums, decay was taken for granted; protection behind glass or in locked drawers was more to prevent theft. But as museums began to treat objects as precious and museum space as sacred, museum artifacts came to seem different from the artifacts of the world. Museums decided their objects needed to last forever, and so they became more protective of them.

Visuality became the proper way to interact with art and artifact. A proposal for a Yale museum in 1845 used a wonderful metaphor: the museum should "lecture to the eye."[20] George Brown Goode began his 1891 essay on "The Museums of the Future" with a paean to vision: "To see is to know."[21] Historian Sam Alberti suggests that attempts to make museums silent and odorless, and therefore more acceptable to "polite" audiences, introduced a much more exclusive focus on the visual.[22] As museums became more open to a wider range of the public, they took on the role of "civilizing" the lower classes to proper behavior: teaching them to control their impulses, to respect the artifacts (and the museum, and the authority it represented), to look and not touch.[23] Historian Constance Classen sums up what visuality meant to visitors. They had to accept several related ideas: "that they were less important than the exhibits on display and thus must behave deferentially towards them . . . that to touch museum pieces was disrespectful, dirty and damaging . . .

and that touch had no cognitive or aesthetic uses and thus was of no value in the museum."[24]

Literary critic Susan Stewart puts forward a related explanation. Touch, taste, and smell entail emotional connections, hidden truths, and an intimate relationship with the past. Sight, on the other hand, offers intellectual understanding. Collections are most easily systematized when laid out so they can be compared visually. Curators wanted museums to be places of rationality, of intellect.[25] The museum, historian Donald Preziosi argues, depended on its visual structure to establish a distinctly modern relationship between the viewer and the object viewed, one that allowed for ease of categorizing. He called the museum "the most extraordinary optical instrument of all; the veritable summa of opticality, of visuality."[26] Museums privilege sight, and to do this they provide an environment that doesn't distract by allowing the use of the other senses.

Benjamin Ives Gilman of the Boston Museum of Fine Arts, in an essay that art historian Kathleen Curran calls the "the birth certificate of the modern American art museum," sums up the work of the museum as "artistic comprehension." That meant attentive looking: "the mind rests in its object, simply and fully beholding it, without deserting it for any other interest whatever."[27] Gilman celebrated the restrictiveness of sight by inventing a device that enforced attentive looking. His "skiascope" narrowed the field of vision, like blinders on a horse, and blocked glare. Gilman wrote that the skiascope would allow museum visitors to see art better. He had a narrow meaning of "better" in mind: a particular kind of aesthetic vision, purer, more abstracted. Looking at only one painting or sculpture at a time encouraged a viewer to appreciate it as a work of art, a pure aesthetic creation, always Gilman's goal.[28]

Gilman's skiascope never became popular, but he was so taken with the idea that he included eight pages of detailed plans for building one in his 1918 book, *Museum Ideals*. I decided to give it a try. I cut and folded cloth, sawed wood, and shaped metal supports according to the instructions, and built my own skiascope—perhaps the first one created in one hundred years. I took it to the Museum of the Rhode Island School of Design to try it out. The Grand Gallery there is a space that Gilman would recognize: a sky-lit room, hung salon style, paintings close together. I held the skiascope to my eyes, and instantly understood why the device appealed to Gilman. It isolated each artwork from everything around it. Because its construction reinforced binocular vision, it made paintings sharper, more three-dimensional. The skiascope purifies visuality. It also separates the user from others in the space, making art viewing a more personal, less social experience. The skiascope renders Gilman's museum philosophy of vision in oak and flannel.

One reason that Gilman's skiascope failed was that Gilman and his successors figured out ways to get the same results without giving everyone a skiascope. They reorganized the gallery—eliminating crowded walls of paintings and improving lighting—to allow museum visitors to interact with paintings in the way that Gilman thought best: one at a time, without distractions. Historian Andrew McClellan argues that "the chief achievement of twentieth-century museology was the refinement of viewing conditions and museum interiors so that the skiascope would no longer be necessary."[29]

Some museum staff fought back against the primacy of visuality. In 1901 the curator at the Whitechapel Museum in London argued that the objects school groups wanted to study "should, whenever possible, be taken out of their cases."[30] Carolyn MacDonald, the first curator of education at the Museum of the Rhode Island School of Design, wrote in 1944 that the museum's "no touching" rules severely limited the learning experience: "Educational work in museums will be jeopardized and seriously hampered until this nonsense about 'precious objects' is discredited."[31] Even Gilman acknowledged that touch could be useful: "If the contents of the public museums of the future were all cheaply and easily reproducible, the loss and damage resulting from their open installation would be as nothing compared with the enhanced interest and comprehensibility of the exhibits."[32]

Some types of museums have been more accepting of exhibit interactions that go beyond sight. When the Brooklyn Children's Museum opened in 1899, it had a hands-on, learning-by-doing approach. Anna Billings Gallup, its curator in chief, wrote that "objects and models are taken from the cases and used in demonstration, living specimens from vivaria and aquaria are shown to the nature study classes." Children built a wireless telegraph station, and learned to collect and mount insects and to identify and sort minerals. They could try on costumes. They could even take objects home with them.[33]

Today, many history and natural history museums have an education collection, separate from the main collection, for handling. Curators at museums of technology, transportation, and industry sometimes operate their machines, arguing that it is the only way to accurately understand and portray the history they represent. Many museums with musical instruments offer musicians the chance to play them. History museums in the age of eBay have begun to rethink the preciousness of some of the mass-produced artifacts in the collections. Why not allow visitors to touch them, to get more out of them, and buy replacements when needed?

The primacy of vision in museums shaped the relationship of the visitor and the objects. "The contagious magic of touch," critic Susan Stewart writes, was "replaced by the sympathetic magic of visual representation." Seeing without being allowed to touch, she suggests, offered "an elaborately ritualised practice of refraining from touch." Museum artifacts became valuable because they were protected.[34] Museum historian Ken Arnold suggests that this "oddly displaced experience" of wanting to touch but not being allowed to lies at the heart of the museum experience with artifacts. Were we allowed to touch the objects, they would lose their mystique. They would become just . . . things. The no-touching rule elevates the museum artifact and creates the museum experience.[35]

Cased, Pedestaled, and Framed

Designers shape exhibitions using tools that take advantage of the museum's sophisticated visuality to make art and artifact tell stories. Objects are displayed on pedestals and in cases. Paintings are framed and hung on walls. Each carries meanings, literal and symbolic.

Consider pedestals. On one level, they're designed to disappear, to quietly do the work of holding objects at the right level. But they do more. They move the object from the space of the viewer into a more rarified space. On a metaphorical level, of course, they do much more. Putting things on pedestals says "This is important." Fred Wilson's 1992 "Mining the Museum" exhibition at the Maryland Historical Society made a point by putting empty pedestals on display, labeled to show what was *not* in the collections.

Museum display cases may seem simple, but they are called on to do many things. In the same book in which he described the skiascope, Gilman wrote that a display case has two functions, "first as guardian and then as expositor of the treasures committed to its charge."[36] As guardian, a case protects the objects within not only from our inquisitive hands but also from thieves, vandals, and environmental threats. As expositor, it shapes our relationships with the things within, encouraging us to look at them in a particular way, in the way the curator thinks appropriate. Like pedestals, cases imply worth and importance.

Building a good case requires skills of design, lighting, and mount-making that are some of the hidden specialties of the museum world. A well-designed case is beautiful in form and function. It maintains temperature and humidity within a narrow band. It allows for lighting with no shadows or reflections, as well as no heat buildup within the case. It opens and closes and locks and unlocks easily, but appears solid and impervious. Ads from case manufacturers claim all these things. One firm notes that its cases avoid "the normal nightmare scenario"—giving a sense of just how complicated cases can be.

Compare museum cases with the similar-looking but differently used cases in collectors' homes. Edmund de Waal writes in *The Hare with Amber Eyes,* a history of his great-grandfather's collection of netsuke, that the collector's vitrine "is for opening." He describes the process—"opening the glass door and the moment of looking, then choosing, and then reaching in and then picking up"—as a "moment of seduction, an encounter between a hand and an object that is electric." The collector removed objects carefully, artfully, "fishing, handing things out to be looked at and handled, to be caressed." Vitrines, he writes, were essential to the "witty and flirtatious intermittencies of salon life." Museum cases, by contrast, "were a sort of coffin."[37]

We are so used to frames that they are almost invisible. We aren't supposed to examine them, writes art historian Ernst Gombrich, but "only to sense them marginally."[38] They serve to separate art from not-art, and to set each piece of art apart from its neighbor. The frames of paintings remind us of window frames, encouraging us to see a painting not as an object on a wall, but rather as something that exists in its own space. The frame, according to philosopher Barbara Savedoff, "encourages the reading of painting as window."[39] Artists and curators fought over appropriate framing. In the eighteenth century, when curators were arguing for a museum organized taxonomically, they wanted uniform frames, to let the art speak for itself. In the nineteenth century, classicists argued for the frame, while romantics argued against it, the better to merge art and life. When twentieth-century artists abandoned the frame, to remind us of painting as object, curators responded with white walls and even spacing, in order to reframe the paintings.

By the mid-twentieth century, the inventions of two hundred years of museum theorists and practitioners had come to seem natural. Art and artifact was isolated in the museum, given a new role to play as work, specimen, or witness. It was there to be looked at: no touching allowed. And it was framed in a way that set it apart. The orderliness of display in a modern art museum—an even line of paintings on a white wall—serves to make art objects only about themselves and their relationships with each other, about their aesthetics and not about the rest of the world.

Art historian David Phillips notes that the entire museum reflects the way it treats the art and artifact within: "Museums and the art they house are framed affairs."[40] These conventions allow a museum to present objects in a new context—cased, pedestaled, and framed—to control the story that they tell. Chapter 11 explains how they do that.

❦ 11 ❦

ORGANIZATIONS AND

JUXTAPOSITIONS

THE JENKS MUSEUM TOLD A STRAIGHTFORWARD STORY. There were three categories: nature, anthropology, and civilized man. Nature was further broken into categories like birds, mammals, and minerals, each with a system of arrangement. Anthropology specimens were organized in one of two ways: by group (Native American tribe, for example) or by type (all of the spears together). Civilized man was represented by local relics and portraits of Brown University donors, faculty, and administrators.

When the Jenks Society for Lost Museums set about reimagining the Jenks Museum, we had to decide how to organize the hundred or so objects we chose for display. We might have arranged them in just the way Jenks did. That was one moment in their history, an important one. We might have arranged them according to their histories after the Jenks Museum closed: these went to the biology department, these to archaeology, these ended up at the university's anthropology museum. Or, most prosaically, we might have organized them by modern category: ethnography, biology, numismatics. This would make the objects useful today, ignoring their history at the old museum. It would also help visitors make a comparison between the Jenks Museum's collecting and that of contemporary museums.

We chose something else. The objects were arranged not by external categories but according to their current condition, their degree of decay. From left to right: "Good condition," "Some wear and tear," and "Com-

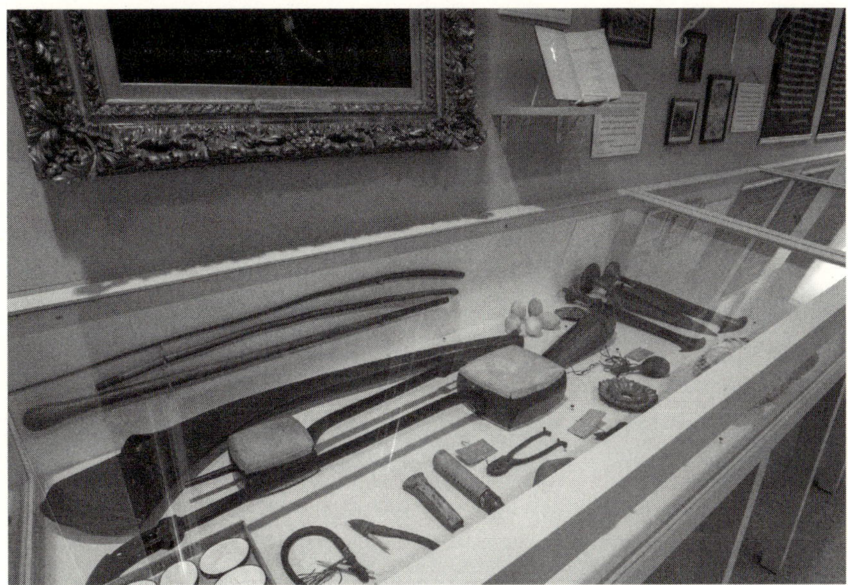

Surviving artifacts from the Jenks Museum on display in *The Lost Museum*.

pletely deteriorated." In the first category were Grand Tour medallions, some odd sticks, African knives and in the next taxidermied birds in glass boxes and Chinese shoes, while to the right lay broken Egyptian amulets, broken bird eggs, a mangled telegraph key, and, all the way at the end, fragile museum labels. From left to right the visitor moved from artifacts in good shape to broken things to fragments to words describing artifacts that had disappeared. That arrangement emphasized the story we wanted to tell: decay, collections beyond the control of a curator, an abandoned museum, change over time. The label on the wall above reinforced the message of arrangement: "Friends, cast your eyes on these shattered remnants and know that all things return to dust."

But while the message was about loss, the loss of both objects and order, the organization in the cases was quite deliberate, and quite beautiful. I watched artist-in-residence Mark Dion arrange things in the case. He was given a most unprepossessing group of objects, and yet he made them fit together beautifully. His artist's eye and hand gave them a coherence that allowed them to tell a story. Design made narrative legible and captivated the viewer. The message of the exhibition came out of the contrast between the ugliness of decay and the beauty of the arrangement.

There's a story here, it said. Pay attention not just to the individual arti-facts but also to the way that, as a group, these objects tell us something about the history of this museum, themselves, and ideas about art and artifact more generally.

Objects, Arranged

"The people's museum should be much more than a house full of speci-mens in glass cases. It should be a house full of ideas, arranged with the strictest attention to system," demanded the Smithsonian's George Brown Goode in 1889.[1] More recently, Philip Fisher argues that what is on view in museums is not only the work of art but also the relations between works of art, what they have in common, and what "in the sharpest way clashes in their juxtaposition."[2]

System, arrangement, and juxtaposition—whether a chronological be-ginning, middle, and end, a taxonomic sequence, contrasts and compari-sons, or a narrative—are the foundation of exhibitions. Choosing a story to tell, and then choosing and arranging the objects to tell it: at the most basic level, that's the art of the exhibition. Curator Ivan Gaskell calls it a "visual argument."[3] There's much more that surrounds the objects—words, images, interactives, environmental settings—but at their most basic, exhibitions are objects, arranged.

The order in the museum mirrors the order that scientists and scholars see in the world. In nineteenth-century museums of natural history, anthropology, and history, it was an orderliness of progress—evolutionary, technological, and racial. In art museums, it was art arranged by chronology and school. Some scholars argue that the orderliness of the museum disciplined the visitors, teaching them how to behave. While that seems to overplay the power of the museum, it suggests the align-ment of the museum with new kinds of order, and new ways of thinking about order.[4]

Museum theorists agree that system is what makes an exhibition more than a collection of things, but they disagree on what system is best. The ordering can be straightforward. The work of an artist, arranged in chron-ological order. European art arranged by school. The progress of invention presented as a history of improvement. The design of exhibitions reflects the underlying order. Philadelphia's early nineteenth-century Peale Museum tamed the world of nature and politics with a rectilinear display. The symmetry of the displays of the mid-nineteenth-century Smithsonian spoke of a world understood and under control.[5]

The history of museum display shows the increased use of system. The earliest museums, the early modern cabinets of curiosities, were, like museums today, collections of artifacts on shelves and in cabinets. But cabinets of curiosities had a different point to make. Early cabinets, in the sixteenth and seventeenth centuries, were heterogeneous, unsystematic displays. The message they sent was one of exoticism and variety. Look at all of these amazing things, they said. What a worldly person the collector must be! Each object stood for itself.

In the late seventeenth century, though, the way those objects were organized began to change. Displays became less about individual objects, and more about the relationships between objects. These museums illustrated the doctrine of signatures—the belief that things that looked similar had some underlying connection. Barbara Stafford suggests that the juxtapositions served to simulate conversation. What did the similarities and differences mean? She notes a "system of sideways looks" that characterized these displays, artifacts that "cacophonously 'chatted' among themselves and with the spectator."[6]

"Sideways looks" might also describe some early art museums. With the rise of the idea of "masters," the role of museum display was comparison. A student might learn about art by comparing one great master with another, one school with another. Variety was key. A good viewer would compare and learn the stylistic traits that made each school of art great.[7]

The way museums displayed art encouraged this kind of use. Seventeenth- and eighteenth-century galleries crammed the walls with paintings. Some treated paintings as decorations, hanging them in symmetrical patterns. Margaret Jackson, assistant director of the Minneapolis Institute of Art, characterized this earlier period with horror in 1917: "Paintings were often cut over to fit some special place; portraits reduced or enlarged to conform to some design; Chinese vases placed on tiny brackets at intervals, even up to the ceiling, simply as part of the decoration; tapestries hung in picturesque folds or used as curtains at doors and windows; nothing mattered so long as the effect of the whole was rich and varied."[8]

In the late eighteenth and nineteenth centuries three new, interlinked ideas began to emerge. Together, they would shape the modern museum.

The first was a new way of looking that changed the interaction of the visitor with the object. Visual studies scholar Jonathan Crary suggests that beginning in the early nineteenth century, the way people thought about vision changed. He calls the new scheme "subjective vision": rather than an observer just seeing what he's shown, he determines what to look at, what he pays attention to. Museums began to consider the visitor as part of the equation.[9]

A second change was a new way of reading objects. Objects no longer were simply there as themselves but rather as tokens, as metonyms, as objects that represented something more than themselves. The objects in the museum stood for things in the larger world, and the museum made the implicit argument that a visitor could understand the world by looking at a few objects out of their original context but in a new, properly organized context. Historian Carlo Ginzburg calls this new paradigm "semiotic" or "conjectural"; in it, objects held clues to the larger world.[10] They were specimens or witnesses.

In natural history museums, this change meant that artifacts became representatives of their type. "Let each object represent so much knowledge," an article in *Hardwicke's Science-Gossip* suggested in 1872, that "the very mention of its name will immediately conjure up a crowd of associations, relationships, and intimate acquaintances, and you will then see what a store of real knowledge may be represented in a carefully arranged cabinet."[11]

In art museums, objects lost their original meaning; they became art, and part of art history, not religious icons or decorations or architecture or furniture. Critic Willem Bürger complained in the 1850s that art in a museum was confusing, out of context: "A Venus is placed side by side with a Madonna, a satyr next to a saint. Luther is in close proximity to a Pope, a painting of a lady's chamber next to that of a church."[12] But by the end of the nineteenth century, this all seemed quite reasonable. The museum created its own context. Objects represented their past, their original context, but were no longer part of it.

The third major change was the notion that organization is the key to museum display. Subjective vision and the idea that a specimen stood for something larger than itself allowed for new ideas about system and order. The juxtaposition of fragments, interpreted by the viewer, would allow him or her to understand the world. Just as the object in the museum stands for the real thing outside of it, so too does the museum display stand for the complex systems of the real world. Art museums viewed art history as a context for all artistic productions, whatever their original purpose. History museums relied on chronology, biography, or filiopietism. Anthropology museums took the work of individuals, and of cultures, and arranged them to show the progress of races, or humanity. Science museums told stories of evolution. Organization and juxtaposition within the museum, every exhibition implied, captured a hidden truth of the world itself.[13]

Museum curators became obsessed with order and arrangement. They slammed museums that lacked good order and argued over the proper order of things. Lambert Krahe, director of the Dussledorf Electoral Palace museum in the 1750s, introduced a new system of organizing paintings. They were to be admired not only as works but also as representatives of their time and place, as specimens, as examples. Krahe aimed to create a pedagogical display that educated viewers in the art-historical principles of the different schools of art. The Louvre followed suit with the goal of "a continuous and uninterrupted sequence revealing the progress of the arts and the degrees of perfection attained by various nations that have cultivated them." The art museum, from this point on, was not solely about individual works but also about art history.[14] A picture collection not arranged by school and artist, wrote art historian Jean-Baptiste-Pierre Lebrun in 1793, is "as ridiculous as a natural history cabinet arranged without regard to genus, class, or family."[15]

Throughout most of the nineteenth century, museums displayed art according to a generally accepted system of art history, organized by time and place. But while the arrangement reflected system, it wasn't making an argument about the relationships of individual works of art. Most eighteenth- and nineteenth-century galleries hung their paintings salon style: the wall painted a dramatic color, elaborately framed paintings floor to ceiling, no individual labels. Arrangement was by fit—larger works were hung nearer the ceiling, smaller works closer to eye level—or by quality, with lesser works "skied," hidden away near the ceiling.

Artists sometimes complained about where their artwork was hung, and some took control of the way their paintings were displayed. Jacques-Louis David showed off his *Sabine Women* with a mirror on the opposite wall of the gallery, so that the viewer could examine details and then turn around to see the entire image. The impressionists also managed the way their work was viewed. They wanted visitors to concentrate on the paintings, so they showed them on neutral walls, well lit, and with generous spacing. They chose simple frames that blended with the wall. In all of this, they wanted to emphasize art as art, not decoration, and make it clear that visitors were expected to pay attention.

The debate over the proper way to organize art on display resumed at the turn of the twentieth century. Should art museums show art or art history? Inspire or teach? Some museums responded to new ideas about visual art, a new belief in art for its own sake. Some artists and critics, writes historian Julia Noordegraaf, focused on the "expressiveness of line and color, the mystical content of the image and the social importance of the work of art" and less on historical orientation. They wanted to privilege the "formal-aesthetic perception of art."[16] The Boston Museum of Fine Arts' Benjamin Ives Gilman was in this camp. He thought the museum's job was to offer the opportunity for aesthetic appreciation, and that the best way to do that was to show art of the finest quality, each piece standing on its own.[17]

The Newark Art Museum's John Cotton Dana blasted that approach, disparaging museums who did it as "gazing museums." He wanted the museum to be of "direct value and service to a city of industries," and urged museums to show not the "old and rare and costly" but rather "objects of daily interest and use."[18] Some museums tried to be comprehensive, art in chronological order. Others, like the Frick Collection or the Isabella Stewart Gardner Museum, showed art in a domestic setting, recommending the taste of their founders as a model for visitors.

In the twentieth century, museums mixed these models with a range of exhibitions that reflected changes in art, curatorial fashion, and visitor and funder interests.[19] Salon-style displays went out of fashion. Robert Fry, curator of paintings at the Metropolitan Museum of Art, explained why: "hanging pictures all over the walls from floor to ceiling . . . is, I consider, equivalent to saying that the pictures so hung are not worth seeing at all."[20] Fry, like Gilman, wanted masterpieces, hung so that they might be fully appreciated. This coincided with the rise of the "dual system" of display, and of the invention of museum storage. All of those paintings no longer good enough for display in the less crowded gallery needed to be moved elsewhere.

The "white box gallery" refined this style. Alfred Barr's installations at the Museum of Modern Art in the 1930s made the white cube famous. White walls; artworks organized by date, or juxtaposed to make a point; presented in a single line; labels explaining what to look at, and what to see; a single path through the exhibition: the strong narrative of the exhibition design suggested that the curator was merely reflecting the truth of the art, a self-obvious art history. By the second half of the twentieth century, this style of display had become standard at most museums.

Brian O'Doherty's 1976 *Inside the White Cube* critiques the white box gallery's ideology. White-walled galleries, O'Doherty writes, deny art's juxtaposition with anything other than art. They move art from the real world into the self-contained world of the gallery. The white cube "subtracts from the artwork all cues that interfere with the fact that it is 'art.' The work is isolated from everything that would detract from its own evaluation of itself . . . art exists in a kind of eternity of display."[21]

The white-walled gallery was the culmination of a long history of museums increasingly isolating art and artifacts from the world. It has come to seem natural. Consider museums that seem, by comparison, unordered. Umberto Eco's *Foucault's Pendulum* describes the Conservatoire des Arts et Métiers in Paris, a church turned into technology museum: "You enter and are stunned by a conspiracy in which the sublime universe of heavenly ogives and the chthonian world of gas guzzlers are superimposed."[22] Or consider private collections that became public. Charles Lang Freer believed that Whistler's *Peacock Room* would be a good place to display his Asian ceramics. Albert Barnes liked juxtaposing Renoirs and Pennsylvania Dutch furniture. In both cases, when they became public museums, they seemed peculiar. Contemporary artists sometimes play with our assumptions of order, using unusual juxtapositions to reveal a deeper truth. Fred Wilson labeled both fine silver work and slave shackles "metalwork" in his 1992 *Mining the Museum* at the Maryland Historical Society.

The debate continues in the twenty-first century. Many museums have continued with white cube galleries, neutral spaces with art hung on the walls in neatly defined categories, paintings separated from decorative arts, everything divided by art historical categories. Some have returned to earlier styles, mixing paintings and decorative arts. Some have gone further. The Tate Modern and MoMA have experimented with thematic hangs (at the Tate, for example, "Poetry and Dream," "Energy and Process," and "Structure and Clarity"). The Detroit Institute of Art has organized galleries around themes chosen for their "broad relevance" to its visitors, showing eighteenth-century European decorative arts objects according to the time of day in which they were used, for example.[23] The Rijksmuseum has combined history, fine art, and decorative art collections into a chronological history. As has been the case throughout the history of museums, many critics find changes like these disconcerting.

The internet has opened up new possibilities for organization, new kinds of juxtapositions, impossible in the museum. French art historian André Malraux set this line of thought in motion with his 1947 publication of *Le musée imaginaire,* usually translated as "museum without walls." His ideas are often represented by photographs of Malraux standing, kneeling, even lying on the floor, surrounded by hundreds of photographs of artworks, ordering and reordering them. For Malraux, the juxtaposition of styles of artwork was the key to the museum. He wanted to move beyond traditional notions of schools and periodization to create a universal museum, albeit an imaginary one, where styles might confront one another. Artworks come to life, Malraux argues, through dialog with other artworks; dialog lies at the heart of our modern response to art. Art belongs to a realm beyond history. "The *musée imaginaire,*" writes Derek Allan, "is the ideal art museum that each of us carries around in our minds; it is our own selection—the works that mean the most to each of us, the works we truly admire."[24]

And that's what online collections promise. We can choose art and artifact from anywhere, reshuffle it as we wish. Unbound from the physical object, choice is personal and juxtaposition is all. Google's Art Project promises "the world's art at your fingertips." It provides three ways to view art, ways that reprise the history of museums of the past few centuries. You can choose to examine individual works of art, close up, as masterpieces, without the museum context. You can choose your favorites and juxtapose them, building a cabinet of curiosities that converse in a "system of sideways looks," that chat among themselves. Or you can visit museums in "Street View," taking the curator's advice on the best way to order your visit. The choice is yours.

Imposing Order

Designer Richard Wurman suggests that there are only five ways to organize information: by location, in alphabetical order, in time, by category, and by hierarchy.[25] While there have been only a few exhibits arranged alphabetically (mostly to make a point about the arbitrariness of order), the others all find regular use in museums.

Time as an organizing principle might seem best for history museums, category for art museums, location for anthropology museums, hierarchy for science and natural history. But in fact, it's more complicated than that. What seems the best story to tell has changed.

Anthropology museums had to decide between two basic kinds of displays. The museum can display artifacts to show something about the way they were originally used, as in a diorama or a display of the tools or costumes of a single cultural group. Or the museum can arrange the objects to tell a larger story of change over time, or human evolution, or perhaps ideas about race and progress. By chronology, or by place, or category? George Brown Goode's 1881 "Scheme of Museum Classification" organizes "the natural history of civilization, of man and his ideas and achievements" both in functional ways—the development of technologies around the world—and ethnographically, showing the development of "races." At one point Goode contemplated putting display cases on wheels, so that they might be moved about easily to tell both stories at the same time.[26]

Chronology might seem the obvious way to organize artifacts in history museums. After all, history is one thing happening after another. Putting things into the order in which they were made or used tells that story nicely. Visitors walk through history as they walk through the exhibit. Philosopher Susan Stewart puts it eloquently. A museum, she writes, is a place where history is transformed into space.[27]

In history and anthropology museums, chronology came to stand for "progress." Technology progressed from stone to steel, from hand tools to machines. The Smithsonian in 1900 used dozens of cases to show "synoptic series" of tools—fire-making, textiles, knives—each in order from the most primitive aboriginal attempts to the most advanced tool in use in American factories. "Each specimen in these series," anthropology curator William Holmes wrote, stands for "a step in human progress. . . . the broader truths of human history."[28] The Field Museum's 1933 "Races of Mankind" exhibition showed human "progress" from "Neanderthal to Nordic."[29]

Art museums faced a similar question: chronology or cross-cultural comparison? Most settled for chronology, or school, a cross between chronology, geography, and technique. Alexander Dorner, newly hired in 1938 to run the Rhode Island School of Design Museum of Art, proposed a simple but all-encompassing chronological scheme. "The entire museum collection needs to be rearranged in historical succession," he wrote to the museum's board.[30] At about the same time sculptor Lorado Taft proposed the ultimate timeline of art for what he called his "Dream Museum." This enormous building, 750 feet long, would showcase copies of the world's most important sculpture and architecture, organized in seven aisles, one for each of the world's seven most important cultures, each arranged in chronological order. You would walk through a timeline of Egyptian, Chinese, or European sculpture and architecture. Interested in comparison across cultures? Walk across the cultural timelines to see what was happening around the world at any given time. Taft insisted that this comparative study would reveal "the meaning of life." Alas, the museum was never built.[31]

In the second half of the twentieth century, many museums reacted against these displays of progress. Historians worried about making the past seem as though it had to happen the way it did, following a preordained timeline. Anthropologists worried about an easy assumption of the progress of all peoples toward current Western technologies and ideals. Explaining the past as an inevitable progression toward today, toward democracy and freedom, faster and better machines, and higher standards of living involves an implicit, scientific-seeming excuse for colonialism and the dominance of Western ideals. It doesn't reveal the lives of people in the past: the choices that they made, the battles they fought, or the challenges they overcame.

Natural history museums faced a different set of options. Arrange the specimens in family-genus-species order and they describe evolution and diversity. Arrange them in dioramas or environmental groupings and they speak to ecology, animal behavior, and the complexity of nature. Put them in a human context, and they address environmental concerns. There are many ways to order exhibitions, many stories to tell.

Biologists looking back at the Jenks Museum criticized it for not telling a story. In 1937, long after the museum had disappeared, a Brown student described Jenks as "a gentle old man, a born naturalist collector," but wrote that as Jenks was "not a believer in evolution," he had no system for organization. He was forced, therefore, "to resort to original and divers methods of classification of his specimens. One month he would deem it best to classify alphabetically and would proceed to arrange all of the A's on one shelf and the B's on the next. Soon tiring of this, he would rearrange his specimens according to the size of the bottle in which they were preserved."[32]

It seems unlikely that Jenks really wandered the aisles of the museum rearranging his collections alphabetically one month, by size the next. But the student's assumption about the importance of *some* plan of organizational—some theory of relationships between objects made clear to the museum's visitors—is correct. Exhibitions depend on organization and juxtaposition.

All Sorts of Impressions and Thrills

In the early 1900s, "museum men"—mostly students and followers of the Smithsonian's George Brown Goode—argued for the importance of popular education in museums. Henry Ward of the Milwaukee Public Museum wrote that museums needed to give visitors "all sorts of impressions and thrills."[33] To do this, natural history museums replaced cases of specimens with habitat displays. Anthropology museums offered life groups and dioramas. Art museums wrote art history labels and created period rooms. History museums began grouping objects together to form cohesive vignettes. Edward P. Alexander, director of the New York State

Historical Association, captured the spirit of this change when he wrote, in 1936, that objects should be surrounded by related objects, images, and text to make them "active," draw visitors back in time, and tell a story.[34]

The most common form of presentation in museums today—objects juxtaposed with photographs and texts—stems from these early twentieth-century roots. Here's a thing. Here's a picture of it being made, or used. Here's what it looked like in its original setting—in history, in its original culture, in nature. Here are some related things. And here's what it all means. Collecting moves things out of context. The first impulse of modern museum display is to restore that context.

Museum theorists have long debated the appropriateness of this effort. Proponents argue that arranging the right objects in the right settings provides a magical experience: educational, entertaining, and engaging, balancing authenticity, verisimilitude, and truth. The object brings authenticity to the setting, and a thoroughly researched context returns the larger meanings that had been lost when the artifact was removed from its original setting. A good exhibit, those in favor of interpretive exhibits argue, captures not an artifact out of time but a moment in time, or change over time. An object in context, they argue, is closer to the truth than the object by itself. It is also more persuasive, telling a story that visitors are more likely to pay attention to and believe. It can explain and interpret and engage. *New York Times* art critic Holland Cotter, lauding the Met's 2015 "Kongo: Power and Majesty" exhibit, called the label text "a model for the kind of truth-telling approach that museums could, and should, be taking to art: factual, incisive, politically astute, connecting the past to the present and inviting argument." Good interpretation "could wake people up; compel them to stop, look and read when they might have passed by; and prompt them to see that art isn't just about objects—it's about ideas, histories and ethical philosophies."[35]

Opponents of the interpretive exhibition argue that museums need to remain a place of art and science, not entertainment or education, and certainly not politics; that they need to double down on the pure power of the authentic artifact. Contextual displays, they suggest, get in the way of appreciating the real thing, replacing it with a moment of fake time travel. A 2016 essay in the conservative art journal the *New Criterion* decries a century of museum interpretation: "From offering an unmediated window onto the real and astonishing objects of history, the contemporary museum increasingly looks to reify our own socially mediated self-reflections."[36]

But all museum display is mediated. There's nothing natural about things in museums. The exhibition is a designed medium. When Mark Dion arranged the collection of surviving Jenks Museum artifacts, he started with a consideration of organization and juxtaposition, the interpretation, the story to tell. But that was only one part of the work of exhibition design. He considered the exhibition case, sketching a plan that captured the style of the Jenks Museum cases but that was also modern. He designed a case that could be built for the funds available, that would protect and secure the artifacts, and that could be installed easily into a modern room not designed as a museum. He thought about the flow of visitors through the space, the ease with which they might read the labels, see the images, and examine the artifacts.

All of these challenges, and more, are the work of the exhibition designer. He or she needs to decide how best to tell the story the exhibition team wants to tell—more than that, how to work with the exhibition team to shape the story—so that it is accurate, dramatic, attractive, accessible, and functional, intriguing but not overwhelming. The design must take into account budget and schedules. It must be a safe home for the objects, keep them secure and environmentally sound. It should be easy to maintain, and look good for the duration of the exhibit. Exhibition design is a collaborative process, with the exhibition designer working together with the rest of the exhibition team to create an environment that meets a complex set of desiderata.

The first set of considerations are those of visitor well-being. Exhibition developer Judy Rand offers a "visitors' bill of rights" aimed at making visitors feel welcomed, accepted, and respected. She lists eleven visitor needs for museums to meet. The first three—comfort, orientation, and welcome— set the tone for a visit. If visitors can't find the bathroom, if they get lost or feel unwelcome, they're unlikely to enjoy the exhibitions.

A second set—enjoyment, socializing, respect—make for a beneficial relationship between the museum and its visitors. Only if they treat visitors as human beings, as whole people, Rand suggests, will museums be able to meet their goals. Acknowledge that visitors are there to have a good time. They most often come in groups, with others, and so it's important to let them find ways to talk with each other, to share the experience. And they come with a wide range of interests and expertise: don't "exclude them, patronize them or make them feel dumb."

A final set of considerations defines the work of the exhibition. Exhibits should communicate accurately and honestly, respect visitors' learning styles, and give visitors the freedom to choose their way in the museum. Exhibits should offer appropriate challenges that fully engage the visitors—Csikszentmihalyi's flow. If all of this works, then visitors will feel refreshed and restored. They'll come back again.[37]

Walt Disney Imagineering president Marty Sklar sums this up in two simple phrases: "know your audience" and "wear their shoes." Don't bore people, talk down to them, or lose them by assuming that they know what you know. Experience your museum as visitors do.[38]

Exhibition design is complex, a three-dimensional jigsaw puzzle of objects, information, emotion, and, of course, the visitor. The challenge of exhibition design is to take the ideas of the exhibition developers—what do you want visitors to see? to learn? to feel?—and make them work in physical space. Designers do that by arranging objects, considering the ways that light, color, and texture shape our interactions with them both close-up and at a distance. They take into account the use and placement of media, interactives, and text. An exhibition is a total environment, and good design takes advantage of that. It must meet the requirements of diverse visitors, including both physical and intellectual accessibility. It must meet the demands of the artifacts—light levels, environmental requirements, and security. And it must meet the legal demands of the building code, the practical demands of maintenance and repair, and the economic demands of budgets.

Small museums often move from scripts and object lists to final exhibits quickly, arranging things on the fly. In many large museums, exhibition designs are worked out in great detail before any object is moved into the gallery.

It starts with the objects. Collections managers measure and photograph each object, update the database, arrange for conservation, and determine any special needs—mounts, lighting, environment, security. Registrars negotiate and secure loans.

The curators and designers work together to determine spatial arrangements, such as where the walls should go and how to shape the ways visitors encounter the art and artifacts. Together they make many sketches, create many versions of files on computers. Some designers print out tiny versions of each painting, make models of each object to scale, and move them about in a model of the gallery space, peering in the doorways and down the hallways to consider views from every angle. Some designers mark the floor of the gallery with tape so that they can walk about the space. Many museums make full-size mockups of the objects to be displayed, moving them around to get sight lines just right.

The sketches and models and walk-throughs are then converted into plans. A large exhibition might have hundreds of pages of blueprints for the space—construction, electrical, lighting, security systems, floor and ceiling treatments, and more—as well as dozens of printouts showing the layout of each wall, including each label. More documents describe interactives, lighting, the details of case design, sometimes the details of object mounts.

The plans, after approvals, are given to the builders. Exhibitions are expensive. Imagine a construction project that's all custom and all detailing. Erich Zuern, an exhibition producer, offers estimates from $200 to $700 per square foot, increasing as the level of media and interactivity increases. Art exhibits tend to be simpler and cost less; aquariums and science centers cost more. Fabrication and installation represents about 68 percent of that total; 8 percent is research and planning, 18 percent design, and 6 percent project management. Conservation, education and public programs, websites and social media, and marketing and publications all add to the cost.[39]

Over days, weeks, or months, skilled construction workers build walls, install display cases, hang ceilings, and finish floors. Art handlers and curators settle the final placement of art and artifact and move them into place. Lighting designers fine tune the lighting. Media experts install monitors, run cables, and check programs. Conservators check light levels. There's always a few final corrections to the text. And then the preview party, and the next day: another exhibit.

EXPLANATIONS AND ENCOUNTERS

THE LOST MUSEUM INSTALLATION did not have many words. The Jenks Society decided not to add new explanations to the artifacts but rather to allow the things and their arrangement to speak. Two large labels, entitled "Life" and "Death," printed in white on black fabric, hung above the display case. Both poetry and explanation, the words offered insight into the lost museum and its founder: "Imagine it! A world of wonder, preserved for the ages." A few quotes from Jenks and his contemporaries provided information about the museum. On the wall next to the storeroom and the re-created Jenks workshop, small plastic rectangles that matched the other door signs in the building read "Museum Storage" and "J. W. P. Jenks, Naturalist," blurring the lines between past and present. In the storeroom, artists' reimaginings of museum objects were described by an accession list, contemporary but done in the style of the 1880s. There were few words but many voices, historical and contemporary. The installation offered a complex context that fit the artifacts and told the story. Labels carried a range of meanings in both content and style.

Chapter 11 considered the organization of artifacts. This chapter and Chapter 13 consider context, the rest of the exhibition—everything other than objects. An object might be accompanied by an explanation: words on a label, an audio with an interview of its creator or the opinions of an expert, a video of the object in use. It might be encountered in a setting, anything from a detailed reconstruction of its original location to a

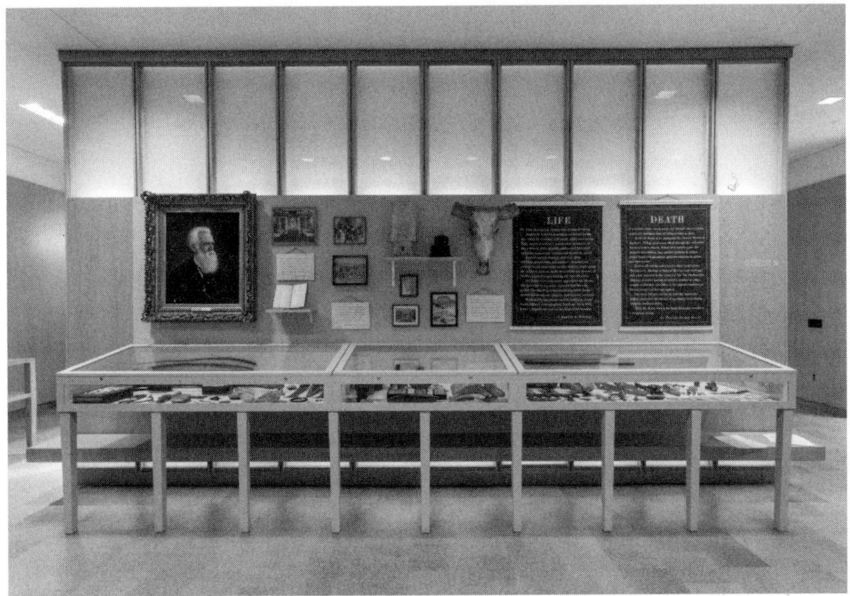

Part of *The Lost Museum* installation. "Friends, cast your eyes on these shattered remnants and know that all things return to dust."

soundtrack and walls painted a color designed to evoke specific feelings. A docent might explain things, or a costumed interpreter might act them out. Objects and explanations work together to capture interest, provide information, tell stories, and evoke emotions.

The Problem of the Label

Cabinets of curiosities didn't need labels. You would visit with the owner, talk about the objects, learn their stories. The objects weren't there to teach order or system, or as specimens, to represent a type. Each stood on its own, for itself. Historian Barbara Stafford writes that these objects offered "garbled messages," "snatches of muttered speech." They were there to provoke conversation, to be read and discussed by experts.[1] One of the earliest pictures of a cabinet of curiosities, Ole Worm's mid-seventeenth century Museum Wormianum, shows labels on boxes ("Metalia," or "Animalum Partes," for example) but no labels on individual things—an invitation to discuss the relationships of bits and pieces and categories.

As Enlightenment ideas of order and system replaced the baroque idea of individual objects fascinating on their own, and as museums turned to face a larger audience, explanations became more important. Louis-Jean-Marie Daubenton, chief curator at the Paris natural history museum in the mid-eighteenth century, labeled each object. The label and the arrangement were designed to "relay an authoritative knowledge from the curator to the visitor." In some ways, writes museum historian Tony Bennett, naming and relationships were more important than the thing itself.[2]

Nineteenth-century American museums offered a wide variety of explanations to their visitors. William Maclure, the director of Philadelphia's Academy of Natural Sciences, believed in the Pestalozzian system of education, which encouraged close observation of nature. He wrote in 1826 that "a careful examination and inspection of the objects themselves [is] calculated to demonstrate their properties and bring them within reach of the senses. . . . Natural history, in all its branches, is learned by examining the objects in substance or accurate representations of them in designs or prints." The object would speak for itself, part of what Bennett calls a "culture of sensory democracy" that rejected expert intervention and encouraged direct sensory connection with objects, so that every observer might judge the truth for themselves.[3]

But while many museum people accepted that they could understand nature through the senses, they increasingly came to believe that the public needed things explained, or, at least, made more interesting. Popular museums like Barnum's supplemented their displays with catalogs, "an illustrated review of the principal objects of interest in this extensive establishment, useful to the visitor for purposes of reference, entertainment and instruction." These gave not just the name of the things on display but also explanations that included details specific to the specimen ("The Black Bear, in this case . . . was one of the largest bears which, up to that time, had ever been taken in the country") as well as some general description ("This species of bear is now very abundant on the Rocky Mountains.") Occasionally the catalog told readers what to think of the objects on display. For example, an automaton is described as "one of the most wonderful and perfect pieces of mechanism ever produced" and a case of Burmese, Chinese, and Japanese idols and images is editorialized thus: "However uncouth these unsightly and miserable attempts at sculpture appear to the inhabitants of civilized nations, they are highly revered and valued at home, where they receive the worship and adoration of millions of ignorant Heathen."[4]

By the end of the nineteenth century most large natural history museums gave the scientific name, common name, location collected, and name of the collector for each specimen, as well as additional information, to make the object interesting and connect it to a larger story. The public needed explanation. "Above all," wrote Sir William Fowler of the British Museum, "the purpose for which each specimen is exhibited, and the main lesson to be derived from it, must be distinctly indicated by the labels affixed."[5] George Brown Goode went even further. An "efficient educational museum," he wrote, was best regarded as "a collection of instructive labels, each illustrated by a well-selected specimen."[6] Labels offered experts a chance to teach. Historian Tony Bennett sums up the triumph of the label writers, the believers in museums as a place to teach visitors to see the world as experts see it: "No matter how much things were said to be able to speak for themselves . . . there was, in museums, an incessant effort to provide a written supplement that would help anchor their meaning."[7]

Art museums started providing labels identifying artists by name at about the same time natural history museums began labeling their specimens, in the late eighteenth century. The change was controversial—"an insult to the eye of the connoisseur" was one reaction to Christian von Mechel's labeling paintings with the name of the artist at Vienna's Belvedere Palace in 1780—and the controversy over how much the visitor should know, and how much the museum should teach, would continue for a long time.[8] The Louvre broke new ground in the 1790s when it provided information about the artist and the subject, and published cheap catalogs to help the general public—not connoisseurs—identify the art on the walls. An 1836 report on London's National Gallery recommended that paintings from each school be kept together, with the name of the school printed high on the wall. Each painting, the report declared, should be labeled with the name (and birth and death dates) of the painter, and occasionally, information about his teacher or most celebrated pupils. This "ready (though limited) information," providing historical context in very few words, would offer the working classes a quick lesson in the history of art.[9] Popular American displays of art offered catalogs, not labels. These give long lists of paintings, most of them with only a title and artist. A few works are singled out for more information: a declaration of its greatness, a notice of the famous people who had previously owned it, obscure characters or symbolism explained. Occasionally the catalog called attention to an artist's technique.

The debate continued in the early twentieth century. "There are those," wrote Frederic Lucas of the Brooklyn Museum, "who consider that objects of art need no labels, that their mere beauty is sufficient justification for their exhibition; but this I believe to be a sad fallacy." On the other side of the debate was the Boston Museum of Fine Arts' Benjamin Ives Gilman. Gilman made his feelings about labels clear in an essay titled, simply, "The Problem of the Label." Some of the words he used to describe labels include "unavailable," "unsightly," "impertinent," "fatiguing," "unsatisfactory," "atrophying to the perceptions," and "misleading." Gilman wanted visitors to the Museum of Fine Arts to appreciate art. Labels, he felt, got in the way, "interposing a rush of abstract historical ideas and even indifferent registration data between the visitor's perception and the artist's intention." Many art museum directors of the twentieth century echoed Gilman's ideas that museums should do nothing to interfere with the visitors' direct engagement with the art.[10]

But the public wanted explanations. A 1929 survey of one thousand visitors to the Pennsylvania Museum (today's Philadelphia Museum of Art) found "that the public desires more information about the Museum, its exhibits, its acquisitions and the significance and meaning of the various objects of art."[11]

Gilman and others struggled to find a solution to the problem. They acknowledged the value of providing visitors with information about art and debated the best way to do that. Gilman suggested that perhaps curators could write "a few sentences of simple information about a room or object succinct enough to be read . . . in the twenty seconds devoted to any one object by the average Sunday visitor." (The museum staff were not interested, and very few of these labels were written.) Visitors might pick up a gallery book. "According to the testimony of a recent Visitor," Gilman told his staff, "the book transforms the netsuke collection from a spectacle mainly indifferent into a mine of fascinating interest." Maybe docents could provide a conversational introduction.[12] (More on that in Chapter 18.)

Gilman's successors also grudgingly admitted that words had their uses. Alfred Barr, founder of the Museum of Modern Art, wrote in 1934 that "words about art may help to explain techniques, remove prejudices, clarify relationships, suggest sequences and attack habitual resentments through the back door of the intelligence. But the front door of understanding is through experience of the work of art itself."[13] Dominique de Menil, cofounder of Houston's Menil Collection, believed that "perhaps only silence and love do justice to a great work of art." Her museum has no didactic labels on the wall or media in the galleries, the better to offer a direct encounter with art. (Though de Menil did continue, "and yet words can be illuminating.")[14]

The debate over labels continues today. G. Ellis Burcaw's widely read 1975 *Introduction to Museum Work* declared that "often the labels accomplish more real education in an exhibit than the objects," but many directors, educators, and curators have pushed back, asking whether anyone actually reads them, and suggesting that they are a distraction from the objects on display or a deterrent from learning from the objects directly for those who do. They also debate what labels should do: identify objects or describe them, help the visitor look at an object or put the object in context? Should labels offer an authoritative explanation or raise questions for the visitor to explore, and if an explanation, should that be an official museum view, the view of the curator, or the thoughts of an invited member of the community whose work is on display? Maybe labels should be signed, so that the visitor knows who is speaking. What knowledge, and what level of interest, should visitors be assumed to have? Should the label be long or short, focus on big ideas, groups of objects, or the individual thing being labeled?

One thing everyone agrees on is that good labels are difficult to write. Gilman, when pushed to write labels, admitted that "to compose them is a task of great difficulty."[15] "There are few forms of literary work," wrote George Merrill, a curator at the U.S. National Museum, "that require greater care than label writing."[16] Label writing, harrumphed Joseph Grinnell of Berkeley's Museum of Vertebrate Zoology in 1921, "is not a chore to be handed over casually to a '25-cent-an-hour' girl, or even to the ordinary clerk. To do this essential work correctly requires an exceptional genius plus training."[17] The museum literature offers endless remarks along these lines.

Stanley Freed, curator of North American ethnology at the American Museum of Natural History, describes the process of writing a technical label: "You have to go to the catalogues, you have to review the labels, and you have to look at all the pertinent literature—at least the major things and the major sources, and read it all and understand it. And then once you've digested it all, then you sit down and present your two hundred or one hundred words or whatever it is, concisely stating what it's about."[18]

But getting the facts right, and stating them concisely, is only the first step. Frederic Lucas at the Brooklyn Museum argues that "nothing is easier to write than a technical label. . . . To write an interesting and popular label is often a difficult matter."[19] Words in museums are not merely about facts. They're about telling a story and reaching an audience. Samuel Langley, Secretary of the Smithsonian Institution, offered a good example when he took on the job of honorary curator of the institution's first exhibit for children, in 1901. He thought carefully about labels and decided to ban Latin names for the birds in the exhibit in favor of popular information, even poems. He left one hummingbird labeled traditionally, writing that it "bears bravely its technical title, *Ramphomicron microrhynchum,* left by the honorary curator as the best explanation of why he has not retained the others."[20] Other museums followed suit. The American Museum of Natural History's bird labels dropped scientific names and range descriptions in favor of lively descriptions of behavior.[21]

It's hard to be thorough, accurate, informative, open-ended, appealing, and useful to a diverse audience, and produce text that is easy to read at the same time. Study after study reveals that museum visitors find labels a challenge. Mostly, they ignore them. "On average," reports one typical study, "about three-quarters of visitors to museums will be unable to pay attention to at least two-thirds of the labels because the vocabulary and sentence structure are too difficult."[22]

The literature on museum labels delights in pointing out their problems. Beverly Serrell lists "ten deadly sins" of labels, including being hard to read, difficult to understand, and hard to match with objects.[23] Rebecca Mileham offers this assessment: "Too wordy, too worthy or too woolly to do their job of communicating."[24] There's something unnatural about the form; after all, what the museum has to offer is not words but objects. Reading standing up, amid many distractions, is far from ideal, and visitors don't spend much time doing it. They seldom have specialized knowledge and often lack confidence in their ability to understand things.

Label-writing experts think about the reception of information. What matters is what the visitor gets, not what the museum presents. The developing consensus on what makes a good label, based on asking visitors what they want, watching what they read, and surveying them about the experience, includes advice on everything from subject and narrative structure to word count and typography.

Labels should connect directly to the objects nearby, answering questions visitors have about them and helping the visitor look at them. Surveys at the Museum of Modern Art suggest that visitors want to know: What is this thing I'm looking at? What does this symbolism mean? How was the work produced? What did the artist think?[25]

Good labels use details and anecdotes to tell a story. They have a narrative arc, a provocative first sentence and a conclusion that connects back to it. Active verbs remind visitors that people made and used the objects on display. Emotions and personal associations enliven labels. Labels should balance what visitors want to know, what they need to know to understand what they see, and what the curator wants them to know. Labels should answer questions and provoke interest and further study, and never tell visitors what to think without telling them why.

Shorter is better. Section and introduction labels should be no longer than 150 words, object descriptions no more than 80 words. Simpler is also better. A label should focus on one thing at a time, one idea per sentence, and not too many adjectives. Short sentences are best. A good label divides information into smaller, readable parts, perhaps with a short summary sentence or two at the start.

Words, once written and tested, are printed and put on the wall or panel. The art of graphic presentation has come a long way since the handwritten labels Jenks attached to the objects in his museum. Recent studies suggest that labels should be part of a unified, cohesive scheme of size, color, typography, and location that signals which labels describe objects, which describe larger themes, and which offer contextualization or other points of view. Good label design attracts attention but doesn't overwhelm the objects. It is consistent and offers the visitor clear signposts. Visual details matter. Good labels should be easy to read: large type, left-justified, no hyphenation, short line lengths, high contrast, and in a color that makes them stand out from the background.[26]

Well-written words in any medium, properly designed and located, offer visitors information when they want it. Most art museums continue to use "tombstone" labels, which offer just the facts for each artwork, perhaps supplemented by general labels offering an overview of an era or an artist's work. Some museums have added the artist's voice, and some provide a critical assessment.

Many museums have moved away from a simple, authoritative voice in their labels. Culture is more than just facts, more than simple assertions. Whose voice is heard is a political decision, so exhibition curators consider carefully whether the tone should be personal, curatorial, official, or institutional.

One option is to present dissenting opinions, so that dueling labels carry on a debate. Such labels can include quotes representing diverse views, as long as the label makes it clear who's speaking. This approach can be especially useful in history exhibitions that address controversial topics. The curators of the National Museum of American History's 1998 exhibition on sweatshops carefully considered whose voice should be heard, and made it clear in each instance who was speaking. An institutional voice—"This exhibition places the current debate on sweat shops in the garment industry in a historical context."—was personalized with the title "Why do museums mount this kind of exhibit?" signed by the director and associate director. The curators signed their own label, stepping back to answer the big question about sweatshops: "Is it getting better?" Six industry and labor spokespeople were offered space to give their opinions. Labels quoted early-twentieth-century sweatshop workers. Workers in contemporary sweatshops told their own story in video interviews. Finally, visitors could join the conversation by writing in talk-back books with the prompt, "Please join the dialogue and share your views on this topic."[27]

Another approach uses historical individuals to bring stories to life—and present challenging issues with the complexity they deserve. For example, at the Museum of the Civil War Soldier, videos presented actors portraying individuals on different sides of the slavery issue in the 1850s. An exhibition of a slave shackle at the John Hay Library at Brown University offered nineteenth-century voices: an abolitionist who had used similar shackles in his antislavery campaign; a quote from a book by a former slave; an abolitionist's recorded testimony of a slave's suffering; and a poem, written by a former slave.

Such voices need not all be from the past. The New-York Historical Society invited politicians and labor leaders to comment on the objects on display in visible storage, as a way of making clear the connections between past and present without the museum needing to take a side in political debates. Contemporary voices are particularly important when the artifacts of indigenous peoples are displayed; there's too long a history of their voices not being heard, their continued existence unacknowledged, their knowledge unappreciated.

Those early twentieth-century curators were right: writing labels is hard work. Writers sweat over their labels, even though they know that many visitors will not read all of them. But every so often, artifact, words, and visitor interest line up. The exhibition works: curiosity is satisfied, interest piqued, information or ideas transmitted, even a life changed. That makes the effort worthwhile.

Beyond Labels

Printed labels are not the only place for explanations in exhibits. Museums also turn to the high technology of their day to tell their stories. "The time may not be far distant," one museum expert suggested in 1904, "when we shall be able, by dropping a cent into a phonograph by the side of interesting objects in the museum, to secure the pleasure of a short discourse on the exhibit."[28] Sound and moving images can provide context and reach visitors who would rather listen and watch than read. When the Museum of Science and Industry opened in Chicago in the 1930s, it promised "one hundred motion-picture projectors" to give "one-man shows of such events and machines as cannot be brought under a roof."[29] A 1940s photograph of the Museum of the Rhode Island School of Design (RISD) shows a visitor sitting on a bench with a speaker to her ear.

But it took a long time for audio and video to fulfill their promise. Not until the 1960s did audio became an accepted part of museum visits. When the Metropolitan Museum of Art offered its first audio tour in 1963, the museum claimed it as "the most important single innovation in the Museum's educational program for the public in many years."[30]

Exhibition design after World War II drew not only on new technology but also on the more practical revolutions in design, entertainment, and information delivery found in stores and at world's fairs. The exhibits modernization program at the Smithsonian aimed to make exhibits "more instructive and entertaining to the millions of school children and casual visitors" who visited, and to compete with "the more attractive methods of graphic presentation of ideas in the world outside museum walls." A curator there noted that visitors were used to "the perfection of colored, talking motion pictures, eye-arresting magazine advertising and store window display," not to mention television. Museums were "polishing the old Rolls Royce," the curator wrote, while their visitors were "vainly looking for some evidence of streamlining."[31]

John Ewers, the curator at the National Museum of Natural History responsible for modernizing the Native American exhibitions in the 1950s, described the new principles for museum display. This was a response to a visitor survey that found that visitors to the present exhibit were confused by the tight-packed displays: there were more than 6,000 objects, many of them similar and of interest only to the specialist. Overwhelmed by the crowded shelves and frustrated by the lack of a logical order of the displays, it was rare that a visitor spent more than 30 seconds at any case. But visitors enjoyed the mannequin groups and wanted to know more about the subject matter and the use of the objects on display.

The new exhibition would have a much clearer structure, with a few primary displays that gave an overview of eight specific cultures. New lighting meant the life groups could be seen better. New small-scale dioramas offered "a graphic synthesis of Indian life." Photographs and drawings showed objects in use. Maps showed locations. The new exhibit was the shared work of the curator and the exhibit designer, one responsible for science, the other for "aesthetic values" and "clarity of interpretation." Together, they had the goal of offering visitors "healthy entertainment and instruction."[32]

Museum curators, once responsible for simply putting collections on display, became members of a team that included label writers, exhibit designers, and preparators. Many large history and natural history museums began to update their exhibits, some of which dated back to the turn of the century, to increase their appeal to a wide audience. Over the next few decades museums would experiment with new ways of presenting information in exhibitions. Some went high-tech. Some offered labels written with diverse audiences in mind, in multiple languages, aimed at different levels of interest and expertise, and in a range of styles of presentation. Some turned to other ways of presenting information, from interactivity to immersion. Exhibitions are not an easy medium, but they are a wonderfully open-ended one. Offering context to artifacts is both an endless challenge and an opportunity for creativity.

Please Touch

At a basic level, every museum exhibit is interactive. Visitors' views change as they move through space. They start and stop, look here and there, discuss objects with friends, snap pictures to tweet or share on Instagram. They interact with objects, spaces, words, companions, strangers.

But some exhibits are designed to let visitors more actively engage with the display—sometimes with the objects themselves, sometimes with devices nearby. The visitor does something and the exhibit responds. Science and technology museums were among the first to make exhibitions interactive, building on a long history of popular science demonstrations at colleges, scientific organizations, and even by itinerant science lecturers who would visit towns to present lively shows of electricity. Children's museums were interactive from their founding at the turn of the twentieth century. Art and history museums were more reticent, fearing that anything beyond quiet enjoyment would debase the museum experience, but by the end of the twentieth century many of them had also embraced a more interactive engagement.

Working machinery delighted viewers to the 1851 Great Exhibition and later international expositions, and by the turn of the twentieth century many technology museums had working models in their galleries. In 1910 the South Kensington museum experimented with interactives. "By pressing a button electric current was applied, and the diminutive ma-chines were at once put into action. . . . It was full of interest to see the thing move."[33] Interactive displays made machines more exciting than static displays and brought them closer to showing their actual use. The Deutsches Museum in Munich, which opened in 1925, set an example many museums of science and technology would follow. Its founder, Oscar von Miller, created interactive experiments to demonstrate science and technology. Visitors could X-ray their hands, see what happened when chemicals were combined, or set a steam engine in motion—all at the push of a button. "Everything was alive, everything moved. In fact, you moved it yourself. . . . You could walk right up to a piece of apparatus, push a button or pull a lever and see for yourself how it worked. There wasn't a 'verboten' sign anywhere," an admirer wrote.[34]

"The case for tactile education in museums grows stronger every day," *Museum News* noted in 1926, and educators encouraged museums to allow visitors to actively participate in the exhibits. Museums became livelier. Natural history museums included ant colonies and bee hives. Visitors could push buttons to take a quiz on electricity, see a model rattle-snake rattle its tail or a chameleon change its colors. They could walk through a butterfly garden.[35] The new Chicago Museum of Science and Industry, founded in 1933 on the model of the Deutsches Museum, em-braced what its director Waldemar Kaempffert called the era's tendency to "vivify teaching." He promised that visitors "will be able to feel and touch, push and pull."[36] Los Angeles' Griffith Observatory, reported *Popular Science* in 1936, offered sixty-four exhibits in which "experiments in physics, chemistry, geology, and astronomy work themselves as if by magic . . . a scientific education at the push of a button."[37] The governor of Massachusetts visited Boston's Museum of Science in 1939 to see the mu-seum's first interactive exhibit—"the first in the world," the museum claimed, to use mirrors "and fancy lighting" to show animals "changing their Summer coat of brown to a Winter coat of white."[38] When the Smith-sonian's Museum of History and Technology (today's National Museum of American History) opened in 1964, it offered "a whole battery of inter-pretive methods" to enliven its exhibits, including "craft and machinery demonstrations, barnyard animals, much appropriate music, and many other lively, mood-producing activities."[39]

Art museums were slower to engage with these new styles of interaction, preferring to offer galleries as places of quiet contemplation and learning. The Metropolitan Museum of Art offered a slightly interactive opportunity to visitors in 1901—they could slip their hands into the side of a display case to turn the pages of an art book.[40] The RISD Museum experimented with push-button interactivity in the 1940s: Visitors could push a button to rotate a sculpture to see the other sides.[41]

Interactives have always had their critics. The curator at London's Patent Museum was the first in a long line of unhappy curators when he told a parliamentary hearing in 1864 that working machines would make for "a more popular museum, and a less instructive museum." He worried about dumbing down: "If you make the machines popular you cease to instruct."[42] The complaint that children in museums were just pushing buttons and turning handles, and not paying attention to what happened next, appeared at the same time as those buttons and handles. Exhibit designers worried about gratuitous interactives, where the visitor gets to push something or move something just for the sake of pushing and moving; fun, but not a way to explore the material on display. Curators worried about the primacy of objects.[43]

The Exploratorium rethought the nature of interactivity when it opened in San Francisco in 1969. Instead of having visitors interact with exhibits to answer questions, let them ask their own questions, creating a constructivist interactive exhibition. The goal was to make science "visible, touchable, and accessible," to get visitors to explore, ask questions, and find their own answers. A classic Exploratorium interactive would ask the visitor to do something, suggest what he or she should notice, suggest ways to think about it, and then ask, "so what?" to encourage further thinking. Directions, diagrams, questions, hints—visitors interacted with the exhibitions as well as with family, friends, and strangers.[44]

Supporters of interactives point to studies showing that when exhibits do things, people pay attention—and they pay the most attention if they can operate the exhibits themselves. Studies suggest that interactive devices encourage visitors to look more closely at the objects near them, to spend more time in the gallery, and to work cooperatively with others. Visitors are also more likely to remember their visits and what they learned when they interacted with a display.[45]

Interactives take a wide variety of forms. At their most basic, they allow the visitor to change something in the exhibit by pushing a button or moving a lever. More interesting interactives allow visitors to gather evidence, make choices, and make a change in the exhibit, and see what happens. In a good interactive, visitors are acted on by the exhibit as much as they act on it.

Consider one of the simplest interactives, letting visitors open drawers, an increasingly common experience at history and anthropology museums. An installation at the Pitt Rivers Museum at Oxford allowed visitors to open drawers and look inside. A glass lid kept visitors from touching the artifacts within, but the sense of exploration—of finding secrets—made the experience very different from the usual visual engagement with collections. The drawers, wrote social anthropologist Sandra Dudley, "give the strong impression of being glimpses into the vast, hidden world of what the museum has in storage." They offered the visitor a surprise (some drawers are locked, the opportunity to open drawers is not advertised), perhaps a feeling of transgression (is this really allowed?), sometimes a revelation (look what I found!), and, most important, a chance to explore, to physically manipulate the display cases to discover something. Some visitors found it disconcerting; many found it exciting.[46]

Reenacting historic activities gives visitors a chance to engage with the past in more visceral ways than most exhibitions allow. A 1960s exhibit at the Farmers' Museum in Cooperstown, New York, provided a historic wooden bowl, the tools used to carve it, and a block of wood on a bench. A label invited the visitor to try making a bowl. "Both the demonstration by the craftsman and the do-it-yourself exhibit," wrote a visiting curator, "rank high in securing visitor involvement and participation."[47]

The Henry Ford museum has long operated the machines in its machine shop as demonstrations, but in the early 2000s it tried something new. Instead of a docent showing visitors how a lathe worked, she took on the role of an instructor, teaching one visitor how to run the machine.[48] One can imagine the consternation behind the scenes—you're going to let a twelve-year-old run that machine? But it worked. The visitor running the machine got a sense of the skills, attention, and effort required to do the work. The visitors watching the lesson saw someone make the mistakes they might make, ask the questions they might ask. They got to vicariously experience running the machine. Everyone was engaged. (And the museum got to charge the visitor for the experience.) Similarly, Colonial Williamsburg opened an educational musket range in 2016, advertising, "Come fire 18th-century firearms." Costumed interpreters taught visitors to shoot flintlocks, with a bit of history on the side.[49] In 2014 the Cooper Hewitt Museum created its Process Lab, a "dynamic interactive space" that gave visitors the opportunity to "brainstorm design solutions," to "play designer."[50]

Most art museums engage visitors interactively through docents, who offer deeper insight into works by guided looking. A new kind of art interactive encourages more physical involvement. "Sculpture Lens" at the Cleveland Art Museum encourages visitors to "Strike a Pose"—to imitate the posture of a sculpture—and "Make a Face"—a computer matches their facial expressions with an artwork in the collection. A letter the Detroit Institute of Art received from a visitor explains the appeal of interactives: "Most art museums focus on teaching you a history lesson with a page of text beside the artwork. The interactive displays at the DIA helped me learn by engaging me: by directing my eyes, by planting questions I wanted answered, and by helping my imagination."[51]

Interactivity—like the other "impressions and thrills" museum exhibit creators aim for—builds on the century-old idea that visitors get more from artifacts if they are presented as part of a story, and even more if the visitors get to help shape that story. It harks back to an even earlier moment in museums when touching and feeling were allowed, when visitors expected to interact with art and artifact. So too does immersion, the topic of Chapter 13.

＞ 13 ＜

SETTING THE SCENE

A VISITOR TO THE BROWN UNIVERSITY archaeology building in 2015 would have been surprised to see a taxidermist's workshop there. Jenks's workshop, reconstructed for *The Lost Museum* project, offered a bit of 1894 in a contemporary space. It is as though Jenks stepped out for lunch, a taxidermied bird under repair on his desk, and never came back. Nothing in the re-created space is authentic, in the sense that none of the artifacts belonged to Jenks. Some are generic: the taxidermy, hunting, and trapping tools, books on nature, and etchings of famous scientists are the sorts of things he probably had in his workshop. Some tell us about Jenks: the two canes on the chair, the ax leaning against the table, the Bible on the desk, and Darwin's *Origin of Species* on the floor under a chamber pot. Yet everything about the room is evocative, capturing the old naturalist's work and life. It is convincing. It can pull the visitor into the past, into another world.

The most dramatic way to restore context to a museum artifact is to set it on a stage. As natural history museums became interested in teaching not only taxonomy but also ethology and ecology, dioramas—taxidermied animals and sculptured plants posed in lifelike groupings before painted backdrops—became appealing, a way to bring nature alive in the museum. When anthropology museums focused on cultural groups rather than racial progress, they set up mannequins wearing authentic clothes, surrounded by authentic goods, in replica settings. In the late twentieth and early twenty-first centuries, museums wrestled with the political and

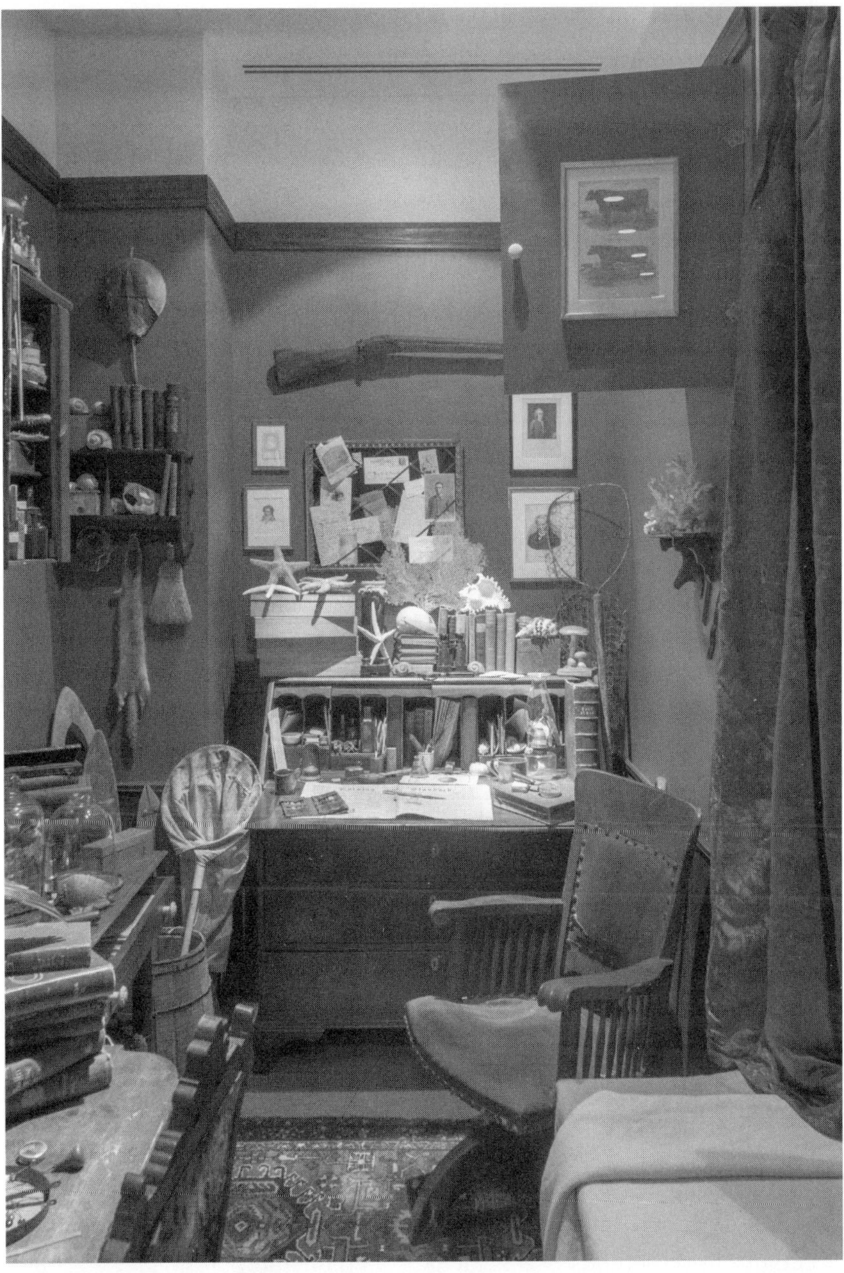

The label on the door reads "J. W. P. Jenks, Naturalist." Jenks's workshop, reimagined by the Jenks Society for Lost Museums.

ethical problems of putting people on display like this, repeatedly reinventing the medium.

History and art museums began to show artifacts in period rooms at the turn of the twentieth century. History museums used them both to show off artifacts and to treat visitors to a vicarious historical experience. Art museums debated whether to make a point of the difference between the current and the original setting for art, as the Louvre had consciously created a new space for art taken out of the church, or to design exhibits that evoked the original setting. Many ended up pulling elements from period spaces to suggest, not duplicate, the places the art was created for. Some experimented with period rooms.

Period spaces came in many formats: with mannequins or without; with real objects or with replicas; a moment in time or a general scene; with sound, light, and video effects or not; open, or closed off behind a velvet rope or glass wall. Each format had its use and a distinct set of meanings.

Museums considering these options drew on a long history of immersive displays of history and exotic cultures. In the nineteenth century huge crowds visited enormous panoramas of battles and voyages for the thrill of traveling back in time or to an exotic part of the world. An 1822 "Exhibition of Laplanders" drew some 58,000 spectators to London's Egyptian Hall to see real Sami people and live reindeer paraded in front of a panoramic painting, as well as clothing and tools on display in a setting designed to evoke a Sami home. A few years later, George Catlin's "Tableaux Vivants Indiennes" included paintings, artifacts, and, first, English impersonators, then actual Ojibwa performers. Expositions in cities in Europe and North America at the turn of the twentieth century put entire villages of "exotic" peoples on display, confirming racial and cultural stereotypes.[1]

Museums experimented with re-created environments, walking the fine lines between a stage set, a world's fair, and a traditional presentation, hoping to find methods of display that seemed appropriate to the art and artifact they wanted to feature and the stories they wanted to tell. They strove for backdrops that would highlight collections, provide context, and capture visitors' attention, but not seem fake or too theatrical. In the last decades of the twentieth century museums, considering the ethics of exhibition, found ways to allow the cultures on display a voice in telling their own story.

The Semblance of Life

In the late nineteenth century, institutions that had been content to serve as storehouses of specimens arrayed on shelves took on the new job of reconnecting Americans with nature and educating city dwellers, especially new immigrants. Museums believed that they could provide more accurate teaching than the increasingly popular lectures and movies offered by explorers and popularizers—if they could attract audiences.

The challenge was how to do this using real things, the specimens that museums were so proud of and which set them apart from other institutions. Some museums tried rearranging their collections, abandoning a strict scientific taxonomy for categories more useful to their audience: "game birds," for example. They rewrote labels to include information about the lives of the plants and animals on display. Some added drawings, watercolors of trees next to samples of wood, and large paintings of dinosaurs next to fossils.

The most important innovation was the "life group," a display of taxidermied animals in a natural setting. Curators worked with department store window designers and theater directors to set specimens into painted and sculpted backdrops, bringing a facsimile of nature into the museum. "The group idea," wrote one curator, was "the highest type of museum installation, for by means of it we are able to present not only the organism but also its environment to the notice of the visitor."[2] "Well-executed groups," a popular science writer noted about the life groups at the 1893 Chicago Columbian Exposition, "give life, reality, and meaning to the objects in the cases around."[3]

Carl Akeley created the first natural history museum life group at the Milwaukee Public Museum in 1890, a display of muskrats, and went on to perfect the medium at the American Museum of Natural History (AMNH). His masterpiece was the 1936 Hall of African Mammals, still one of the museum's most popular exhibitions. Akeley's displays put taxidermied animals and artificial plants into dramatic scenes and an environmental setting. Each was based on meticulous field research that set it in an actual place at a precise moment.

Life groups seemed true not only because of the taxidermied animals and preserved plants on display but also because of the deep research on which they were built. They had a focused coherence that argued for their authenticity. "It is Africa. Not only the animals, but the trees, the leaves and grass, and the earth itself were brought from the place," wrote the director of the American Museum of Natural History.[4] The public loved these presentations—the visitor, a Field Museum curator told his colleagues in 1916, "wants life, vividness, and action, or the semblance of it"—and so museums across the country competed to create ever more spectacular displays.[5]

In recent years many natural history museums have rethought these displays, which often said as much about the people who made them as the animals displayed. Many showed scientifically incorrect "nuclear families" of animals, for example. Their makers favored the largest animals, not typical ones, as more impressive. They featured charismatic megafauna at their rare moments of activity rather than more common scenes. They separated nature from culture to show nature as unchanging, stable, pure. In 1990 the Smithsonian's National Museum of Natural History added "dilemma labels" to some of its life groups to point out the racist, sexist, and colonialist attitudes embedded in the displays.

Museums ran into even more difficulty with the contextual settings they had built for their ethnographic exhibits. Ethnographic dioramas have their politically fraught roots in the imperial displays of nineteenth-century world's fairs, which carried messages of a benighted primitive people, the triumph of civilization and technology, and the benevolence of colonialism. Early dioramas were staged displays of primitivity and progress, with peoples arranged on a scale from barbaric to civilized.[6]

The Smithsonian's Otis Mason and William Henry Holmes built dioramas of American Indian groups at the 1893 Chicago exposition. These displays moved to the National Museum, where, at the turn of the twentieth century, visitors could see "primitive peoples" in dioramas based on "geo-ethnic" units. Each diorama showed a family of realistic mannequins, dressed in authentic garments and using authentic artifacts, engaged in activities of daily life or ritual as curators imagined them prior to European contact. Anthropologists liked dioramas for the way they contextualized objects. Margaret Mead wrote in 1965, "The feeling of being in the midst of life is due to the care with which the habitat groups show total scenes—Indians making pottery, Eskimos using their dogsleds, villagers on a Pacific island working in the shade of the houses built up on stilts. In every scene the trees are right and the color of the sky, too. The relationship between men and other living things is maintained."[7]

While anthropologists turned to dioramas as a way of providing context, they kept the context narrow, for the most part maintaining the "ethnographic present" that froze non-Western people in time, outside history, and within outmoded racial categories. Anthropologists Susan Harding and Emily Martin note that the African dioramas at the AMNH, created in 1968, present an ecological and functionalist account, placing African people in a zone between nature and culture, an "ensemble of human adaptions to natural and social environments largely uncompromised by Euro-American slave-trading and colonial incursions."[8] Cultural studies scholar Pauline Wakeham calls the ethnographic diorama "a spectacle of otherness permanently paused for the fascinated surveillance of the white spectator," turning Native people into objects of a "fetishistic colonial gaze."[9]

A protest at the AMNH in 2016, "Decolonize this Museum," noted the politics that underlay many of the exhibitions. It decried the way that "The Hall of African Peoples" misrepresented and oversimplified "the vast multiplicity of African social and cultural life": "Africans are depicted as pre-modern, bearing curious instruments and colorful costumes, instead of as present-day people." It connected the museum's presentation to contemporary concerns, arguing that "discrimination against African diasporic peoples is everywhere reinforced by these primitivist stereotypes." The protesters demanded that "the museum's display arrangements and classifications be reconceived by curatorial representatives of the 'exhibited' populations."[10]

The history of African ethnographic exhibits at the Smithsonian shows the challenges of this kind of display and some of the ways they might be improved. The exhibition label for the diorama of the Zulu-Kaffir tribe, created in 1915, describes a typical scene: "The group shows a section of the house with doorway; a fireplace on which a woman is cooking mush;

a woman dipping beer from a large pottery jar." The settings were inside cases, viewable from all sides, reinforcing the sense of a vanishing people, museum objects. There were also miniature dioramas showing building types and various arts, and individual mannequins showing off native attire. Like the natural history displays, these seemed real because they were so meticulous, so strange, based on a scholarship of detail.[11]

The National Museum of Natural History's Africa Hall, which opened in 1967, took a new approach to ethnographic exhibits. It was organized by ethnic group and topical theme. Short text labels described how things were used. Though it was set in contemporary times, it insisted on unchanging tradition: "This is rural Africa, traditional Africa, most of Africa. Here, where outside influence is only beginning to penetrate, most Africans still follow their traditional cultures or ways of life." Mary Jo Arnoldo, curator of African ethnology, wrote in 1999 that this was a "romantic and indefensible notion in the 1960s." The exhibition's "outdated and pejorative nomenclature," she continued, "reinforced stereotypes of Africans as primitive, exotic and savage," "an Africa outside of time."[12]

Protests closed the hall in 1992, and a new Africa hall, a collaboration with a wide range of stakeholders—academics as well as local African and African American communities, a core group of some sixty advisors—struggled with the challenge of presenting a more complex story. When it opened in 1999 it used a new kind of immersive environment. A diorama of a marketplace used life-size photographic cutouts, not mannequins; visitors read photographs as contemporary. They were photographs of real women, who told their story in their own voices. A Somali house diorama used photographic, not painted, backdrops. Labels include Somali proverbs and poetry, to present a Somali perspective. Two videos situated the house in the museum. In one, a Somali American family visits the diorama; the father discusses what the house means to him, and how the family can build on the culture it represents as they start a new life in the United States. The second shows Somali women directing the reconstruction in the museum. The curator hoped that by showing the house as both "a significant object of memory and as a museum object" the house would be not a "static diorama" but a "more dynamic dialogic display."[13]

The National Museum of the American Indian also reinvented the diorama. It was important to the curators that the museum not freeze Native Americans in one period but rather show their continued survival, a distinct change from the diorama tradition of the natural history museum. And so the museum worked closely with native communities to tell their stories, and set up dioramas of contemporary life, not only a historical native past.

Perhaps the most impressive ethnographic immersive environment was created at the Mashantucket Pequot Museum in 1998, a 22,000-square-foot sixteenth-century Pequot village. It dealt explicitly with the challenges of earlier dioramas. Visitors could walk into the scene, look inside of structures, and examine the dozens of mannequins engaged in daily activities. An interactive audio device let visitors hear native commentary on the historical display and its connections to contemporary native life. The environment showed change over time; the final part showed the village after European contact. The whole exhibit was alive with sounds and smells. Authenticity was assured not only by thorough research, presented in great depth in the audio guide, but also by native voices on the audio guide and even the making of the mannequins: "All the figures were life-cast from Native American people; the traditional clothing, ornamentations, and wigwams were made by Native craftspeople."[14]

From Period Rooms to Immersive Environments

History museums experimented with their own version of immersive display, the period room. Its roots were in historic house museums and in European national museums.

American historic house museums date back to the 1850s. Mount Vernon, preserved by the Mount Vernon Ladies Association, set the model. Women were responsible for most of these, part of what historian Patricia West calls the "aesthetic moralism" of the mid-nineteenth century women's movement, a way for women to leverage their power in the domestic sphere into a larger political sphere. Mount Vernon gave the country "a shared ancestral home and sacred heritage."[15] Historic house museums featured a famous individual and a house furnished with a mix of original furniture and similar pieces, set in a moment in time. They were shrines, a way to visit a domesticated, patriotic past.

Women also turned to historic domestic settings when they organized sanitary fairs in cities across the North to raise money to support Civil

War soldiers. Many showcased colonial kitchens and relic rooms that combined history, domestic morality, and patriotism. Expositions continued this tradition. Visitors to the 1876 Philadelphia Centennial could take tours of a "Connecticut Cottage" and an "Old Log Cabin New England Kitchen." In the "Old Log Cabin," guides "dressed in the costume of a hundred years ago" showed off "the wonderful articles of furniture and cooking utensils, whose very simplicity made them incomprehensible to the victim of modern improvements."[16] An engraving in *Frank Leslie's Illustrated Weekly* newspaper shows an impossibly large room with antiques scattered about. No one would have found this a convincing re-creation; it was more a set for old furniture than an authentic place.

Artur Hazelius, a Swedish curator, invented the history museum period room for the Nordiska Museet in Stockholm in the 1870s. Rather than organize works of art by material or type, he pulled together material from the same time and place and re-created the setting it might have come from, complete with mannequins in period clothing. In 1891 he went further and opened an outdoor museum where guides in costume re-created life in the "olden days." This was an era when European museums were eager to show off—and, in some cases, create—the roots of a national culture, and period settings allowed the public to imagine a national past. Many European national museums showed off dozens of period rooms.[17]

The rooms at the Essex Institute in Salem, Massachusetts, which opened in 1907, presented an American version. Curator George Francis Dow designed period rooms "to present a truthful picture of an interior." He had observed the splash dioramas made in natural history museums and wanted to bring that experience to the history museum. And, he noted, "paintings, furniture, objects of household utility or adornment never look so well as displayed in surroundings approximating those for which they were originally intended." His notion of authenticity was "atmosphere of liveableness," and so he felt free to use reproductions, dressing the set to bring it to life: "An effort was made to heighten the illusion of actual human occupancy by casually placing on the table before the fireplace in the parlor, a Salem newspaper printed in the year 1800 and on it a pair of silver-bowed spectacles, as though just removed by the reader." Dow's rooms seemed real because they told stories. A visitor could imagine that the inhabitants left just moments, not centuries, ago.[18]

Like the historic house museums, museum period rooms were often tinged with nostalgia and a political conservatism. So too were living history villages.

Colonial Williamsburg, founded in 1928 is a restored and re-created town used as a stage set for reimagining and re-enacting eighteenth-century history, the better to understand the "lives and times" of early Americans and appreciate their contribution to the "ideals and culture of our country." The authenticity of the setting was there, Williamsburg historian Cary Carson wrote, to lend credibility to "the great patriots."[19] Henry Ford created Greenfield Village, an outdoor living history museum, by moving a collection of historic structures—and sometimes even the dirt beneath them—from across the country to Dearborn, Michigan, in 1933. "When we are through," he wrote, "we shall have reproduced American life, and that is, I think, the best way of preserving at least a part of our history and our tradition."[20] In some "living history" museums, the politics was explicit. The American Iron and Steel Institute rebuilt the seventeenth-century Saugus Iron Works in the 1950s to celebrate it as a "living illustration of what individual initiative and American freedom can do."[21] Period settings evoked a nostalgia for "the good old days" to serve political ends.

The Smithsonian turned to immersion and interactivity, in a modest way, in its 1957 "Everyday Life in Early America." "For the first time in an exhibition of American cultural history in a large museum," the Smithsonian bragged, "it introduces the revolutionary new exhibit techniques first applied to exhibits of natural history." The exhibition presented objects "not only to exploit their own intrinsic qualities but also to illustrate ideas and tell a narrative history." It used new methods: illustrations, diagrams, models, maps, narrative labels, colorful backgrounds, dramatic lighting, pushbuttons to change the lighting from daylight to candlelight, and a tape-recording triggered when a visitor entered the space.[22] Curator Malcolm Watkins believed that showing artifacts in this way let the museum show "how it was made, who made it, and who the people were who used it, and why they used it, and how they reflect cultural manifestations."[23] Edward Alexander, Colonial Williamsburg's historian, explained in 1968 that placing "hundreds of small objects" in period rooms organized "myriads of small facts into meaningful wholes." "Authentic details" make these exhibits convincing, he said, "especially when the visitor supplies his built-in feeling of nostalgia to bring it all to life."[24]

Some of the rooms from *Everyday Life* moved into the new Museum of History and Technology when it opened in 1964, and many were still there when the building was renamed the National Museum of American History in 1983. Most are gone now. Period rooms seem old-fashioned in the era of ubiquitous video and interactivity. New research revealed inaccuracies in many of them. And they seemed inflexible, trapped not only in the past illustrated by the setting, but in the past history of the museum. A colonial room at the museum, wrote Spencer Crew and James Sims in 1991, looked authoritative, but it was inaccurate, serving merely to reinforce colonial revival ideas and what visitors already believed. Perhaps thinking of the "dilemma labels" at the natural history museum next door, they wrote, "No interpretive text disclaimer can outshout the room, with its rightness of scale and warmth of recognition."[25] Period rooms worked best, perhaps, as Alexander had suggested, on the level of nostalgia.

Crew and Sims wrote in an era when history museums turned to social history to tell the stories of workers, people of color, and other underrepresented groups. They were less interested in showing off elegant furniture than in teaching history, and they argued for a new style of period rooms. What makes them "real," they suggested, is an accurate telling of a moment in history. Artifacts add only "corroborative power" to the "historical expertise of the exhibition team." The insistence on museum practice that holds the "authentic" object to be inviolable, they wrote, meant that a period room couldn't be interpreted to tell a different set of stories, of a different moment. What seems to be important changes over time, and period rooms don't.

A few of these period rooms were reinvented, changed dramatically as views of history changed, and as new techniques become available. In "After the Revolution" (1985) eighteenth-century period rooms were repurposed to show not an idealized past but rather to tell the stories of specific men and women at moments in their lives when they had to make decisions about politics and culture. "We spent lots more on the research than on the installation," boasted Roger Kennedy, the museum's director. "The ideas arranged the objects."[26]

In their 1987 exhibition "Field to Factory: Afro-American Migration, 1915–1940," Crew and Sims tried a new approach. They assembled a "constellation of objects . . . not authentic in any sense to this particular house . . . in such a way as to create an image of an authentic event." That image was helped by theatrical techniques (Sims's background was in theater design): scrims, mannequins in dramatic poses, careful attention to sound and light.[27]

Another exhibit, the same year, created from scratch a cabin at a Japanese-American internment camp and made it "real" by re-creating one family's home, basing it on the original plans and outfitting it with "authentic" furnishings and artifacts; almost everything in the room was actually used in the camps. But what brought it to life for visitors was a video projection, life sized, of an actor playing a survivor of the camps visiting them with his daughter, and reminiscing about them. In this new kind of exhibition, the story—the social history, based on archival research and oral histories—provides its own authenticity.[28]

The U.S. Holocaust Memorial Museum, which opened in 1993, combined authentic things and deeply researched history to create compelling museum experiences. Leon Wieseltier, reviewing the museum shortly after it opened, noted the way it mixed history and memory, facts and things: "the vividness of recollection joins the sturdiness of research. The stinging subjectivity of the testimonies of the survivors is met in these galleries by the tart objectivity of photographs, films, maps, statistics and objects. . . . Feeling must be annotated by fact."[29] The use of objects reflected this. As cultural historian Alison Landsberg writes, "The experiential mode complements the cognitive with affect, sensuousness, and tactility." Objects—the museum called them "object survivors"—were both evidence and environmental cues. The cobblestones visitors walk on are real, from the Warsaw ghetto. The boxcar they walk through is one that transported Jews to Treblinka. The display is immersive. "There is no longer a clear distinction," Landsberg notes, "between your space and the exhibit, your body and the history of the objects all around you. Gone are the display cases." The "seductive tangibility" of the real things, displayed immersively, "draws you into a lived relationship with them."[30]

A new emphasis on story and new ways of thinking about authenticity define some of the best period settings in history museums and historic houses today. The Lower East Side Tenement Museum, like the "Field to Factory" room, places an assemblage of appropriate-but-not-from-the-building artifacts in authentic spaces to serve as an armature for storytelling. "I think of it as writing a visual historical novel," Pamela Keech, the museum's curator of furnishings, told the *New Yorker*.[31] "Open House: If These Walls Could Talk," a ground-breaking exhibition at the Minnesota Historical Society, tilted almost entirely to stories. Neither the space nor the artifacts were authentic to the history told. Rather, they served as an evocative space for exploration, provoking conversation and letting visitors connect on a personal level.[32]

Museums have also experimented with opening the fourth wall and allowing visitors into re-created spaces, to give them a more immersive experience. "America on the Move" (2004) at the National Museum of American History allowed visitors to walk into and through settings of trains, trolleys, and automobiles, and even inside of a rapid transit "L" car from Chicago to take a ride (with sound, light, motion, and video of actors) on a winter morning in 1959.[33] The Indiana Historical Society let visitors walk through photographs projected onto mist screens into a stage set: "Step into another era in 'You Are There.' Historic photographs are brought to life three-dimensionally, right down to the actors playing the people in—and around—the images. Become part of the story and see, hear and touch history because . . . You Are There."[34]

Perhaps the most fully immersive museum experiences are found at living historical farms. They allow visitors to transport themselves to a different time, to walk through historic spaces and connect with people and nature. From the visitors' point of view, their reality stems from the physicality of the experience: the smell of wood burning in a fireplace, the taste of bread made in a historic oven, the feel of a just-sheared sheep.

But behind the scenes of many historical farms and other reenacting experiences is a heated debate over competing notions of authenticity, over what makes them good history and good teaching, over how to produce and take advantage of the power of the immersive experience. "To function properly and successfully, a live museum should convey the sense of a different reality—the reality of another time," wrote archaeologist James

Deetz about his work at Plimoth Plantation in the 1970s. He pioneered "first person" interpretation there, using costumed interpreters (he called them "informants") to speak in the voices of people from the era. There were no authentic objects—Deetz sold at auction the antiques that had been donated, replacing them with replicas that could be used by the interpreters—but there was a feeling of authenticity based on deep research not only into material culture, but also language and culture. Deetz wanted to re-create not a place but a culture.[35]

Like Plimoth, many historic farms turned away from authentic artifacts to accurate reproductions and an authenticity of minutia. Get the details of the costumes and animals and furniture right, and learn the skills, some reenactors believed, and they would capture the reality of a time. That made the experience authentic for the reenactors but rarely for the visitors, who didn't have the knowledge needed to appreciate those details. The challenge for many living farms is to connect the history to the visitors and to the history beyond the immediate scene. Most living history sites use third-person interpretation, guides that may dress in period clothes but speak of historical figures as "them," in the past tense.

Second-person living history experiences, a fairly new kind of presentation, lets visitors pretend to be part of the past, a chance not only to go back in time, but also to engage with the spaces, objects, and people of the period. Most of these offer fairly traditional opportunities—helping with chores, for example. In the most adventurous programming, the visitor can cocreate the experience. The "Follow the North Star" program at Indiana's Conner Prairie Interactive History Park, created in 1998, put visitors in the very uncomfortable position of being a freedom-seeking slave on the Underground Railroad—"an intense, living drama where guests become actors on a 200-acre stage." This kind of immersive living history allows visitors to make strong emotional connections to the past, but requires careful design; without historical context, it can trivialize history and traumatize participants.[36]

Picturesque Treatment

Art museums, with their focus on the authentic object, moved more slowly than other museums to embrace immersive settings. Designers and curators worried about confusing the real and the fake, or even suggesting that the art was just decoration. Contextuality, worries architectural historian Victoria Newhouse, can become kitsch, a "fake stage set."[37] But done well, a contextually appropriate setting can introduce drama and bring art to life.

The trustees of the Boston Museum of Fine Arts (MFA) debated "Current Theories of the Arrangement of Museums of Art" at length in 1904 as they were considering their new building. On the one hand, they were told, German museums claimed that "artistic and educational truth may be attained by the suggestion if not by the reconstruction of this original environment." Designing rooms "in the style of the objects which they contain" was thought to help the visitor experience "more vivid emotion."[38] Wilhelm von Bode, the director of Berlin's Kaiser Friedrich Museum in the early twentieth century, set this style, arguing that "the chief aim should be the greatest possible isolation of each work and its exhibition in a room which, in all material aspects, such as lighting and architecture, should resemble, as near as may be, the apartment for which it was originally intended."[39] He displayed art, furniture, and *objets d'art* in the same gallery.

Other advisors gave the MFA trustees contrary advice. In this new style of exhibition, they warned, the room itself would become the exhibit and "the contents of the room become subsidiary to the room as a whole." Perhaps it would be acceptable for domestic artifacts. But "paintings and sculpture cannot be submitted to picturesque treatment, for they may not be subordinated to their surroundings. . . . A work of art . . . does not need the help of a background, but only to be unhindered by it."[40] Benjamin Ives Gilman advised the board that some backgrounds were necessary—after all, "a museum is a collection of fragments"—but that they should be "abstract . . . restricted to certain traits of dignity and charm." That would allow "the imagination of the beholder" to do its work.[41] The MFA listened to Gilman, designing its new museum to highlight the art.[42]

The Metropolitan Museum of Art embraced contextual settings more than the Boston museum did. It started slowly with what it called "period walls," walls from historic houses as backdrops for furniture. (Above the display, to make clear that this was not about lived experience, a sign read "American Art / XVIII Century.") The American Wing, which opened in 1924, had nineteen period rooms that showed off furniture from before 1815 in settings designed to provide historical context. They also provided an ideological context, showing off the good taste of early Americans—as well as that of their descendants who donated the rooms. "Traditions," read the catalog, "are one of the integral assets of a nation. . . . The tremendous changes in the character of our nation, and the influx of foreign ideas utterly at variance with those held by the men who gave us the Republic, threaten us and, unless checked, may shake its foundations." Immigrants visiting these spaces would be inspired by fine furniture to abandon their alien ideas.[43] The Met added European period rooms, parts of a Jain temple from India, and even, in 1938, an entire medieval stage set, the Cloisters. (The museum used more respectable language: "an ensemble informed by a selection of historical precedents."[44])

Art museum professionals had mixed feelings about the period room. In the 1920s Paul Sachs staged a debate in his Harvard museum studies class. On one side, some students found period rooms useful to the average viewer. They might not appreciate art, but rooms were interesting and related to their everyday life. Others worried that period rooms were lowbrow. They "cheapened the museum by too evident a display of showmanship. . . . The museum curator should not be forced to become an interior decorator." Sachs, true to his elitist perspective, hated period rooms. He even disliked the Isabella Stewart Gardner Museum, where paintings blended into the rooms. The Cloisters was more to his style, because of the way it highlighted the art.[45]

But Sachs, and many art museums, were out of touch with the new ways that their visitors were thinking about art and display. The Met's president, Robert de Forest, though he didn't fully approve of period rooms—they seemed too much like department store displays to him—acknowledged their value. "Period rooms," he wrote, "have proved themselves to be among our most attractive exhibits, and period rooms which are helpful in suggesting tasteful modern arrangement are highly educational."[46] Indeed, a 1929 survey at the Pennsylvania Museum showed an "overwhelming preference" for period rooms. Forty-four percent listed a period room as their "first preference." Women especially liked them. Only 26 percent mentioned paintings.[47]

The survey reflected wider trends in display, the excitement of the world's fair, the theater, and the department store. Newark Museum director John Cotton Dana insisted that people could learn more from shop windows than from museum displays. Department store magnate Samuel Wanamaker explained why. "In museums," he wrote, "most everything looks like junk even when it isn't because there is no care or thought in the display. If women would wear their fine clothes like galleries wear their pictures, they'd be laughed at."[48] A museum's job, theater designer Lee Simonson declared, "is not to chronicle art as a fact but to enact it as an event and to dramatize its function. Its role is not that of a custodian but that of a showman." Simonson urged museums to use "a few principles of stage setting and architectural arrangement."[49] A few took him up on it; most did not.

Some museums took a different approach altogether. Building on ideas coming out of the surrealists and the Bauhaus in 1920s Europe, a few art museums experimented with multimedia, hoping to produce a "dynamic interrelation" between art and the visitor through multiple large images and an "extended field of vision." A 1938 MoMA show on the Bauhaus surrounded visitors with visual information: photographs and diagrams hung by strings from the ceiling; the floor painted in geometric shapes; footprints on the floor; graphic hands on the wall. The goal, historian Fred Turner writes, was to allow visitors to "synthesize the elements presented to them into coherent internal pictures of their own."[50] Many critics and visitors were mystified, but this immersive style would be repeated in MoMA exhibitions during World War II.

Alexander Dorner's "atmosphere rooms" at the Rhode Island School of Design museum in the 1940s also built on European ideas. These exhibitions weren't re-created spaces but attempts, through sound, color, and illustration, to capture the essence of a culture. Dorner wanted rooms that would display each artwork in context so as to "exert the greatest possible emotional effect," and the essence of the period to "be mediated to the visitor emotionally."[51] To do this, he painted walls colors he thought appropriate to the period of the art. Ancient Greek art was displayed in a room with blue walls, Early Christian art with a beige background. Backlit transparencies showed scenes of buildings and cities. The exhibition of Greek art, Dorner wrote, tried to "bring out in color and arrangement" its "joyous character," the "confidence in physical life of the individual," and the "colorful atmosphere of the surrounding landscape." The Babylonians didn't have frames, and so a Babylonian fragment was installed embedded in a wall, and the patterns of the fragment extended on the wall around it. A drawing reconstructed what the original might have looked like, for context. Loudspeakers enriched rooms devoted to Egyptian, Babylonian and medieval art with music Dorner thought appropriate to each of those periods.[52]

Although no other museums followed Dorner's model, many began to use settings that suggested the period of the art. An exhibition designer wrote in 1949 that his job was to "visualize the atmosphere, shall we say, of the collection and try to bring this to the visitor's eye by choosing the right background."[53] In the later twentieth century, Washington's National Gallery of Art became famous for blockbuster exhibitions that put art into context. Its 1985 "Treasure Houses of England" offered seventeen elegant settings that made visitors feel like they were touring an imaginary British country house. The goal of the exhibition was to show the country house as "more than an accumulation of objects" but rather as "a container for a way of life." Gil Ravenel, the exhibition designer, described the process: "You begin with a group of objects and then you build a room, like a glove, to hold them."[54]

In the late twentieth century, period rooms posed a quandary to art museums. New research often proved them incorrect, telling visitors— like the history museum period rooms—more about the era of their curation than of their creation, more revival than reconstruction. All too many turned out to be composites from several houses—acceptable practice when the rooms were created in the 1910s, but something of an embarrassment a century later. Museums took a range of approaches to fixing the problem. Winterthur kept a set of period rooms as a memorial to the taste of its founder. The Met redid its rooms several times, including with a computer interactive that not only made it easy to identify the furniture, but which also told the visitor something of the social history of the room, how it came to be in the museum, and what about the display was less than accurate.

Some art museums found new ways to use the old spaces. "Dangerous Liaisons" at the Metropolitan Museum of Art made a big splash when it

4. Rooms of Louis XV and Louis XVI furniture, previously
st to specialists in French decorative arts, were suddenly
sets for dramatic installations, mannequins showing off not
t's fine collection of clothing from the era, but also the era's
leals of "beauty and pleasure." The focus, at least in the reviews,
was on pleasure. The Met was coy in its description of the scenes: "The
artfully composed scenes include: a woman sitting for her portrait while
her husband flirts with her friend; . . . a girl receiving more than a harp
lesson from her teacher, while her oblivious chaperone reads an erotic
novel; a woman giving up her garter as a memento of a very private dinner."[55]
Adding people made the scenes come alive, added interest. The success of
the show brought an encore, two years later, "Anglomania: Tradition and
Transgression in British Fashion."

These Met exhibitions suggested new ways of bringing life to the
period room. Artists like Yinka Shonibare MBE would take that even
further, using the rooms to create new art. (See Coda.)

$\Rightarrow$ 14 $\Leftarrow$

TURNED INSIDE OUT

PROFESSOR JENKS MOVED THINGS about in his museum, searching for the best way to display the wonders of nature. One rule seems clear: put everything you have on exhibit. Jenks packed his display cases full, using them both as exhibit and storage.

There is much appealing about this kind of combined storage and display, though it went out of fashion even before the Jenks Museum closed. The few museums that fought off the waves of exhibition design ideas and kept this style—among them Philadelphia's Wagner Free Institute of Science—have come to love and protect their old-fashioned presentations. This kind of display is based on the ideas that artifacts speak for themselves and that groupings of artifacts tell stories. Simply show off the collections in large numbers, and—in an echo of the "system of sideways looks" that historian Barbara Stafford noted in the cabinet of curiosities—artifacts might chat among themselves and with the visitor.

In the past few decades, visible storage has returned to fashion. It furnishes what packed display cases offered Jenks: a chance to show off the entirety of a collection. It also provides something new: a way of allowing the curator to slide out of the picture, to suggest that the objects and visitors, working together, create their own meaning. It fits well with the expectations of an era of easy online access. The contrast of visible storage with the storytelling of most modern exhibitions provokes us to contemplate the relationships of collection and exhibition.

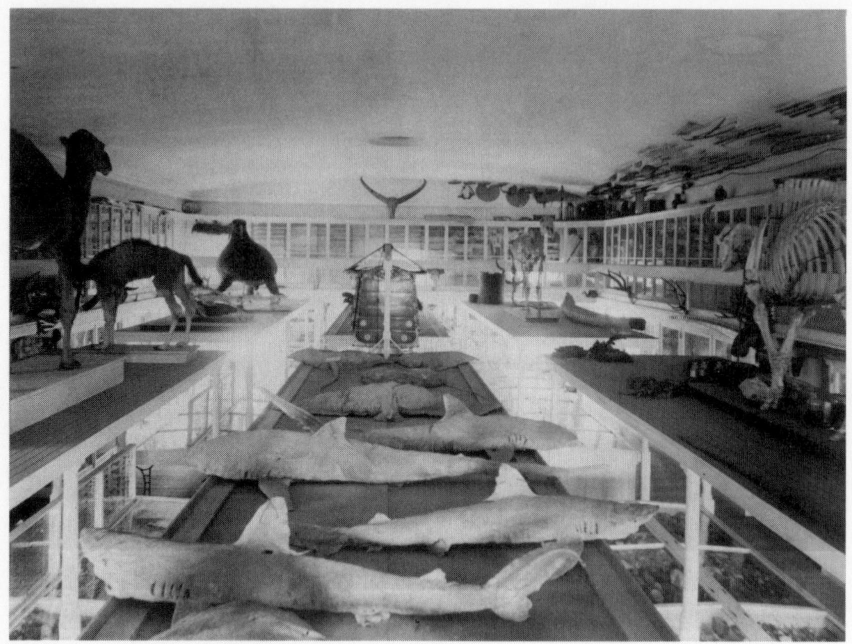

Sharks stored atop cases in the Jenks Museum.

Unmediated Access

Bring visitors to a museum storeroom and almost without fail they say, "This is better than the exhibits! Why don't you just open the storeroom for display, and let us see all of the good stuff!"

That was the reaction of the students in a Rhode Island School of Design (RISD) interior architecture course I critiqued in 2011. The assignment was to reimagine Brown University's Haffenreffer Museum of Anthropology. No holds barred. Don't worry about practicalities, budget, or what curators wanted. What would be most interesting?

The result, in almost every case, was a museum designed to reveal collections, not to tell stories. One student, Amy Boyle, came up with these goals:

1. As MUCH of the collection as possible should be on display at all times.
2. As MUCH of the space as possible should be used to support mission goals of research and learning.

3. As MUCH of the display and organization of the space as
   possible should be used to support the close-range observation
   and study of objects.[1]

Inspired by cabinets of curiosity, recent displays in natural history museums, and a visit to the Haffenreffer's Collections Research Center, Boyle's design filled the exhibition space with cabinets. Shelves and drawers of artifacts were open for exploration. The cabinets were on wheels, to allow them to be moved into place for individual study, classes, and temporary exhibitions. The walls were tightly packed with artifacts to create visual interest, provoke comparisons and close looking, and, most importantly, to display "as much of the collection as possible."

Almost all of the students wanted to let each object stand on its own so visitors could enjoy a sense of discovery. These were museums designed for an era of internet searching: songs, not albums; browsing and linking, not predetermined paths. Visitor choice trumped curatorial narrative. The students would agree with artist Mark Dion, who writes, "The museum needs to be turned inside out—the back rooms put on exhibition and the displays put into storage."[2]

The modern era of visible storage began about forty years ago, at the Museum of Anthropology (MOA) at the University of British Columbia. MOA was the first to use the term "visible storage," in 1976, although in some ways the idea of displaying everything is a much older idea, a throwback to the nineteenth-century museum before the dual system of storage and display became fashionable. Visible storage differs, though, in how things are organized. In MOA's visible storage, the collections were arranged according to a cultural logic—first by geography or culture region, and within those areas, by use, of which there were fourteen categories, including dress, masks, music, toys, and games. Reproductions were included along with originals, for hands-on teaching.[3]

The museum's project manager described the new system as a "parking garage for objects," but Michael Ames, the MOA director who introduced the system, made larger claims. It was, he said, "a radical departure" in museum display. He wrote that "if visible storage is carried to its logical conclusion, it could do for the study of material culture what the printing press did for literary culture."[4] Open collections, and the unmediated access to knowledge that represents, he argued, would allow museums to reach a larger audience. Access to heritage was a democratic right, and the best way to do it was to make museum collections available to all.

Limited access, Ames contended, not only made research difficult for scholars but also presented a distorted view of the collections and the cultures they represent. It was the museum's way of giving itself airs. By hiding away duplicates, copies, and "junk," museums presented themselves as "treasure houses" and their collections as "fine art," keeping from the public the wide ranges of style and quality that exist. That might be fine for an art museum, Ames implied, but it was wrong for a museum of anthropology. Visible storage was democratic and nonhierarchical, a way of sharing authority.

The Strong Museum, in Rochester, New York, was one of the first history museums to turn to visible storage, in 1982. A floor of visible storage—319 cabinets!—showed off some of its half million toys and other objects of everyday life, arranged by type of object, which appealed to collectors. In 2005 the Senator John Heinz Pittsburgh Regional History Center adopted visible storage, with some objects arranged by ethnic group, to appeal to local audiences.

The New-York Historical Society turned to visible storage in the late 1990s for a different reason. The organization was in trouble, running out of funds and accused of inappropriate deaccessioning. It was spending $500,000 a year on off-site storage, with little to show for it. The *New York Times* reported that the society felt "pressure to take artworks and collectibles out of warehouses and to put them before the public," which it did with an $11 million visible storage area that could house 75 percent of its art and artifact collection. The new facility aimed at not only improving collections management and public access to the society's holdings, but also at demonstrating its commitment to caring for its collections.[5]

The new space, some 21,000 square feet, housed 31,000 objects, arranged by categories including paintings, tools, decorative objects, and maritime artifacts. It was more than storage. Staff helped visitors find things. Audio tours provided paths through the forest of artifacts. Exhibit stations offered an overview of the collections. Famous New Yorkers shared stories about their favorite things. Computers gave visitors access to the museum's collections database. There were originally no individual object labels, but the public wanted to know what things were, so some were added. A review of the space validated the result: "It creates a delicate balance between storage and exhibition, using storage as a platform for programs and interpretation."[6]

Technology museums adopted visible storage too. It fit nicely with a sense that technology was about progress, and it appealed to their audiences of collectors and buffs. Automobile museums traditionally showed off their vehicles in rows that looked like sales rooms or parking lots. Some have played with the presentation, putting vehicles onto more museum-like visible storage shelves. The Barber Vintage Motorsports Museum in Alabama has done that with some 600 motorcycles, stacked high. Dozens of automobiles decorate a "car wall" and bicycles a "suspended velodrome" at Glasgow's Riverside Museum. Touch screens on the floor allow visitors to explore them virtually. The National Air and Space Museum's Udvar-Hazy Center has two hangars with hundreds of planes and thousands of other artifacts on display—part storeroom, part exhibition.

A few art museums adopted the idea. In 1983 the Metropolitan Museum of Art used exhibit space to store its Egyptian collections. That was a surprise success, and in 1988 the museum opened the Henry R. Luce Center for the Study of American Art. Forty-four cases showed off ten thousand items previously in storage—two-thirds of the museum's American collection, everything that wasn't on display elsewhere, or light sensitive. Artifacts were organized first by medium or material, then by form and chronology. Most of the objects were labeled only with their accession number; visitors could take that to a nearby computer to find out more. They could also query the computer for objects of interest and their locations.[7]

Some art museums made their visible storage look beautiful, and couldn't resist exhibition-style interpretation. The Brooklyn Museum's visible storage cases glowed with blue light, and included "focus objects" and small exhibitions. Its goal was to deliver a sense of "the great depth and richness of the museum's collection and a fascinating, exciting peek into what's behind the scenes." In a 2016 interview, the museum's chief curator argued in favor of the flexibility of interpretation visible storage allows. "By massing things together, you learn things just from the sheer quantity, which you don't learn when you look at one or two of the very best examples."[8]

Others designed their visible storage not simply to display art but also to reveal something of the museum's inner workings. The Smithsonian American Art Museum (SAAM) "wanted to provide visitors with a genuine behind-the-scenes experience—an opportunity to walk through art storage and see thousands of artworks that are normally kept off-site."[9] Next to its visible storage area was an open conservation laboratory. At Los Angeles' new Broad Museum, Joanne Heyler, the director and chief curator, turned to visible storage to "make it clear to visitors that storage and conservation are a core part of the museum's function." There, visitors could look in the paintings storage area and see storage racks. Of course, that led to thinking about how to make storage look good. "That's the funny thing," Heyler noted. "You end up curating your storage."[10]

Peering In

Visitor studies suggest a mixed response to visible storage, and some museums have abandoned the experiment. The director of collections at the Strong Museum noted that "we were wowing people with the number of objects on the shelves. . . . You would get the people with an intense interest in Flow Blue china, who wanted to see every pattern, every shape. But for the average person, all they saw was case after case of blue-and-white china." The Strong backed away from visible storage.[11]

The Met was surprised by the way the public used its Luce Center. It had hoped that visitors would ask curatorial questions. Carrie Barratt, the center's manager, told the *New York Times*, "What visible storage does is put things out there and lets the visitor shake it out. Now suddenly you aren't quite so sure why this is on a pedestal and this isn't. It may be baffling, but it causes people to think."[12] But visitors weren't asking curatorial questions. Instead, Barratt noted, they were exploring personal interests: "The centre is, contrary to our expectations, most popular with people who notice that a particular work of art is related to something that they own, such as a glass dish, a porcelain vase or an upholstered chair."[13]

Does visible storage demystify the museum or further mystify it? Some argue that it shares authority with the public, letting them ask their own questions and tell their own stories. But without a curator's narrative to guide them, visitors are left to their own devices, imagining a narrative that makes sense of a hodgepodge of material—or simply feeling overwhelmed and moving on. At the Smithsonian American Art Museum, a survey found that visible storage was visited less and rated lower than other exhibitions.[14] The New-York Historical Society noted that very few visitors found their way to its visible storage area. It first experimented by adding more thematic and narrative elements, and is now reimagining it, looking for ways to increase public access and engagement.[15]

Abigail Glogower, who worked in visitor services at the Open Depot Collection Display at Chicago's Spertus Institute of Jewish Studies, found that for many viewers, visible storage presented a "maddening inscrutability." The objects, some 1,500 items of Jewish art and material culture, were intended to speak for themselves; what that meant, in practice, was that they privileged "whatever preexisting knowledge . . . viewers bring to the exhibit." Visitors are supposed to have "an aesthetic and associative" encounter with the exhibits; instead, many were confused. Some wondered if the items were for sale. Many visitors felt they didn't know enough to ask interesting questions and so asked no questions at all. Some felt that the space was haunted by the memory of the Holocaust: Jewish objects with only the ghosts of Jews. Glogower used this metaphor: "Seeing the veiled machinations of the museum laid bare can be like entering the kitchen of one's favorite restaurant: there is excitement, to be sure, but also discomfort." Too many objects can impress or overwhelm. Some visitors know how to navigate the collection, most don't. It can work for experts, but not for a typical visitor.[16]

Visible storage means different things to different kinds of collections. As Glogower notes, Jewish artifacts are particularly complex. Visitors to the Jewish Museum Vienna find two displays of artifacts. One, the gift of a Jewish donor, is organized according to Jewish rituals; the objects are intended to be read as "the materialization . . . of a fundamental idea." The second, the *Shaudepot* ("open depot," or "viewable storage area"), stores objects expropriated during the Holocaust or taken from synagogues that had been burned during Kristallnacht. For historian Margaret Olin, the *Shaudepot* evoked a sense of disquiet, sadness, and melancholy, reminding her of the Jewish possessions seized by the Nazis. Olin writes,

"A chill ran down my spine. . . . Entering the room felt like coming upon a body." The room seemed like a tomb. The objects had "the aura of mysterious relics from a vanished culture." They "are condemned to haunt the present only as a ghost."[17] Edward Rothstein, the *New York Times'* exhibition reviewer, saw the same display but found it "one of the most moving exhibits of religious objects I've seen." For Rothstein, the relics were "monuments to a world of belief and practice wrenched not only from its habitat but from the earth."[18]

For some types of objects, like those at the Jewish Museum Vienna, visible storage allows the museum to sidestep a complicated history. For others, like automobiles and airplanes, visible storage plays up objects' aesthetic dimensions and makes it difficult to interpret more complex social and cultural histories. Visible storage seems to work best for those who know and care about the art or artifacts on display, who know how to look and what questions to ask. It doesn't work so well for those who don't bring the knowledge to supply the missing context. Some visitors are drawn in. Others are confused or overwhelmed. Visible storage—exhibition without interpretation—reinforces existing ideas and beliefs.

Open to Many Meanings

Visible storage has always been controversial. Michael Ames, who devised that first visible storage display at the University of British Columbia's Museum of Anthropology, reported on the resistance he faced. "Curators complained that it was wrong to show artifacts without explanation, or cultural context. Visitors, they said, were confused by all of the things, or bored by them." Curators claimed that they were experts at knowing what was important; it was dereliction of duty not to offer that expertise to the public. With the collections out in the open, without interpretation, the public would just draw their own conclusions. Or—even worse, from some curators' point of view—education staff, not curators, would explain them. Ames summed up the complaints drily. The main drawback of visual storage, he wrote, was that it "threatens some prerogatives and assumptions of the curatorial profession."[19]

Others suggest that visible storage only pretends to share authority, that it merely hides the curating. Nicky Reeves, a curator at the Hunterian

Museum at the University of Glasgow, argues that the openness of visible storage is false, pointing out that the "behind-the-scenes" shown to the public was replaced with a behind-the-visible-storage-scenes. Curators were still at work, but they were invisible, and thus the choices they were making were invisible. Further, only some of the behind-the-scenes was shown: the curator's work, not, say, the director's or the janitor's, misrepresenting what actually happened at the museum.[20]

Visible storage bares the intersection of collections and exhibition, and reveals that neither is simple. Both depend on hidden assumptions about appropriate categories of taste and meaning and history. Both are supported by not only the machinery of the museum but also the machinery of scholarship; hiding that machinery, pretending it's not there, can reinforce its power. Neither collections nor visitors are blank slates, and to suggest that there's a simple, honest, unmediated interaction that occurs by allowing the two to meet underestimates both the complexities of artifacts and the subtleties of museum visitors. Curators bring their knowledge and object-love to the storeroom. How might open storage offer some of that same depth?

Several museums are experimenting with a new generation of visible storage that wrestles with these challenges. The MOA reconfigured its visible storage, which had been successful in many ways but had come to feel both overcrowded and old-fashioned. The museum decided that "visible" wasn't enough.

Native peoples had not been consulted about the display of their objects at MOA's visible storage, and many found it cold and abstract. A selection of masks, for example, was shown arranged in neat rows; a community member described the display as "like cordwood." Ruth Phillips, the director of the museum in the late 1990s, noted that the display "conformed to standard anthropological classifications according to culture, area, and type"—practices now considered "artifacts of Eurocentric and cultural evolutionist premises." Visible storage, she argued, offered only the pretense of neutral, unmediated display.[21]

In the years since the original installation, the notion of access had broadened. Source communities didn't want to merely see their objects as anthropologists saw them. They wanted to go beyond visual access, to reintegrate objects into ceremonial and cultural use, to reconnect with them. The revised space, called the Multiversity Galleries, provided "multiple ways of knowing, categorizing, and organizing tangible and intangible culture." Its goal, wrote one of its curators, was "to decolonize older systems of museum classification" and "reflect ways of knowing that are most relevant to originating communities."[22] The new space offered many ways to explore the collections. Community members could use objects, handle them, learn from them, and teach the museum about them.[23] In the Multiversity Galleries, collection objects were accessible in more ways than just being visible: they were open to many meanings and many uses.

A proposed redesign of the University of Washington's Burke Museum is another model of openness. The plan, called "Turning the Museum Inside Out," starts with visible storage as the physical representation of transparency and moves beyond that to open up all of the work of the museum to the public. Visitors will be able to see into labs. Collections managers will talk with visitors. And, as at MOA, openness would work both ways. The museum set up community committees to engage not only potential visitors but also source communities. Openness and access would be not only visual but through engagement of every kind.[24]

The Los Angeles County Museum of Art is also considering a redesign that would use visible storage in a new way. Exhibits would be on the second floor, supported by glass "stems," densely packed with art, visible from the museum grounds and illuminated at night. Visible storage here presents the collection as both the foundation of the museum and an advertisement for it. Director Michael Govan uses a commercial metaphor: he calls these stems "storefronts on Rodeo Drive."[25] The storeroom is for shopping.

The rise of visible storage paralleled the rise of online collections access, and the two have much in common. The collections database on a museum's website is a version of visible storage, but one that is more flexible than any actual storage could ever be. An endless display of pictures of objects that can be instantly reorganized by category or date or location means the visitor may virtually stroll through collections. Online has advantages: the same object can be in two places at once, letting the visitor toggle between categories, or between a database of artifacts and a carefully curated path through them. What it lacks, of course, is the real thing, the physicality of the artifacts, their presence, and the social space of the museum.

The RISD student assignment to redefine Brown's Haffenreffer Museum of Anthropology, mentioned at the beginning of this chapter, envisioned a museum shaped by both delight in the real thing and the instant access promised by the digital world. The assignment shaped the museum's CultureLab installation, a combination of visible storage, seminar room, and hands-on exploration area. We moved museum storage cabinets into the exhibit space and put museum storage shelves inside exhibition cases. Artifacts were arranged by topic, using not the museum's categories but the categories defined by the faculty teaching courses that used them. We provided the tools of museum research: instruments like microscopes, as well as reference books, catalogs, and online access. Staff supervised hands-on work with selected collections. We tried to make it possible for visitors to use artifacts in the way museum staff did, giving them access beyond just viewing. Trying to share some of the object-love that the curator feels in the storeroom, we rediscovered some of the hands-on teaching values of the Jenks Museum.[26]

Open storage means easy access to some aspects of museum knowledge, the visual and categorical, and at least the possibility of revealing what goes on behind the scenes of the museum. But it plays down the secrets that expert intervention can reveal, the meaning added to artifacts by expertise. That's what we hoped would happen in the CultureLab when we offered artifacts along with the knowledge and tools to understand them and their context. And that's what museums do when they combine their expertise and artifacts in being useful to a wide range of audiences. That's the topic of Part IV.

IV

❧ Use ☙

WHAT USE IS A MUSEUM?

PROFESSOR JENKS THOUGHT HIS MUSEUM should be useful. The artifacts were tools for research, teaching, and learning. The museum itself was an essential part of the Brown community and a resource for the Providence community. He wanted his students to learn about biology and the wider world by working with the collections. He hoped museum visitors would grasp a much more profound truth, a religious one, about the glory of God's creation.

The Jenks Museum was not the only museum on campus or nearby. Within a few blocks there were many opportunities for students and others to see art and artifacts, to learn from them, to make use of them. Let's take a quick tour of the varieties of art and artifact on display at the turn of the twentieth century, a key moment in the debate on how a museum should be useful.

There were many exhibition spaces on the Brown campus. In the same building as the Jenks Museum a room was set aside as a museum of fine arts, serving both for the display of paintings and as an office for the museum's director. Across the campus green was the student union, where dozens of photographs and photogravures of art from museums around the world decorated the dining room and reception spaces.

A few hundred feet north of the Jenks Museum, on the first floor of Brown's Manning Hall, was the Museum of Classical Archaeology. This was a small museum of plaster casts, comprising about two dozen busts and a few full-size reproductions of classical sculpture. The casts served many purposes, including as models for art students and a lesson in

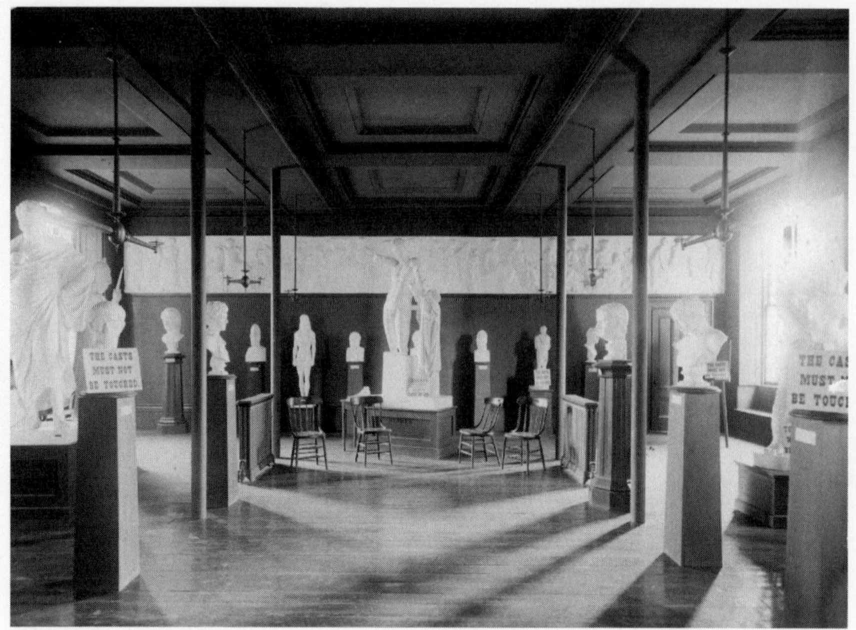

Brown's Museum of Classical Archaeology, about 1893.

beauty: Brown's president Elisha Andrews declared that the casts presented "the principles for all the aesthetic developments of modern times."[1]
They were also avatars of classical ideals. A professor at Johns Hopkins
wrote that "many advantages can be gained for the higher Hellenic hu-
manism and classical culture by a contemplation of casts of the noblest
treasures of Greek art."[2]

There were two libraries with exhibitions within a block of the Jenks
Museum. Directly across the university's green was the John Carter
Brown Library, its collection focused on Americana. It had a large and active exhibit program. In 1906, the year after it opened, students could see
displays of letters written by the signers of the Declaration of Independence and "ancient and modern bookbindings in connection with talks
given before the local Society of Arts and Crafts." The university's new
John Hay Library also had an Exhibition Room where students might admire Civil War autographs, a loan exhibit from the Newark Public Library
on the making of printed books, and selections from a recent bequest of
paintings, busts, vases, and a large amount of china, silverware, and bric-

a-brac. It was necessary, according to the university's annual report, "to keep good examples of art within daily reach of our students, to minister to the esthetic needs of the University and to aid all, the School of Design included, who are seeking to educate the public in the significance of art."[3]

A few blocks to the west of the Jenks Museum was the museum of the Rhode Island School of Design. RISD was founded in 1877 with three objectives: the instruction of artisans, the training of artists, and "the general advancement of public Art Education, by the exhibition of works of Art and of Art school studies, and by lectures on Art." A museum seemed essential for all these goals.

A few blocks south of the Jenks Museum was another art museum, the Annmary Brown Memorial, a museum of paintings by "ancient and modern masters" and a library of early books. Its founder, General Rush Hawkins, established it "for the purpose of honoring and perpetuating the memory of a beloved wife and woman." He wanted to "share a space of beauty with the public for reflection and edification." His museum, he declared, would "be of use to those who love the beautiful, and who may desire to study their interesting historical features—in the belief that there are interests worth living for other than those of the materialistic side of existence." The memorial was also a place for scholarly research on the history of printing.[4]

A block north of the Jenks Museum was the Cabinet Building of the Rhode Island Historical Society. It included an art gallery hung with portraits of great men (and a few women) from Rhode Island and beyond "who have done honor to themselves, their State, and their Country," as well as a display of "engravings, badges, medals, flags, swords, and relics of various kinds . . . that illustrate local history." The display was intended to "perpetuate the memory" of the state's "founders and benefactors."[5] Historical artifacts provided moral and practical lessons, and a sense of place, community, and history. Brown had its own "Hall of Fame" in a building not too far away, "a notable portrait gallery of former presidents and other officers of the University, as well as of distinguished alumni."[6]

Within a ten-minute walk of the Jenks Museum there were at least nine places where visitors might see art and artifact on display. They served scientists and scholars, artists and mechanics, students and the general public. They memorialized significant men and women, offered lessons in art history, morality, and beauty, and provided opportunities for contemplation and inspiration.

But this was a moment when the usefulness of physical things in research and teaching was beginning to be questioned. Historian Steven Conn argues that the waning of what he calls "object-based epistemology" at the beginning of the twentieth century marked the decline of many museums. Scientists and scholars thought they could learn more from books than from things. Museums seemed to have lost their utility.[7]

The Jenks Museum began to seem outmoded even before its founder died in 1894, and after his death it quickly came to seem a holdover from another era. Professor Albert Davis Mead summed up attitudes toward the museum in the biology department in 1914: "The reasonableness of spending money for the dusting and rearranging of miscellaneous curios of a university junk shop for the gratification of a few straggling sightseers is, we readily admit, not obvious." In the next year the museum was closed, the remaining specimens stored in attics and basements across the university. In 1945 most of what was left—ninety-two truckloads, mostly taxidermy—was carted off to the university dump.[8]

The Case for the Museum

Professor Mead's derisive language—"miscellaneous curios" and "junk shop"—reflects a long history of poking fun at museums as guardians of useless accumulations. One of the most cynical descriptions of museums appeared in an 1806 poem, "The Age of Frivolity":

> Some spend a life in classing grubs, and try,
> New methods to impale a butterfly;
> Or, bottled up in spirits, keep with care
> A crowd of reptiles—hideously rare;
> While others search the mouldering wrecks of time,
> And drag their stores from dust and rust and slime;
> Coins eat with canker, medals half defac'd,
> And broken tablets, never to be trac'd;
> Worm-eaten trinkets worn away of old,

And broken pipkins form'd in antique mould;
Huge limbless statues, busts of heads forgot,
And paintings representing none knows what;
Strange legends that to monstrous fables lead,
And manuscripts that nobody can read;
The shapeless forms from savage hands that sprung,
And fragments of rude art, when Art was young.

This precious lumber, labell'd, shelv'd, and cas'd,
And with a title of Museum grac'd,
Shews how a man may time and fortune waste,
And die a mummy'd connoisseur of taste.

Part III of this book considered how museums use their collections in exhibits. Part IV considers all of the other ways that art, artifacts, and specimens—and the museum itself—have been and are put to use. Museums and museum collections have value to more than the "mummy'd connoisseur of taste" of the poem.[9]

As museums tackle new missions, their artifacts, expertise, and cultural clout find new uses that their founders never imagined. Scientists make use of that "crowd of reptiles—hideously rare" and the other specimens in natural history museums to investigate taxonomy and ecology. Some of the artifacts collected by anthropologists and explorers to document what they imagined were vanishing races, and to prove the superiority of the white race—the poem's "shapeless forms from savage hands that sprung"—have found new uses when they are made available to the communities that created them. Communities turn to museums and collections to reconnect with traditions, to stake their claim to history and their place in society. Historians and archaeologists learn things about the past from piecing together those "worm-eaten trinkets . . . and broken pipkins." Artists use museum collections to learn to be artists and create new art. At each stage of their history, museums and museum artifacts have found new utility.

The very first treatise on museums made a case for their usefulness. Samuel Quiccheberg's 1565 *Inscriptiones* imagined the museum as a place for collecting, research, and display, for self-knowledge and power.

Quiccheberg, a Flemish physician, was a librarian and curator for Hans Jakob Fugger, a wealthy Bavarian art patron, and Duke Albrecht V, Duke of Bavaria. The book's very long subtitle sums up Quiccheberg's advice on what to collect and its purpose: "Repository of artificial and marvelous things, and of every rare treasure, precious object, construction, and picture. It is recommended that these things be brought together here in the theater so that by their frequent viewing and handling one might quickly, easily, and confidently be able to acquire a unique knowledge and admirable understanding of things."[10]

Quiccheberg argued that a museum had practical value. By collecting the right things, putting them in the right order, and examining them, one gains not just knowledge, but also skill and power. Over the next 450 years, museums would prove him right, though not quite in the way he imagined. Museums would prove useful in the service of empire, nation, and community. They would support scientific, medical, and technological advances. And they would also serve, in the most general way, as sources of inspiration and education. Collections of artifacts turn out to be enormously versatile tools.

Useful or Proudly Useless?

Nineteenth-century British museums set the tone for museums' multifaceted, even contradictory, nature. They were part of the social reform agenda, aimed at elevating (and shaping) the working class. At the same time, they were elitist in their view of art and science. They focused on both general education and sophisticated scholarship. They were cosmopolitan and imperial, nationalistic and local.

American history and natural history museums have always been comfortable with a balance of useful work, both educational and economic; educational outreach, both ideological and practical; and scholarly endeavors, both pure and applied research.

Art museums, though, were the scene of a long-running battle about purpose that came to a head in the early twentieth-century United States. On the one side stood John Cotton Dana, founder of the Newark Museum in 1909, and on the other Benjamin Ives Gilman, longtime secretary of the Boston Museum of Fine Arts. Dana spoke for those who demanded that museums be useful by meeting society's needs. Gilman spoke for those who argued that art is its own reward, that it is useful in its uselessness.

Dana, a librarian by training, created the Newark Museum for the people of the city to use, to "find pleasure and profit therein." "The one and obvious task" of every museum, according to Dana, was "adding to the happiness, wisdom, and comfort of members of the community" and thus "produce beneficial effects on their respective communities." Dana wanted to attract a wide audience, not just those who felt comfortable in museums. A museum didn't need a fancy building; that would just scare people away. It didn't need a great collection. Instead, Dana would show off reproductions of famous paintings, hobbyists' collections of local history, regional flora and fauna, and the products of nearby factories, in a utilitarian building in the center of the city, and at department stores and industrial sites. Dana wanted to move from an old model of museum-as-storehouse to a new model of museum-as-workshop. His goal for museums: to entertain, interest, and instruct—even if that meant breaking museum rules and traditions.[11]

Dana and his followers interpreted his argument for usefulness broadly. The museum delivered many things of value to its visitors, not just "happiness, wisdom and comfort," but also enjoyable hobbies and useful knowledge and skills. Dana wanted to increase his visitors' "zest of life by adding to it new interests."[12] He also argued for museums' civic value. Rossiter Howard, a colleague of Dana's and curator of educational work at the Cleveland Museum of Art, summed it up in one bold sentence. He imagined the museum "lightening the life of the poor, chastening the tastes of the rich, vitalizing the work of the schools, improving the output of industry, creating more efficient salesmen in the stores, increasing the value of real estate, a possible community centre for music, drama, and all the arts which go to make the city a better place in which to live."[13]

Gilman, in Boston, argued not for usefulness but uselessness, for museums as ends, not as means. He believed that museums were places for enjoying the beauty of fine things. "A work of fine art," he wrote, "is something of which the simple contemplation is worth while; and the artist creates it that this contemplation may take place." "We plant flowers for their own sake; grain for a harvest. In the first case a thing has value in itself; it is an end. In the second case it has value through other things, to which it is the means." Art, for Gilman, was an end in itself, and the museum's job was to "collect and exhibit the best obtainable works of genius and skill" then get out of the way and allow visitors to engage with it. And so, he argued, museums should not teach art history or technical skills. Rather, their goal should be to "bring about that perfect contemplation of the works of art it preserves." His goal for museums: to behold, appreciate, and love.[14]

Dana wanted the museum to be part of and useful to society; Gilman imagined the museum as a place of retreat, useful to its individual visitors. Dana found beauty in everyday things; Gilman, in the finest art. Dana imagined an active museum; Gilman, a place for quiet contemplation. Dana was eager to attract everyone to the museum, especially the working-class men and women of Newark and recent immigrants. Gilman also wanted to attract a wide range of visitors, but he imagined visitors who—no matter what their background—were, by virtue of their ability to appreciate art, members of an elite.

The Dana-Gilman debate on just how museums might be useful has continued for the past century, changing to reflect the times. In the late 1930s and early 1940s the debate echoed concern over the degree to which museums should serve the individual, as educational institutions, or the state, as instruments of propaganda. Thomas Adam, a professor of political science at Occidental College, weighed in on Gilman's side: "Direct sensory perception" of the carefully selected art in museums, he said, would allow the "ordinary citizen" to "build up his cultural tastes."[15] Others supported Dana's cause, arguing the case for education—"the purpose and the only purpose of museums," wrote Theodore Low, director of the Metropolitan Museum of Art. Low believed that museums should strengthen "that thing which we like to call 'the American Way of Life'."[16] But whose education, and how that should be accomplished, was still a matter of debate. Low aimed at "the intellectual middle class," charged Blanche Brown, a lecturer at the museum, in a stinging critique of her boss's book. Along with other museum educators, she wanted to reach out to a broader audience. Museums, she wrote, should "learn that a public institution must not only build itself a house in a community, but must penetrate the lives, the ideas and ideals of the members of that community, so that the good things of cultural experience also may be democratically disseminated."[17]

During World War II, many museums were eager to show themselves both practically and ideologically useful. The Museum of Modern Art announced that it would "serve as a weapon of national defense," and mounted exhibits like "Wartime Housing" and "Road to Victory," which used new multimedia techniques to promote patriotism and raise morale. The Brooklyn Museum offered shows like "The Consumer Front: Essential Wartime Economics" and "Inventions for Victory." "The Homemaker and the War" at the Chicago Museum of Science and Industry told visitors "the best ways to do all that is expected of them as patriotic housewives."[18]

Scientists at museums also served military goals. Anthropologists and zoologists at the National Museum of Natural History prepared twenty-one "war background reports," most of them focused on Pacific islands suddenly of interest to the U.S. military. They wrote *Survival on Land and Sea* for downed airmen. The collections were useful, too: specimens of disease-bearing insects were shipped to field hospitals to help with identification. In the first year of the war, some 10 percent of the scientific staff was working full-time on military projects.[19]

Museums retreated to a less active role after the war, but in the 1960s found themselves with new duties, as participants in Great Society social programs. In 1968 the American Association of Museums (AAM) issued the *Belmont Report,* a call for federal support for museums. It focused on the many ways that museums served the public, "the astonishing range of their educational and cultural activities." Like all the reports the AAM would issue over the next half century, it was the work of a committee, and it argued in both Gilman's and Dana's voices. On the one hand, it noted the "pleasure and delight" a museum offered. On the other, it made the case for museums' practical and educational value: some eight million school children and five million adults participated in museum programs each year. The *Belmont Report* also made a tentative economic case for museums. "We are not accustomed to think of a museum as having an economic value to a city or to the nation," the report declared before arguing for museums as engines of tourism and economic development.[20]

The AAM continued to weigh in on the purpose of museums every decade or so. In 1984, in *Museums for a New Century,* it declared that museums "represent certainty in uncertain times." On the one hand, museums honor tradition; on the other, they help "fuel the growing pride in community and ethnic history." The key, the AAM wrote, was the recent democratization of museums. While in the 1960s and 1970s they had been ivory towers, in 1984 museums were committed to a public role, "deeply aware of their public obligations."[21]

The AAM's 1992 *Excellence and Equity* balanced the two kinds of usefulness, arguing that the goal of a museum is "to nurture an enlightened, humane citizenry that appreciates the value of knowledge about its past, is resourcefully and sensitively engaged in the present, and is determined to shape a future in which many experiences and many points of view are given voice." Channeling Gilman, the report urges museums to "take pride in their tradition as stewards of excellence." Channeling Dana, it asks museums "to embrace the cultural diversity of our nation, to foster "the ability to live productively in a pluralist society and . . . contribute to the resolution of the challenges we face as global citizens."[22] In 2002, in *Mastering Civic Engagement,* the AAM counsels museums to "revisit the power of community and consider what assets museums contribute to the shared enterprise of building and strengthening community bonds."

Impact

The authors of the 1968 *Belmont Report* might have been shocked that anyone thought of museums in economic terms, but starting in the 1980s, facing cuts in government support, museums began touting economic impact as an essential part of their argument for support. Museums began to make the instrumental argument that they were not only good for people, but also essential community anchors and engines of economic development.

This continues to be an important theme. The AAM website lists "Museum Facts" selected to show that "museums are educational, trusted, beloved and economic assets to communities everywhere." Museums, the site declares, are "economic engines." They employ more than 400,000 Americans and contribute $21 billion to the U.S. economy. They attract big-spending leisure travelers. For every $1 government invests in arts and culture, $7 is returned in tax revenues.[23] The AAM argument was part of a larger economic case for arts and culture. Americans for the Arts, an arts lobbying group, adopted "the arts mean business" as a slogan and issued a series of reports on "Arts and Economic Prosperity."

In 2014 the Boston Museum of Fine Arts (MFA) hired an economic consulting firm to outline the good the museum did the city. It published the results in a thirty-six-page illustrated report, "The Economic and Community Impacts," with an equally long technical appendix. (The city's attempt to increase the museum's payment in lieu of taxes no doubt encouraged the report.) The report pegged the economic impact at $337 million, or, if construction spending were included, $748 million. Expand that to the state as a whole and the impact was over $1 billion. The museum was responsible for 1,313 jobs; 3,872 if you counted "indirect and induced effects." The museum spent $25 million in the state, and generated $22 million in public revenue. Tourism was another way the museum strengthened the economy: 448,000 tourists came to Boston primarily to visit to the MFA, pumping some $168 million into the city's economy.[24] The Met in New York presented even more detail, issuing press releases on the economic impact of individual exhibitions: "Three Metropolitan Museum Exhibitions Stimulate $742 Million 2013 Economic Impact for New York." In that year, the museum claimed, visitor spending in New York by out-of-town visitors to the museum in fiscal year 2013 reached $5.4 billion. Almost half of the visitors were from other countries, and 77 percent of them were from outside the city. The Met claimed credit for their presence: 54 percent of the travelers "cited visiting the Met as a key motivating factor in their visit to the city."[25]

Some cities turned to museums as anchors for new development, even complete renewal. The best-known and probably most successful example is the Guggenheim's outpost in Bilbao, Spain, which opened in 1997 in a declining industrial city and attracted visitors from around the world. "The Bilbao effect," it was called. The regional government put up a great deal of money for the museum, but quickly earned it back, some $110 million in taxes in the first three years. And so other cities signed up for Guggenheim franchises, including Abu Dhabi, Helsinki, and Guadalajara. To date, though, none have come to fruition. It's hard to pin down what failed: the projects all faced financial and political challenges. Some blamed the recession, some a reaction against globalism, some a sense that the projects were too ambitious.

Cities around the world built new museums in hopes of economic and cultural expansion. The Louvre Abu Dhabi—the name alone cost $520 million, loans and management an additional $747 million—is just one element in the city's Saadiyat Island cultural development. In Doha, Qatar, the Qatar Museums Authority is building eight museums. The country's goal is to become "a cultural destination in its own right. . . . Our museums and galleries play a vital role in helping us fulfill our unique, long-term ambition." H. E. Sheikh Hassan Al Thani, Qatar Museum's vice president, saw museums as essential to future creativity. "Art can only truly flourish," he wrote, "if the artists, their collectors, curators, and audience are able to connect to yesterday's achievements." Qatar's museums aimed to engage "local audiences with the rest of the world."[26] The number of museums in China doubled to over four thousand between 2000 and 2013, part of a plan to make culture a "pillar industry," the "spirit and soul of the nation," and an essential element in economic development. Some cities hoped for the Bilbao effect. Some saw museums as a tool for real estate development. Many new museum buildings were signature works by "starchitects," destined to become a key piece of a city's brand identity.[27]

Museums argued for their presence not merely as economic development, but as something more profound: anchors of a new kind of economy. Richard Florida's *The Rise of the Creative Class* (2002) helped set off the trend. The future of American cities, he claimed, was the "creative class," the one-third of workers who "create for a living." Cities needed to attract them, and one of the best ways to do that was by supporting the kind of cultural organizations they liked.

Museums eagerly claimed this role. The Museum of Fine Arts report argued that "the MFA is a very important part of the Boston creative economy," not only as a cultural institution, but also as a school for artists and a venue for events that attract "creatives." At the other end of Massachusetts, in the depressed industrial city of North Adams, the Massachusetts Museum of Contemporary Art (MASS MoCA), which opened in 1999, made a similar argument. It pointed to the many new "creative" tenants in its once abandoned factory buildings. MASS MoCA, one report proclaimed, "has become an anchor to the North Adams economy, and sparked a burgeoning arts and creative culture that has made the city a more desirable place for people to live, work and play."[28]

Economic impact has become one of the ways that museums make the case for support. Not everyone is happy with it. Activists have argued that cultural economy arguments are an excuse for gentrification and real estate development, that museums have become more interested in their finances than in their collections and educational responsibilities. By focusing too much on their economic impact, museums, they say, have lost sight of their true mission. *New York Times* critic Holland Cotter sums up the concern. Museums, "once considered standoffish, genteel and politically marginal . . . are now viewed as being emblematically engaged players within the power network of global capitalism. And some are seen as using that status badly."[29]

MUSEUMS MAKE COMMUNITIES

In 1884 Jenks visited the World's Industrial and Cotton Centennial Exposition in New Orleans. He wanted to see one thing: Samuel Slater's forty-eight-spindle spinning frame, built in 1790 in Pawtucket, Rhode Island, which was on display in the Rhode Island exhibit.

Slater had been a Rhode Island industrial pioneer, the founder of the first mechanized spinning mill in America. By 1835 the first spinning frame was no longer in use but was displayed at the mill as a curiosity. A visiting textile expert called it a venerable relic "which ought never to be destroyed." He suggested it be given to a museum as a memorial to Slater and to "exemplify the rapid improvements" in technology.[1] This came to pass: In 1856 it was given to the Rhode Island Society for the Encouragement of Domestic Industry for display in its downtown Providence office. A member saw it there, writing that it and another Slater machine "stand in state in the rooms of the society." In 1876 it was proudly shown at Philadelphia's Centennial International Exposition, and when it returned to Providence the society donated it to the Jenks Museum, along with hatchets and spinning wheels of "ye olden time."[2]

It didn't stay at the Jenks Museum for long, if indeed it was ever moved there. Four years later Jenks told the society that he believed "these valuable instruments marking the beginning of cotton manufacture in America" should be sent to the Smithsonian. The letter proposing to donate the machine to the national museum argued that "this relic is most valuable—the first spinning frame ever started in America (1790)—the rude beginning of all that vast New England industry—cotton spinning

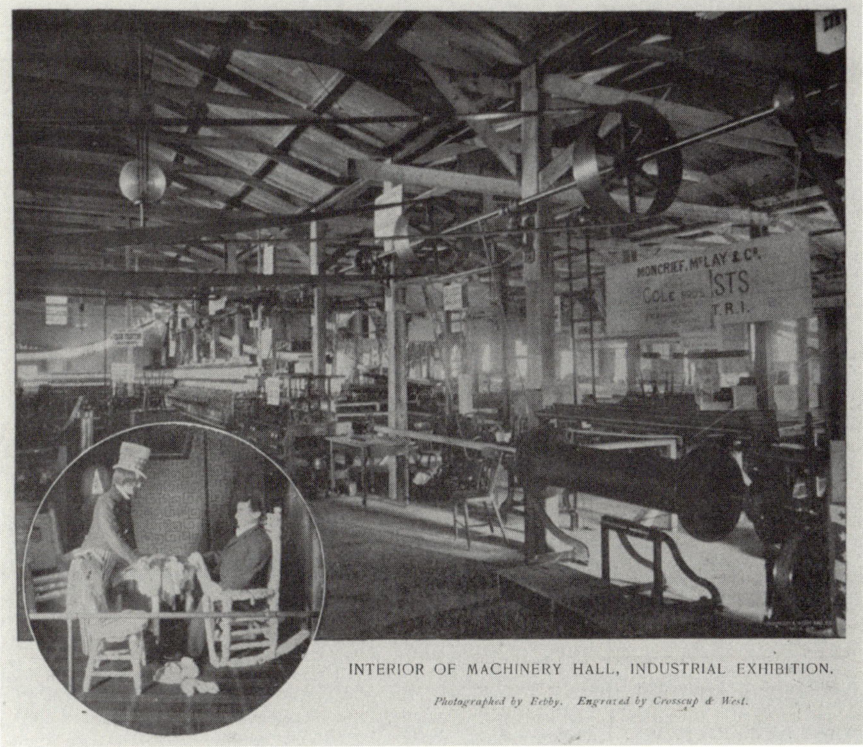

The Cotton Centenary in Pawtucket, Rhode Island, 1890, showcased the 1790 Slater spinning frame. The inset shows Samuel Slater and Uncle Sam.

and weaving." Although the society regretted seeing the spinning frame leave the state, it believed that the Smithsonian was the place "where the citizens of our Common Country may view it and learn of its history. No place offers a superior claim to your Department for the preservation of this valuable historical relic."[3]

The Smithsonian accepted the gift, displaying it first in Washington and then in New Orleans, where Jenks stopped by to see it on his way west. In 1890 the spinning frame would return to Pawtucket for the Cotton Centenary, operated by an elderly man who had tended it when he was a boy. There was also a display of modern textile machinery and a tableau featuring Uncle Sam congratulating Samuel Slater on his success.

For Rhode Island industrialists at the end of the nineteenth century, the Slater spinning frame was a museum artifact to be put to use, and not just by the elderly spinner. It was a symbol of the state's pride in its industrial history, a way to highlight Rhode Island's place as the "birthplace of the American industrial revolution." The displays at the Jenks Museum, the Smithsonian, New Orleans, and Pawtucket made a claim about the importance of technology and industry in America's past, present, and future, as well as the importance of Rhode Island in that story.

The society's donation of the Slater spinning frame to the Smithsonian showed an appreciation of the political and cultural work that museums could accomplish. Putting an object in a museum can serve a range of purposes.

Museums Strengthen Communities

Museums strengthen communities by providing a shared history and usable past, saving things and telling stories that seem important. They make political arguments. Technological museums might seem above this, but as the story of Slater's spinning frame shows, they embody the dual purpose of museums. By showing off historical machines, museums make heroes of inventors and promote the ideology of technological progress. They support inventors not only in a practical but also in a political way.

Quiccheberg made this point in his 1565 *Inscriptiones,* the first treatise on museums. His ideal museum included models of machines so that "larger ones can be properly built and, subsequently, better ones invented." Inventors were important too: the portraits of "men . . . preeminent in learning" and "those famed for their craftsmanship" were included in his museum, alongside portraits of kings, princes, and aristocrats.[4]

Francis Bacon, the influential English philosopher of the scientific revolution, also acknowledged the museum's dual purposes of spreading knowledge of technology and acclaim for inventors. In his 1627 *New Atlantis* utopia, Bacon imagined a scientific institution he called Solomon's House. Here would be gathered together "the sciences, arts, manufactures, and inventions of all the world" and "books, instruments, and patterns in every kind." "We have two very long and fair galleries," the head of Solomon's House said. "In one of these we place patterns and samples of all manner of the more rare and excellent inventions; in the other we place the statues of all principal inventors."[5]

Many early museums included technological artifacts, just as Quiccheberg and Bacon recommended. Models and machines seemed the best way to teach technology and encourage inventors. Philadelphia's Peale Museum included both inventions and portraits of inventors. It had "the beginning of a Collection of Models of useful, foreign, and domestick Machinery." Peale placed inventors and scientists Robert Fulton, Joseph Priestley, and David Rittenhouse among his portraits of great Americans.[6]

America's most important early technology museum was at the Patent Office in Washington. Inventors submitted a model with their patent application. The Patent Office 1837 annual report called itself a "museum of American ingenuity" that would stir "interest and patriotism." The tens of thousands of models on display were a tourist attraction and a symbol of the democracy of learning as well as a useful resource for inventors.[7]

Many of the patent models were transferred to the Smithsonian, and there too they served both practical and ideological ends. J. Elfreth Watkins joined the Smithsonian as its first curator of railroads in 1885. He owed his job to congressional lobbying by the American railroad industry, and had a clear sense of the usefulness of collections to the industry. He collected historical objects to show historical developments and to raise inventors into the pantheon of American heroes, making the case that "the Railway and Steamboat" were "the greatest civilizers of the century."[8]

The Smithsonian of the early twentieth century was eager to be useful to industry. The 1924 *Annual Report* noted that the museum was "more and more appreciated" by industrial organizations, who offered donations and helped create exhibitions.[9] Curator of engineering Carl Mitman's "highest priority," writes historian Arthur Molella, "was to provide the American engineer a recognized and respectable place in history." In the 1920s Mitman proposed a new museum, the National Museum of Engineering and Industry, to demonstrate that "engineers, inventors, and scientists have played as important a role in the nation's history as diplomats, politicians, and generals."[10]

The museum project of framing inventors and engineers as heroes continues today. Jerome Lemelson, a wealthy inventor, established the Smithsonian's Lemelson Center for the Study of Invention and Innovation in 1995 with the institution's largest cash donation to that date. His goal was "to boost appreciation of the nation's innovators." Lemelson lamented that "American youngsters know more about Michael Jordan than Thomas Edison. . . . Yankee ingenuity was the talk of the last century and is the talk of this century—but we don't know the role models." His hope, he told the *Washington Post,* was that inventors would become the new American heroes.[11] The Lemelson Center would collect the archives of inventors, present historical exhibitions showcasing inventors, and host hands-on activities "where museum visitors become inventors"—all goals that would have been appropriate for Bacon's Solomon's House.[12]

It's not only inventors and businessmen who use museums to build their community and support community goals. Every museum makes an argument that something—a thing, an idea, an ideal—is worthwhile, important, and worth saving, sharing, and celebrating. A museum can define, strengthen, even create a community. It can testify to the community's glory, extol its history, and eulogize its heroes.

One of the most curious items in the Jenks Museum's collection served this role. "Chief and foremost" among the articles in the museum that "would delight the heart of a curiosity seeker," wrote the *Providence Journal* in 1893, was "the twisted apple tree root which bears the legend: 'The apple tree root that ate up Roger Williams.'" This root, now on display at the Rhode Island Historical Society, was found at the supposed site of Roger William's grave, in the shape of his body. The paper explained: "This root has become familiar to all Rhode Islanders. Its story has for generations interested the inhabitants. . . . Historic lore has cast its air of sacredness about it until it has even become an object of veneration, perhaps of awe."[13]

Communities define themselves in part by their "objects of veneration," the things that seem sacred, even an apple tree root, if it has a good story, and especially if it is preserved in a museum. Museums give communities things to be proud of. John Adams, in Paris as a diplomat during the American Revolution, visited the Cabinet d'Histoire Naturelle and wrote to his wife Abigail that the new United States needed a similar museum: "Our Country affords as ample materials, for Collections of this nature as any part of the World."[14] Thomas Jefferson and Charles Wilson Peale both displayed mastodon fossils as part of an argument that animals in America were larger than animals in Europe.

Local and state historical societies created the earliest American community museums. The first was the Massachusetts Historical Society, founded in 1791 "to collect, preserve and communicate, materials for a complete history of this country, and accounts of all valuable efforts of human ingenuity and industry," and to organize a museum to inform the public about American history.[15]

Americans established historical societies throughout the country in the nineteenth century—including in the West, often not long after settlement—to celebrate both the significance and the potential of their communities. These societies also served to suggest, by collecting documents and artifacts, the deep roots of settlers who had in fact just recently claimed the land. They often had a strong genealogical interest, and celebrated local heroes. James Ramsey, founder of the East Tennessee Historical and Antiquarian Society in the 1830s, was typical in his aims: to collect archives and historical collections, or as he put it, to mine "the gems of the past in our state which should sparkle & illuminate the pages of biography & illustrate the virtues of our ancestors."[16] Historical organizations created a usable past.

Halls of fame promoted a different kind of community. The Hall of Fame for Great Americans, the first American hall of fame, was established in 1901 with the goal of promoting a vision of a national American culture. The selectors, mainly university presidents, defined American culture narrowly, in one way—native-born and male—but, broadly, to their minds, in another way: not just generals and politicians, but also artists, business men, scientists, missionaries, and (no surprise) authors and teachers. Those selected were honored not only with busts at the Hall of Fame, located on what was then the campus of New York University in the Bronx, but also with artifacts in the associated museum, to be filled with the "thousand-and-one memorials which vividly call to mind the departed great."[17]

The hall of fame idea spread. The first community-specific one was the Baseball Hall of Fame, founded in 1939. Like so many since, its goal was to define and strengthen a community. The Baseball Hall of Fame aimed to establish baseball as the national sport by proving that baseball was an American invention, to increase the sport's popularity, and, not least, to bring tourists to Cooperstown by making the (dubious) claim that it was the site of baseball's invention. It combined a public selection process and induction ceremony with familiar museum features. Visitors could see "funny old uniforms," "baseballs thrown out and autographed by presidents," and the "bats of baseball's greatest players." Museum artifacts and an official list of "greats" stabilized the community by giving it a history.[18]

Hundreds of halls of fame were created over the following decades, from sports to entertainment to industries, defining and strengthening communities by honoring their great and good. The Rock and Roll Hall of Fame (1983) made rock 'n' roll respectable with its I. M. Pei-designed museum, library, and archives. The Rhode Island Heritage Hall of Fame aims to "combat the state's negative self-image and to foster pride in our state."[19]

Museums by and for racial, ethnic, and cultural groups stimulate another kind of community building: community cultural development. Hampton Normal and Agricultural Institute (today Hampton University), founded in 1868 to educate freedmen, built a museum of African artifacts. Cora Mae Folsom, the museum's curator, explained why in 1917: "Collections should and do have a large mission in stimulating race pride."[20] Toward the end of the twentieth century, building a museum became a way to say "We are a community; we have a history; our history is important, and useful." This was especially valuable for those who had been left out of, or treated poorly by, history textbooks and mainstream museums. Museums built pride in accomplishments and portrayed a past that suggested a brighter future. They also created an opportunity for a broader, more inclusive understanding of a diverse society.

Andrea Burns traces the history of African American museums from the nineteenth-century intellectual tradition known as vindicationism— "we, too, were here"—to their role in the second half of the twentieth century in shaping black identity and serving the needs of the black community. In the 1960s black museums used black art and history to convey a message of identity and self-worth. Margaret Burroughs founded Chicago's Ebony Museum of History and Art in 1961 (today's DuSable Museum of African American History) as a place where "art and history should teach racial self-appreciation." Dr. Charles Wright, who established Detroit's International Afro-American Museum in 1965, argued that the museum's job was "to erase 350 years of dehumanizing brain washing." The new museum, he believed, would "correct the false and distorted history of the Black Man, and . . . give him a symbol of his own identity and worth." (The museum was renamed after Dr. Wright in 1998.)[21]

Other community museums focused on art. The Studio Museum in Harlem, founded in 1968, supported community artists with an artist-in-residence program and by showing their work, and provided art education for school children.[22] The director argued for programs that grow "organically out of the needs and desires of the people we serve." El Museo del Barrio opened in East Harlem in 1969 to offer materials on Puerto Rican culture for schools and soon expanded to include what some of its founders called "an empowering and cosmopolitan vision" of "alternative avenues of validation for local artists." Both museums faced controversy as they tried to balance the needs of local communities and artists of color, connections with the international art world, and a permanent collection. The history of these museums charts changes in the community, the meaning of community, and the purposes of museums.[23]

In the 1980s a new kind of museum—what Jack Tchen, founder of the Chinatown History Museum, called a dialogic museum—suggested a new way to collect and tell the story of a community. Tchen argued that a community museum must be organized not for the community, but by the community. It should present not history from the top down, but a place to "facilitate the collaborative exploration of the memory and meaning" of the past, to share authority with the people whose story is told. He wrote, "We want to fashion a learning environment in which personal memory and testimony inform and are informed by historical context and scholarship."[24]

The Wing Luke Asian Museum in Seattle pushed the model of the dialogic museum even further. Its means was a museum, but its goal, according to director Ron Chew, was to be "a vehicle for community organizing and empowerment." One of Chew's first exhibits was on the Japanese-American internment, to tell "in the voices of the former camp inmates and their children, the painful story of the wartime injustice." That and other community exhibits "affirmed the restorative power of the oral tradition and the first-person voice of ordinary people. It gave sanction, in a museum setting, to the notion of students, non-professionals, and elders as scholars and lead decision-makers, rather than token advisors." The process, exhibition developer Cassandra Chinn wrote, was to "work closely with community members in spurring a broad-based dialogue, then creating a format for accurately and sensitively representing the community's vision."[25]

A group of high school teachers created Chicago's Mexican Fine Art Center Museum (now the National Museum of Mexican Art) in 1982. They tried to balance several kinds of community building, combining art, history, and community service. The museum reminded the Mexican community in Chicago of Mexican artistic traditions, both fine art and folk art, from both Mexico and from Mexicans in the United States, but it also served the community's practical and political needs. Indeed, Carlos Tortolero, one of its founders, thought of its museum function as a social service. The museum was a "first voice" museum, speaking for the community not only to the community, but beyond it as well.[26]

The Queens Museum in New York has reimagined its role along community museum lines. José Serrano McClain, a community organizer employed at the museum, sees his job as using "art strategies in service of community organizing efforts that are already underway in the community. . . . We lend our creative services to help social campaigns in the community that we do not lead but participate in." The museum supports the community's artists and cultural institutions, hosting local art installations, and opening up the museum for community meetings. Its paradigm is not outreach, but partnership, collaboration, and reciprocity.[27]

The Smithsonian's National Museum of the American Indian (NMAI), which opened in 2004, is both a community museum and a national museum. Richard West, its director, was clear on its mission: to celebrate the cultural accomplishments of the native peoples of the Americas by using the voices of native people themselves. That would be accomplished through community curation, a collaboration between professional curators and native communities, and would support "an indigenous way of knowing." The museum featured exhibits based on five principles: 1) community—our tribes are sovereign nations; 2) locality—this is Indian land; 3) vitality—we are here now; 4) viewpoint—we know the world differently; and 5) voice—these are our stories. Amanda Cobb, writing in

American Indian Quarterly, sums up the message: "We are still here—isn't that amazing and beautiful? We are still here—we have something lovely to share. Welcome."[28]

Not everyone heard that message. Some native reviewers expressed concern that the challenges that native communities face were played down: the museum should have been more forthright in explaining the horrors of conquest and colonization. Native scholar Amy Lonetree calls the museum a "missed opportunity," declaring that it "falls short in telling the hard truths of America's treatment of Indigenous people."[29] Others wished for more information on historical events rather than an emphasis on unchanging culture.

Many non-native reviewers criticized the museum equally harshly, but from the other direction. Edward Rothstein, the *New York Times*'s history museum reviewer, blasted the museum, calling it "a self-celebratory romance" that betrayed "a studious avoidance of scholarship." He accused the NMAI of looking through "a gauze of romance" to portray "an impossibly peace-loving, harmonious, homogeneous, pastoral world that preceded the invasion of white people—a vision with far less detail and insight than the old natural history museums once provided."[30]

Rothstein's criticism of the NMAI was part of a larger critique of what he calls "identity museums." He takes this view of their goal: "Me! Me! Me! That is the cry, now often heard, as history is retold. Tell my story, in my way! Give me the attention I deserve! Haven't you neglected me, blinded by your own perspectives? Now let history be told not by the victors but by people over whom it has trampled." He accuses community museums more generally as "celebrations of hyphenated existence," advancing "a grievance narrative" instead of what he sees as "a fuller understanding of both sides of a hyphenated identity."[31]

The Smithsonian's National Museum of African American History and Culture (NMAAHC), which opened in 2016, took a different approach to combining community, national, and even international history. Its curators thought carefully about two audiences, visitors particularly interested in black history and culture, and a more general audience, volunteering its exhibits as a place where "all Americans can learn about the richness and diversity of the African American experience, what it means to their lives and how it helped us shape this nation." The African American experience was a lens to view the nation's history and culture, "to understand what it is to be an American."

By connecting African American history to American culture more generally, the museum hoped to transcend "the boundaries of race and culture that divide us," instead focusing on "a story that unites us all." It does this by showing how American values like resiliency, optimism, and spirituality are reflected in African American history and culture, and how the culture of all Americans is shaped by African and African American culture. Lonnie Bunch, the museum's founding director, emphasizes the museum's inclusivity: "This is America's Story and this museum is for all Americans."[32]

The NMAAHC straddles the boundary between community museum and national museum, supporting the identity of the African American community at the same time it reshapes national identity. In that it merges the traditions of the community museum and the traditions of the national museum.

Just as communities built museums to reinforce identity, so did nations.

Serving the Nation

The British Museum, the first national museum in Europe, opened in 1759, but most national museums had their start after the Napoleonic Wars, when many of Europe's modern nations were created. These museums were created to showcase a national history and culture, to make the case for the nation. New nation-states found museums particularly useful: Germany, Belgium, and Bulgaria, among others, built museums soon after their formation. The world's newest nation, South Sudan, is already planning a national museum: "a place where the people of South Sudan can tell their stories, discover their common heritage, and find a forum for a constructive dialogue between meaningful pasts and desirable futures."[33]

Stefan Berger, in his history of national museums, assesses their purpose bluntly: "to protract a national storyline into the dim and distant past. The further the national history can be traced into the past, the more distinguished the present-day nation is."[34] Archaeological museums made this point well. Denmark's Nordic Ancient Museum, founded in 1819, was "a temple for the remains of the spirit, language, art and power of our past, where every patriot can study the successive advances of the nation's culture and customs."[35] History museums extended the narrative by demonstrating the nation's continuity to the present day. Folk history museums, popular at the end of the nineteenth century, defined the nation's past by highlighting preferred ethnic groups and peasant traditions. Skansen, Sweden's open-air folk museum, was originally created to justify Sweden's union with Norway and to include the Samis as part of the nation. Art museums, often focusing on national art, aimed to reflect both the importance of a country's artistic tradition and its national character. National museums of technology and industry showed off a country's inventive and economic prowess.

National museums serve both a national and international audience. When President Lyndon Johnson dedicated the Smithsonian's Museum of History and Technology in 1964 he hoped that American children would "see, with their own eyes, yes, even touch with their own hands, the ripe fruit of America's historical harvest . . . [in] this treasure-house of our inheritance." And he hoped that "visitors from the newly emerging nations" would visit to learn "that their labors are not in vain—for the future belongs to those who work for it. Let them go back to their home secure in the knowledge that from coarse and barren beginnings come the fulfillments of hope. They will have seen the evidence here."[36]

A more recent use of national museums is to remember moments of trauma, to "never forget" moments that shaped the nation, whose memory binds the nation together. Some remember moments when the nation did wrong. Holocaust memorials and museums in Germany present in-depth histories and analyses of the horrors of the Nazi era. In the United States, National Park Service sites preserve the story of the internment of Japanese Americans during World War II to stimulate "dialog and greater understanding of civil rights, democracy, and freedom."[37] Other museums remember when the nation suffered. In 1958 the USS Arizona Memorial was established to commemorate military personnel killed in the Pearl Harbor attack. In 2008 its mission was changed. Now the World War II Valor in the Pacific National Monument, the memorial's goal is to "preserve, interpret, and commemorate the history of World War II in the Pacific from the events leading to the December 7, 1941 attack on Oahu, to peace and reconciliation."[38] The National September 11 Memorial & Museum has multiple missions. It "bears solemn witness to the terrorist attacks," honors the victims and those who risked their lives to save others, and recognizes the survivors. It attests to "the triumph of human dignity over human depravity," and endeavors to "reaffirm respect for life, strengthen our resolve to preserve freedom, and inspire an end to hatred, ignorance and intolerance."[39]

As nations became empires they used museums in new ways. Imperial museums not only served to show off the extent, grandeur, and ambition of the empire by displaying the art and culture of the colonies but also connected the nation to the glories of the ancient world by acquiring Egyptian, Greek, and Roman art and artifacts. Natural history museums were useful in straightforward economic ways, supporting the agriculture that European tastes and industry demanded, as well as in cultural ways. They made the rest of the world an object of Western science.

Colonial powers used museums to shape their colonies. Political theorist Benedict Anderson suggests that three devices—the census, the map, and the museum—allowed colonial states to conceive of themselves as "imagined communities." The census defined racial and ethnic categories. The map defined the borders of a place. And the museum provided a history. It legitimized the state's ancestry. Old sacred sites became archeological and tourist sites, "repositioned as regalia for a secular colonial state." Antiquities became museum artifacts. A museum offered the state a genealogy, making historical images and artifacts into "an album of its ancestors."[40]

Some national (or imperial) museums, like the British Museum or the Louvre, have rebranded themselves as encyclopedic or universal museums. James Cuno, director of the Getty Trust, has been most eloquent on the importance of bringing together artifacts from around the world. "Antiquities," he writes, "are the cultural property of all humankind. . . . We have much more to gain by seeking similarities between us than by relying on crude, reductive, falsely unifying cultural identities." He argues for "the Enlightenment ideal of the museum as a force for understanding, tolerance, and the dissipation of ignorance and superstition about the world, where the artifacts of one time and one culture can be seen next to those of other times and other cultures without prejudice."[41]

Of course, the countries from which some of those objects were taken have a different point of view. The Parthenon Marbles, the Benin Bronzes, and the Rosetta Stone are only the best known of the British Museum's objects whose countries of origin have requested repatriation. Battles continue in the courts, and in scholarly, public, and political arenas.

Philosopher Kwame Anthony Appiah argues against return partly on practical grounds—how can a modern nation-state claim as its cultural heritage something from long before the state existed?—and partly on the grounds of cultural connection. Rather than return the objects looted by the British to the palace museum in his home of Ghana, why not send "a decent collection of art from around the world." Yes, people connect to cultural things from their heritage. But equally important, he says, is the human connection: "My people—human beings—made . . . those things. . . . The connection through a local identity is as imaginary as the connection through humanity."[42]

Appiah imagines not an imperial museum, nor a universal museum, but what might be called a transnational museum, one that connects cultures rather than separating them. Just as the museum was an essential piece of imagining community and nation, so it is in imagining a global culture.

Orhan Pamuk, Turkish novelist and creator of the Museum of Innocence in Istanbul, offers a critique of national museums that expresses a different, complementary alternative to the focus on the nation. He suggests that they should turn from "the narrative of nation, History with a capital H," to the stories of individuals in the nation. "The great museums invite us to forget our humanity and to accept the state and its human masses," he laments. They are epics, he writes, and they should be novels. They should "reveal the humanity of individuals. . . . Ordinary and everyday stories are richer, more human and above all more joyful." These museums should not "represent the state but [rather] re-create the world of individual human beings."[43]

The past few years have seen new kinds of museums that begin to address these concerns. They focus on connections between countries, on global issues, on individual stories, and on new ways of defining community. The Canadian Museum for Human Rights, the Center for Civil and Human Rights in Atlanta, and the National Underground Railroad Freedom Center address contemporary issues of human rights in historical perspective. The U.S. Holocaust Memorial Museum builds from historical events to address contemporary issues: one of its goals is to "encourage its visitors to reflect upon the moral and spiritual questions raised by the events of the Holocaust as well as their own responsibilities as citizens of a democracy."[44] Many countries now have museums of migration, addressing the movement of people between nations.

Most ambitious is the United Nations Live Museum for Humanity, planned for Copenhagen in 2020. Among its goals are sharing "knowledge, experiences and solutions that encourage recognition of cultural diversity and cultivate freedom of thought" and encouraging visitors to act, both locally and globally, on important issues through "concrete collaboration and individual action."[45] It wants to connect visitors to the museum with people around the world, allowing them to "not only co-create with participants the ongoing exhibition in real-time, but also to co-create new exhibitions, bottom-up, with initiative and contributions from participants around the world."[46]

Museums reflect and reinforce community. That's what the Slater spinning frame did for the industrialists of nineteenth-century Providence, the reason they were eager for its display at the Jenks Museum, the Smithsonian, and at expositions. At their best, museums can also shape and even create communities, using artifacts to tell important new stories. The Slater spinning frame could today tell a story of immigrant success, or of a community reshaping its economy. Community, national, and transnational museums bring people together by offering new and useful narratives that connect past and present.

‌⇒ 17 ⇐

LEARNING FROM THINGS

WHEN THE SMITHSONIAN INSTITUTION ISSUED a call for natural history specimens in 1850—in particular for "SMALL QUADRUPEDS, as field mice, shrews, moles, bats, squirrels, weasels"—Jenks was one of many naturalists who responded.[1] Jenks sent the Smithsonian's Spencer Baird hundreds of mice, voles, shrews, weasels, muskrats, and skunks, along with one rat and two foxes. "I interested my pupils and others to bring them into me till he cried enough," Jenks wrote in his autobiography.[2] (Jenks paid them six cents per mouse.) The Smithsonian's *Annual Report* thanked him for his work: "One of the most important contributions to the geographical collections of the institution has been the series of mammals of eastern Massachusetts received from Mr. J. W. P. Jenks of Middleboro. Large numbers of all the species from about Middleboro have been collected and forwarded by Mr. Jenks amounting to over eight hundred specimens and with the result of adding several species to those known to inhabit the State."[3]

Baird analyzed the specimens he received in *The Mammals of North America: The Descriptions of Species Based Chiefly on the Collections in the Museum of the Smithsonian Institution*, published in 1857. The book shows the importance of a large number of specimens. Baird singled out Jenks's specimens for particular praise: "I have taken the white-footed mouse of Massachusetts as my type in describing the species on account of the very large series of specimens on hand received from Mr. Jenks, and representing all of the variations of age, sex, and season."[4]

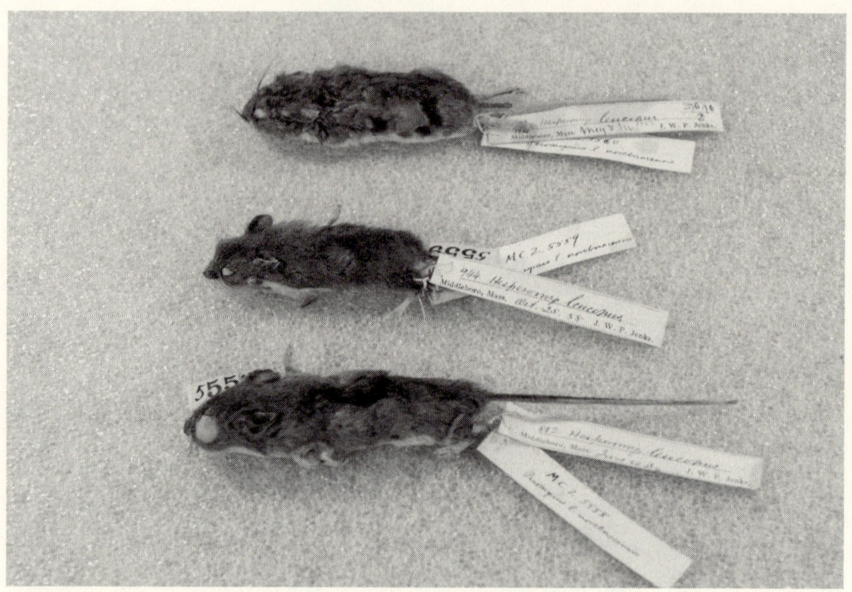

Some of the mice Jenks caught in the summer of 1855 for the Smithsonian, now in the collection of Harvard's Museum of Comparative Zoology Mammalogy Department.

When Baird finished looking at and measuring Jenks's "varmints," they were stored at the National Museum along with all of the other animals Baird had used for his *Mammals*. They were also made available for other scientists to use for their work. In 1866 Joel Asaph Allen, a curator at Harvard's Museum of Comparative Zoology (MCZ), began work on his *Catalogue of the Mammals of Massachusetts*. This 1869 catalog was based mostly on Allen's own collecting in Springfield, but Allen knew about Jenks's collections at the Smithsonian from Baird's book, and he wanted to examine them. On June 24, 1866, the Smithsonian shipped them to the MCZ, not too far from their first home in Middleboro, for Allen to work on. Allen learned new things from Jenks's mammals and offered this appreciation of his work: "No one has done more to increase our knowledge of their history than Mr. J. W. P. Jenks, of Middleboro."[5]

The next revision of the Rodentia—Allen and Elliott Coues's 1,100-page 1877 *Monographs of North American Rodentia*—was based on Darwinian principles, so it required more specimens and a different way of looking at them. Allen and Coues wanted to show the range of variability within a single species. The "liberal policy pursued by the authorities of the National Museum and of the Museum of Comparative Zoölogy of Cambridge," they wrote, made this work possible. Jenks's collection of specimens came in handy once again.[6]

Jenks's mice would continue to show up in taxonomic texts, but they would also serve another purpose. In February 1876 the MCZ received a shipment of rodents from the Smithsonian, among them several Jenks specimens. These were "Types of Dr. Coues's Monograph North American Muridae"—some of the specimens included in Coues's book. In its role as the national museum, the Smithsonian distributed identified sets of specimens like these to museums across the country. Jenks's mice found new homes at, among other places, the University of Michigan, the Chicago Academy of Sciences, and "Women's College, Baltimore."

Jenks's mice were useful. Scientists examined them and measured them—a dozen or more measurements for each mouse—built taxonomies with them, and used them in other types of research. That's why they were collected, and that's why they have been preserved. Many of Jenks's mice are still at the Smithsonian and the MCZ and other museums across the country, awaiting further use. I wanted to see them, so I went to visit the MCZ.

I visited the specimens online first. Museums have put enormous effort into making their catalogs available online, searchable and usable in new ways, as well as connecting them to other online resources. The metadata—the records—have become important in their own right. VertNet is one place to start: 19 million vertebrate specimens from 171 collections in twelve countries. The MCZ has its own database, MCZbase, of almost 2 million specimens. There are about fifty fields one can search: a vast range of identifier codes, any part of the scientific name of the species, seventeen varieties of location, fourteen varieties of individual relationships ("eaten by," "littermate of"). I kept it simple, typing in "John Whipple Potter Jenks" under "collector." Jenks's collections appear, 107 rows of blue type. The first specimen's identification is *Blarina brevicauda talpoides*. (That's a shrew.) Collection location for all of the specimens is Middleboro. The date of collection is either 1856 or "date unknown," with a few other more complicated dates scattered here and there. There are twenty-four northern short-tailed shrews, one northern red-backed vole, two meadow voles, and eighty white-footed mice. Twenty-eight shrews and a few mice are noted as "not found," and forty-eight *Arvicola riparia* (water voles) have been discarded.

The online catalog lets me jump from the single-line description to a full page of information about each specimen and then to a photograph of the page of the museum's register where the specimen's accession was originally recorded. Even though the information has been carefully converted to digital form, it's still useful to see the original page. Details of handwriting, drawings, and other notes can tell important tales.

I went to visit Jenks's mice in August 2015. The mammalogy collection is two floors underground, in a new building. Mark Omura, the curatorial assistant, showed me around. Brand-new and filled with gleaming white storage units, its first appearance is one of orderliness, rather like the lines of type of the online database, or the rectilinear rules of the register pages. Charismatic megafauna, mounted, decorate the room: a black bear, several deer, an antelope.

Omura showed me the catalogs first. They are large, heavy, cloth-bound volumes, about fifteen inches high by twenty-four inches wide when they're opened up. A few dozen volumes trace the entire history of the collection. Omura calls them the gold standard for documentation, even more so than the database. The specimens, described in beautiful handwriting early on, and in neat printing in the twentieth and twenty-first centuries, were once living things, members of family groups, parts of ecosystems; but here they are summarized into a single line, given a species name and a number that links the data to their earthly remains in the cabinets nearby.

The storage units are arranged in taxonomic order. The cabinet we want is labeled RODENTIA, and then Cricetidae, Cricetinae, and then, in smaller type, "*Peromyscus leucopus ammodytes* through *Peromyscus leucopus noveboracensis* to MCZ 40025." These are mice. There are maybe a dozen drawers in the storage unit. I check the database later. The MCZ has 36,497 specimens of Rodentia, almost 8,000 *Peromyscus*, and 1,603 *Peromyscus leucopus noveboracensis*. Eighty of those came from Jenks, and those are the ones I want to see.

The mice have been turned into skins and skeletons. One drawer might have fifty mouse skins and almost as many small bottles, each containing a mouse skeleton. The skins are flat, about three inches long, their tails straight behind them, adding another two inches. They look remarkably similar to the mice I occasionally see in the traps I set in my basement; they've aged well. Jenks was a good preparator. They're stuffed with cotton, and you can see the cotton through their eye sockets. Each has a tag with a name and a number.

Omura picks up one of the mice and hands it to me. It looks like the others, but it's special: it's a Jenks mouse. There are three tags: two attached to a hind leg, each almost exactly the length of the mouse's tail, and one to its ear, smaller, with just the number: 5558. One of its leg tags reads "802 *Hesperomys leucopus* / Middleboro, Mass. June 20 '55 J. W. P. Jenks." That's Jenks's original tag. The other reads "MCZ 5558 / *Peromyscus l. noveboracensis.*" This mouse went from a quiet summer in Middleboro to Jenks's home in Middleboro to Washington. It was measured and described by Spencer Baird, examined by Joel Allen and Elliot Coues, distributed to the Boston Society of Natural History, and then transferred to the MCZ in 1876, after twenty-one years of traveling and service to science. There it joined the other mice Jenks had collected that quiet summer. It moved a few hundred yards to the new building last year and now rests in the subbasement, its skin and story intact, still useful to anyone who cares to seek it out.

The mice preserved in alcohol are next door, in the "wet" storerooms of the old MCZ building. We head over, passing through rooms filled with jars of pickled creatures. Omura opens a storage unit and points to a jar. It's a new jar, maybe a foot tall, maybe ten inches in diameter, with a white plastic lid. It's filled with fluid, and mice. At first glance, I can just see tails and feet, but on closer inspection I can make out the entangled mouse bodies. They're intact, as far as I can tell. Each has an embossed metal tag attached, with its number. And the jar is labeled on the outside with the scientific name, Jenks's name, and the location caught: Middleboro, Massachusetts.

The big jar of mice stopped me cold. John Whipple Potter Jenks had collected these mice 160 years ago. He had probably followed Spencer Baird's instructions from 1850: keep a small keg handy, partially filled with liquor, and throw the mice in alive; this would make for "a speedy and little painful death" and "the animal will be more apt to keep sound."[7]

The mice had been transferred to a new jar and they had been retagged. But here they were. I had been following Jenks's trail for several years, and suddenly felt that I was, oddly, in his presence.

Harvard's mice are part of a larger universe of mouse collections. Hopi Hoekstra, Harvard's Alexander Agassiz Professor of Zoology and Curator of Mammals at the MCZ, and a mouse expert, counts more than 120,000 *Peromyscus* specimens in natural history museums. The Smithsonian has more than 38,000 mice, the Museum of Vertebrate Zoology at Berkeley more than 34,000.[8] Large numbers are useful. Wilfred Osgood looked at more than 27,000 specimens for his 1909 *Revision of the Mice of the American Genus Peromyscus*—including some of Jenks's mice from Middleboro.[9]

But what use are they today? That's the question museums need to answer. Hoekstra considers their value in a 2015 article. "Our knowledge of the distributions, home ranges and habitat preferences of deer mice comes primarily from the trapping data and field notes of early natural historians," she writes. The mice they collected are invaluable, documenting "more than a century of dynamic relationships between deer mice and their environment." That allows scientists to undertake studies of population genetics, and to answer questions about climate change and how mice respond to changing environments.[10] Hoekstra continues to collect mice for the Museum of Comparative Zoology, over 1,200 *Peromyscus* to date. They sit on the shelves not far from *Peromyscus* that Jenks collected for the Smithsonian a century and a half ago.

A Great Reference Library of Material Objects

Jenks's mice tell a traditional story of scientific collections. They weren't collected for display, have never been on display, and probably never will be. Neither will 99.9 percent of the world's 3 billion natural history specimens.

But that doesn't mean they aren't useful. Natural history specimens, as well as the art and artifacts at other museums, are essential for research and teaching, and have practical, commercial, and political uses. Jenks's mice, collected as part of the Smithsonian's project to document the animals of the United States, have been used by successive generations of scientists to determine the taxonomy of the Rodentia and to answer other scientific questions. Alexander Kellogg, the director of the U.S. National Museum, was unequivocal about the place of artifacts in research. In his 1959 *Annual Report* he wrote, "In the National Museum research naturally enough starts with the assembling of the great collections." He outlined the process of research as he saw it. First, the objects are described and classified to provide a basis for "further theorizing and evaluation," and to lead "to the establishment of broad general facts, or to the development of fundamental laws or conclusions." The Smithsonian was "a great reference library of material objects."[11]

Look behind the scenes, and you see them put to use.

Anthropologist Margaret Mead led a virtual tour of the American Museum of Natural History in her 1965 *Anthropologists and What They Do.* "Up here, on the curators' floor, the long halls are lined with tall wood and metal cabinets and the air has a curious smell—a little stale, a little chemical—a compound of fumigating substances and mixed smells of actual specimens, bones, feathers, samples of soils, and minerals." You might get the idea that a museum is "a place filled with specimens smelling of formaldehyde, all rather musty and dated and dead." But then you open a door into a curator's office: "A curator's office is a workshop. Here he spreads out new specimens to catalogue or old ones to study. Here he makes selections for exhibits, comparing his field notes and his field photographs with objects collected on a recent field trip or perhaps a half-century ago."[12] The researcher gives the specimen new life.

Richard Fortey, a paleontologist at London's Natural History Museum, leads us on another behind-the-scenes tour. He shows us "the natural habitat of the curator," the "warren of corridors, obsolete galleries, offices, libraries and above all, collections." There are endless drawers of fossils, arranged taxonomically, like the mammals at the MCZ. Each is labeled with its Latin name, the rock formation from which it was recovered, its geological era, location, and the name of the collector, and, sometimes, where it was published. This is where Fortey does his work, assigning names to new species, comparing examples to understand systematics (the relationships between species), and generalizing about evolution and geological and climate change. "The basic justification of research in the reference collections of a natural history museum," writes Fortey, "is taxonomic."[13]

The earliest uses of scientific collections were practical. European universities' botanical gardens included plants from around the world with medicinal value, arranged to show their botanical and medicinal relationships. Apothecaries accumulated natural history specimens for their medicinal values as well. As European countries began to acquire colonies around the world, they collected specimens both to understand their new world and to create new commercial opportunities. Kew Gardens and places like it, writes historian Lucile Brockway, "played a critical role in generating and disseminating useful scientific knowledge which facilitated transfers of energy, manpower, and capital on a worldwide basis and on an unprecedented scale." The first director of the Calcutta botanical garden, founded in 1787, described its purpose as not curiosity or luxury, but rather as "establishing a stock for disseminating such articles as may prove beneficial to . . . Great Britain, and which ultimately may tend to the extension of the national commerce and riches."[14]

Natural history collections have been the basis of the most important biological breakthroughs from Georges Louis Leclerc Buffon's 1749 *Histoire naturelle, générale et particulière* to Georges Cuvier's theories of animal anatomy in the early nineteenth century, and from Darwin's 1859 theory of evolution to Ernst Mayr's mid-twentieth-century evolutionary synthesis.

Gathering together and ordering specimens in museums made it easier to learn from them. It became simpler to compare and to build theories from them. "How much finer things are in composition than alone," wrote Ralph Waldo Emerson after a visit to the Muséum d'Histoire Naturelle in 1833. Emerson saw there "the upheaving principle of life every where incipient," the organization of the universe.[15] Similarly, scientists could find principles of organization useful to their work. Science historian Bruno Strasser writes, "When objects become accessible in a single place, in a single format, they can be arranged to make similarities, differences, and patterns apparent to the eye of a single human investigator; collections concentrate the world, making it accessible to the limited human field of view." As Buffon put it in 1749, "The more you see, the more you know."[16]

Collecting for scientific ends has always been central to American museums. The goal of Charles Wilson Peale's Philadelphia museum, established in 1786, was the promotion of useful knowledge. That was also the goal of the nearby American Philosophical Society, the Smithsonian when it was founded in 1846, and of natural history museums across the United States in the nineteenth century. They built collections for researchers. They published volumes of scientific papers. Outreach—exhibits, lectures, popular education—was a secondary goal for much of their history.

Museums were the home of an object-based nineteenth-century science. Collecting and observing things seemed central to understanding. William James wrote that his teacher Louis Agassiz, founder of Harvard's Museum of Comparative Zoology, "used to lock a student up in a room full of turtle shells, or lobster shells, or oyster shells, without a book or work to help him, and not let him out till he had discovered all the truths which the objects contained." Museums were places to discover the truth about things.[17]

Taxonomy and systematics—the identification and classification of plants and animals—was, until the twentieth century, the most important work of biology, and put natural history museums at the center of the field. Taxonomy, explains Harvard's Edward O. Wilson, another denizen of the museum storeroom, "is a craft and a body of knowledge that builds in the head of a biologist only through years of monkish labor. . . . A skilled taxonomist is not just a museum labeler. . . . He is steward and spokesman for a hundred, or a thousand, species."[18]

But by the middle of the twentieth century, biology based in the museum seemed less important than biology based in the laboratory. Experimental and analytical sciences—genetics, biochemistry, crystallography, and eventually molecular biology—made natural history seem old-fashioned. Function seemed more important than form, chemistry more important than taxonomy, behavior more important than appearance. Collections were out of fashion. The museum biologists fought back. Harvard's Museum of Comparative Zoology was one of the places this battle—Wilson called it "the molecular wars"—was fought. He wrote, "The molecularists were confident that the future belonged to them. If evolutionary biology was to survive at all, they thought, it would have to be changed into something very different. They or their students would do it, working upward from the molecule through the cell to the organism. The message was clear: Let the stamp collectors return to their museums."[19]

Wilson did, indeed, return to the museum, but to regroup, not to retreat. "What I desired most," he wrote in his autobiography, "was to emigrate across the street to the Museum of Comparative Zoology, to become a curator of insects, to surround myself with students and like-minded colleagues in an environment congenial to evolutionary biology." There, he and his fellow curators would start a revolution in evolutionary biology and ecology that drew on the collections—as well as the new ideas about molecular genetics—in new ways.[20]

Bruno Strasser points out that the natural historians who worked in museums had always collected more than just specimens of animals and plants. They had also collected, starting in the nineteenth century, seeds, blood, tissues, and cells. More important, they had also collected data: locations, descriptions, drawings. All those measurements of Jenks's mice were part of a vast database that included not just the collection of skins and skeletons but also information about the creatures. This proved useful for answering new questions.[21] Joseph Grinnell, founding director of Berkeley's Museum of Vertebrate Zoology, emphasized the importance of this data for the new biology of the early twentieth century: "The museum curator only a few years since was satisfied to gather and arrange his research collections with very little reference to their source or to the conditions under which they were obtained. . . . The modern method, and the one adopted and being carried out more and more in detail by our California museum, is to make the record of each individual acquired."[22] Grinnell's California collection included not only 100,000 specimens but also 74,000 pages of field notes and 10,000 images. "These field notes and photographs are filed so as to be as readily accessible to the student as are the specimens themselves." Grinnell thought that this data might end up being more important than the specimens.[23]

When scientists like Wilson became interested in theoretical questions of population ecology in the 1970s, the collections and the data about them proved essential. When issues of pollution and environmental contamination became important in the 1980s, or climate change in the 2000s, the collections were useful. Museums have pivoted from a focus on systematics to biodiversity as they look for new ways to take advantage of their hard-won collections. Biodiversity research relies on systematics; you can't know what's going extinct unless you know what you have. The 1998 Presidential Panel on Biodiversity and Ecosystems called for digitizing collections data as a vital first step—a call that was answered over the next twenty years with systems like the ones that allowed me to find Jenks's mice scattered across the country.[24]

Over the past decade there have been many arguments for the practical value of natural history collections. Collections are useful in tracking invasive species as well as documenting, for example, the presence of DDT (measuring the thickness of eggs from museum collections) and mercury contamination (using bird and fish specimens). Collections are useful in the study of pathogens and disease vectors; millions of mosquito specimens collected over the course of a century provide information on the spread of malaria, West Nile virus, and other diseases. The invasive Asian long-horned beetle was identified from a specimen in the Cornell entomology collections.

The molecular revolution of the 2000s unlocked even more information from the collections. It's possible to extract DNA from some specimens, not only to improve taxonomy but also to learn about diseases and even the evolution of viruses. Researchers have used material from collections to trace the history of the 1918 influenza virus. An analysis of the 1990s hantavirus outbreak using museum rodent collections was useful to public health officials in predicting new outbreaks—and researchers argue that had there been good collections from Africa, the recent Ebola outbreak would have been easier to understand and control. Natural history museums continue to serve as a "great reference library of material objects." Pulled from across time and space, they pose—and answer—old questions and new.

Objects as Evidence

Art and history museum collections, like natural history collections, contain hidden evidence of the past. John Keats's "Ode on a Grecian Urn" expressed this most longingly. The poet addresses an urn at the British Museum:

Thou still unravish'd bride of quietness,
  Thou foster-child of Silence and slow Time,
Sylvan historian, who canst thus express
  A flowery tale more sweetly than our rhyme . . .

Objects, Keats claims, are better historians than we are. They hold truths; the poet—the historian—can only ask questions. The object as keeper of the truths of the past, as the most truthful historian, appeals to the museum curator. The curator can reveal those truths and make the objects speak.

Historians and archaeologists have worked hard to use objects as evidence. Connoisseurs read objects to determine style, authenticity, location made, and date. Technical art historians use infrared and X-rays to look at paintings in new ways. Material culture historians have developed a range of frameworks of analysis. Edward Fleming suggests starting with materials and then focusing on design and classification. Jules Prown recommends starting with description, then considering what it would be like to interact with the object, both physically and emotionally. Archaeologists use the details of provenance to understand, in James Deetz's phrase, the "small things forgotten," and new technologies to examine old things.[25]

Most curators combine methods like these with a visceral understanding of artifacts. Nicholas Thomas of the Cambridge Museum of Archaeology and Anthropology describes the museum method as more about "discovery" than critical enquiry: "it often involves finding things that were not lost, identifying things that were known to others, or disclosing what was hidden or repressed." He continues, "The encounter with arrays of objects" is usefully destabilizing. "For example, the search for a 'good' or 'representative' piece may put at risk one's sense of a genre or place. One may be distracted by another work or by some aspect of the provenance or story of an object that is not good or not typical. This is, in one sense, entirely unremarkable; it reflects the contingency of dealing with things, but, in another sense, it represents a method powerful because it is unpredictable." Object-love as method![26]

Looking closely at museum things and asking the right questions—more accurately, conversing with them—can raise new questions and surface new evidence about the past unavailable any other way, complementing the history revealed in oral histories and books and archives. Historian Leora Auslander calls attention to the evidence available only from objects. They offer us not only "rich sources for grasping the affective, communicative, symbolic, and expressive aspects of human life," she writes, but also, because "they act, and have effects in the world," they "structure people's perceptions of the world, thereby changing that world." We can understand something about history and culture by examining things—and there is much we miss by not examining them.[27]

That notion comes easily to museum curators, who are close to artifacts in all their work. The familiarity that comes with closely examining and handling objects provides insight into the objects' previous lives. Sometimes taking objects outside the museum and seeing how they work precipitates an understanding impossible to come by any other way. And in recent years, scientific studies—everything from CT scans to dendrochronology to chemical analysis—have been used to determine just how old things are, where they're from, how they're made, and what they did. All of these techniques can tell us not just about the objects, but about history.

Historian Laurel Thatcher Ulrich demonstrates, in *The Age of Homespun: Objects and Stories in the Creation of an American Myth*, how close looking can lead to new understandings. Ulrich chose fourteen objects from small museums—baskets, woodwork, textiles—and figured out what they meant to the people who made, used, and preserved them. An Algonquian basket from 1676 at the Rhode Island Historical Society offers insight into the conflicts that led to King Philip's War and how English and Native American cultures changed in its aftermath. She reads the label on the basket, from 1842, as an object too, using it to understand the ways in which the basket was made useful as a museum-worthy artifact, the ways that it connected the preindustrial past to the rapid industrialization of the day and helped reimagine King Philip's War in a way that exonerated English settlers.[28]

Hands-on work with historical objects sometimes provides direct information. David Jeremy, a historian of the textile industry, discovered important insight into early nineteenth century spinning frames by examining one closely—almost too closely. To understand the mechanism that stopped the loom when a thread broke, he accidentally triggered it "and the curator nearly lost a finger."[29] Curator Edwin Battison of the Smithsonian discovered that the supposed interchangeable parts of early American muskets—a key claim of Eli Whitney and an important question in the history of manufacturing—were not interchangeable, simply by trying to interchange them. His work raised new questions about skill and management in nineteenth century industry.[30]

Examining tools and machines closely for wear marks and other indications of use allows us to reconstruct the lives and skills of the men and women who operated them. Machine design reveals managerial oversight, labor relations, and technological skills and knowledge. Operating historic lathes, looms, and other machines brings a new appreciation of the levels of skill and attentiveness they demanded of the men and women who operated them, often revising notions of "deskilling" based on other kinds of sources. I once worked with machinists and conservators for days to get an 1830s pin-making machine to run, and in the process gained a great respect for the factory operatives who had what historians called "unskilled" jobs as machine operators, as well as the ways in which skilled machinists understood mechanisms and materials.

In 1981, in honor of its 150th anniversary, the Smithsonian's National Museum of American History decided to repair and operate the *John Bull*, the oldest operable steam locomotive. Bill Withuhn, one of the museum curators who repaired and ran the engine, writes that "in the course of firing up and operating the engine, we gained several fresh and important insights into the history and design of the locomotive. These insights could only have come from the actual, hands-on experience of running and operating the engine." The curators running the engine were surprised at how easy it was to get up a head of steam, and how well the engine tracked, leading them to reevaluate the history of leading-truck design on early locomotives. They discovered, by accident, that one of the wheels was deliberately loose, to allow it to round curves more easily. "A live specimen," Withuhn concludes, "is a far different thing than a lifeless one."[31]

Historical musical instruments are very different when they are alive, being played, than when they sit on a museum shelf. Playing an instrument brings a new understanding and appreciation of the music written to be played on it. Johnny Gandelsman, a violinist, comments about his experience of playing Bach with a Baroque bow: "It was as if the bow was telling me how to play the music."[32] Similarly, wearing historical clothing reveals truths about posture and movement. One reenactor writes that you can "understand the physicality of another time and place when it literally swathes you." The clothing shapes the way you behave. Reenactors say that they can do historical work better when wearing historical costume.[33]

The Association for Living History, Farm and Agricultural Museums sponsors plowing matches, an opportunity to actually use some of the re-created historic equipment in the museum. Deborah Arenz, a curator at the Nebraska State Historical Society, argues that every curator should give it a try. "Using a plow," she writes, "was the most complex interaction I've had with an artifact. . . . I was full of questions and found myself thinking of the plows in our collection and what it would be like to use them." After using the plow, she could relate to history differently: "We had a shared experience: we all stood behind a plow."[34]

Close examination of works of art can reveal how a painting was painted, how furniture was constructed, how a sculpture was carved; all of these are questions that can only be answered by investigating the actual thing. Was Jackson Pollock's *Blue Poles* painted in a single drunken session? No, it turns out: if you look closely, you can see that the paint dried in five separate sessions. That seventeenth-century chest shows, in the grain of its wood, the landscapes of the town where it was made. Look closely at a Greek sculpture and you can find evidence of the paint that once covered it, the holes where swords and jewelry were fastened. Conservators and materials scientists contribute a different kind of detailed analysis using scientific tools. Scientific analysis of pigments, for example, can definitively give a date before which a painting could not have been painted. X-rays and infrared reflectography can reveal underpainting, earlier versions, or changes the artist made to the work. Ultraviolet light can help date varnishes. Raking light shows brushstrokes, revealing the way a painting was created.

Archaeologists have a different set of tools. They use instruments that measure radiocarbon, thermoluminescence, and archaeomagnetism, among others, to date artifacts. X-ray fluorescence can reveal the elemental composition of an object; knowing what an object is made of can reveal when and where it was made, which in turn provides information on trade routes and skills. X-rays, CT scans, and other imaging technologies let archaeologists look inside objects and learn more about them.

Metallurgist and historian Robert Gordon of Yale carried out what might be called forensic analyses of historical artifacts. In the 1990s he analyzed the residue at the bottom of the "Kelly converter," long on display at the Smithsonian as the first device that produced steel using a pneumatic process, and supposedly proof that an American, William Kelly, and not the British Henry Bessemer, had invented the process. Gordon tested the slag that adhered to the firebricks and found no trace of metal. He was able to prove that the device had been used in an attempt to make steel and that it had failed. The Kelly invention was a useful myth, Gordon explained, not only for American steelmakers wanting to get around the Bessemer patent, but also for dismissing the skills of practical ironworkers who had scoffed at Kelly's work. The converter came off display.[35]

"Museum exhibits and stored material are provocative of research," claims the 1935 *Manual for History Museums.*"[36] "Provocative of research" is a good phrase; museum objects raise questions and offer innovative ways to answer them. The examples above show the possibilities. But curators are aware that most historians do not use material culture in their work. Cary Carson, of Colonial Williamsburg, states the problem bluntly: "No matter what standard measure objective scholars use they can hardly avoid the conclusion that the study of artifacts has contributed to developing the main themes of American history almost not at all."[37] In part, that's because written documents seem more direct. But it's also because there's so much else to do in museums: The 1992 *Manual of Curatorship* notes that "collections research has often been regarded as something a Curator did if he found time," a luxury.[38]

Professor Jenks would be pleased, but not surprised, to find that many of the "small quadrupeds" he collected more than 160 years ago have served science, and that they are still in a museum, and still useful. Though he might be disappointed that collections are stored away in a subbasement, he would take delight in the new ways that they are put to use, the new questions they raise, and the new answers they suggest.

≫ 18 ≪

TEACHING WITH THINGS

PROFESSOR JENKS MUST HAVE BEEN pleased when Brown University hired Hermon Carey Bumpus as assistant professor of zoology and assistant curator of the museum in 1890. Bumpus had been one of Jenks's students. He went on to receive his PhD at Clark University and work at the Marine Biological Laboratory in Woods Hole, Massachusetts, where Jenks spent many summers.

Bumpus, who would become the director of the American Museum of Natural History, was a member of a new generation of biologists. He learned from Jenks that biology was best taught hands-on, but he took that precept much further than Jenks had. "With his coming," wrote Bumpus's son, "the methods of teaching biology underwent a drastic change."[1]

Bumpus, like Jenks, believed in teaching with things. But he used those specimens in a different way, to teach different lessons. The first course Bumpus offered was vertebrate anatomy, and he put the museum collections to use: "Specimens majestically installed in the museum were ruthlessly withdrawn from their alcoholic containers and used for class demonstrations." Bumpus took his students out collecting. He undertook biological surveys. When an animal at the nearby zoo died, the class met there, to observe the autopsy. The museum collections seemed less useful to him than laboratory work: "A ponderous stuffed walrus, the memorial gift of some early class, was denounced as occupying space that might be better used for a laboratory table." One of Bumpus's students, later a

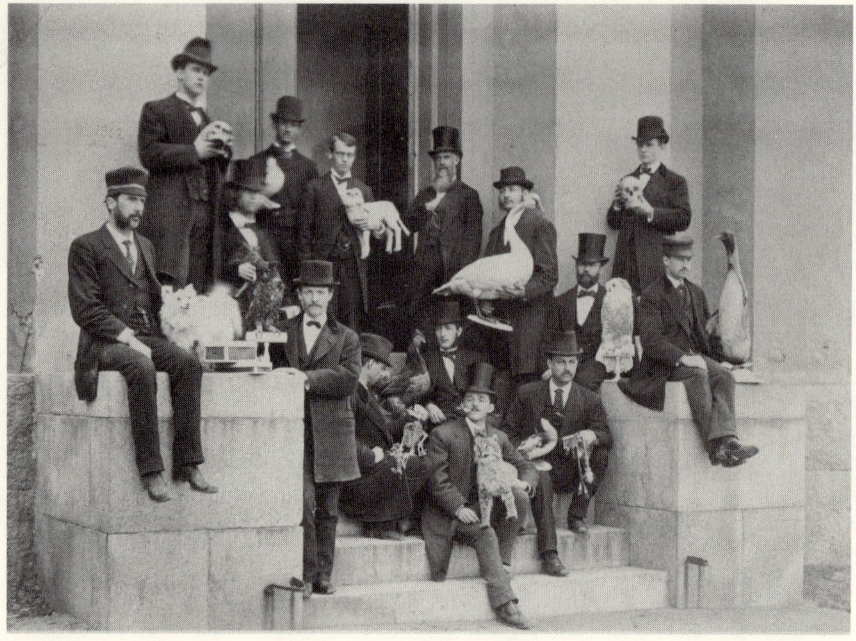

John Whipple Potter Jenks with his students, showing off their taxidermy on the steps of the museum building.

biology professor, remembered nothing of textbooks or lectures but recalled with pleasure "studying specimens of all sorts at first hand."[2]

Those who work in museums insist that museums are educational places and that objects are of educational value—while at the same time worrying that museums are not appreciated for what they do and that they have not reached their full educational potential. That has been true from at least the mid-nineteenth century. Adele Silver, editor of a 1978 survey of museum education, sums up the conundrum this way: "Those who run American museums have never fully agreed on what museums should teach, to whom, or for what ends, but from the outset museums have undertaken to teach someone something."[3]

What Good Is Art?

Benjamin Ives Gilman contended that art was art, not a tool for learning something else. He railed against what he called the "didactic fallacy" and argued that it was wrong for art to be treated as the means for teaching something else. It was simply to be enjoyed and appreciated. Art was good for people. *Harpers Monthly* had written along similar lines much earlier, arguing in 1876 for the creation of art museums because "there can be no question as to the ennobling and refining influence of art upon personal character and upon the community."[4]

But while many art museum directors and curators believed in art for art's sake—its ennobling influence—most also made the argument that art was useful. Museums have used art in many ways, from evidence in political arguments to exemplars for improving the quality of manufactured goods. Most importantly, art is used to teach topics from art and design to art history to taste and style.

A primary purpose of art museums for much of their early history was to teach artists and artisans. Artists visited museums both to be inspired by masters and to learn by copying work from "the most outstanding works of art." An eighteenth-century artist's guide advises: "In copying these the student receives a first insight into the higher spheres of art." To do this, an art student needed "easy access to a picture-gallery."[5] In its early years, half of the Louvre's opening hours were reserved for artists. Its galleries were filled with copyists. In 1868 *Harper's Weekly* described the Louvre as "a splendid studio" for American artists, illustrating the article with a wood engraving showing a gallery made almost impassable by young artists copying paintings.[6]

Many early American art schools were associated with art museums. The Pennsylvania Academy of Fine Arts, founded in 1805, started a museum school in 1813. Artists' associations like New York's National Academy of Design (established in 1825 "to promote the fine arts in America through instruction and exhibition") and the American Art-Union (founded in 1839 for "the cultivation and diffusion of correct taste in the Fine Arts, by the exhibition and distribution of good works of Art, and the encouragement of Artists") maintained large galleries that showed off contemporary art, for sale and appreciation. At the turn of the twentieth century, the Museum of Art at the Rhode Island School of Design (RISD) was an essential part of teaching at the school. It opened its galleries to students mornings and evenings, "in order that the day and evening classes may study the different collections on exhibition during the year."[7] An 1897 review of training in the arts listed Boston's Museum of Fine

Arts (MFA), the Metropolitan Museum of Art, and the Art Institute of Chicago, among others, as institutions with "fine Art schools attached to them, and the excellent facilities afforded by the collections of paintings, statuary, and art works, are of inestimable value to the students."[8]

Colleges, too, used art museum collections for teaching. College and university museums were sometimes directly tied to art departments. The Fogg Art Museum opened at Harvard in 1895, part of the Division of Fine Arts. It served as a resource for the division (providing slides, photographs, and plaster casts of sculpture), and a site for studio courses in drawing, painting, and design. The Department of Fine Arts at Indiana University created a museum in 1898, its collections "selected with direct reference to the university courses in painting, sculpture, drawing, architecture, and engraving."[9] A museum seemed an essential part of the training of a modern artist.

As art history became an academic subject, university museums began to seem necessary in a new way. William Carey Poland, who was both professor of art history at Brown and president of RISD, took his art history students to the cast gallery at Brown and the RISD museum, as well as to private art collections. John Shapley, a Brown University art historian, wrote in 1918 that "art objects form the foundation of art teaching. Instruction on any other basis is as obsolete as alchemy or astrology."[10] Louis Earle Rowe, director of the RISD Museum of Art, went further. Museum collections, he claimed, "are as essential to the study of history, literature and design . . . as a laboratory is to the study of chemistry or physics."[11] A fund-raising pamphlet for a new building for the Fogg Art Museum at Harvard in 1924 made the connection more explicit. Titled *The Fine Arts in a Laboratory*, it put forth the museum as a place for "educational experimentation." Instruction in fine arts, Harvard's president wrote, "can be most effectively given as scientific instruction is given,—by a study of specimens; in short, by the laboratory method."[12]

Many museums focused on training not artists but artisans. In 1872 Louis J. Hinton, a stone mason and an English immigrant, shared with New York State's Commissioner of Education what he called "after-hours-of-labor-thoughts" on the education of artisans. Museums, he declared, were a necessity. "The good such institutions effect is immense," he wrote, not only because of the delight they bring, but because they offered workers the opportunity to imagine new, finer things. Museums teach observation. They allow workers to create beautiful things. "It is evidently to the interest of the public to increase the number of such ingenious and useful workers," he wrote, and museums were the democratic way to do it.[13]

The museum that the New York mason had in mind was London's South Kensington Museum, today's Victoria and Albert Museum. It was not a museum of fine arts but a practical museum, founded in 1852 as the Museum of Manufactures with the mission of boosting British industry. In the late nineteenth century, many American museums copied this model. The Maryland Institute's proposed museum offered a vision of utility:

> Surrounded by these objects, the offspring of a knowledge of practical and theoretical Art, the artisan, mechanic, manufacturer, architect, designer and decorator would breathe an atmosphere of industrial knowledge. Here lessons would be taught which would educate the eye, train the hand, inculcate taste, and expand the intellect and which would open the door of a respectable livelihood to those who profited by them.[14]

Today's Philadelphia Museum of Art was founded as the Pennsylvania Museum and School of Industrial Art in 1876 with the goal of developing "the Art Industries of the State." The Met, whose charter called for "encouraging and developing the study of the fine arts, and the application of arts to manufacture and practical life, of advancing the general knowledge of kindred subjects, and, to that end, of furnishing popular instruction," established an art school aimed at designers and artisans in 1880, soon after it was founded. The school didn't last long, closing in 1894.

In the early twentieth century, many museums reached out to industry in a new way, promoting their collections as models for good design. The "progressive connoisseurs" running art museums, historian Jeffrey Trask writes, linked "art and beauty to citizenship." They believed that better cities and better homes would create better citizens, and that museums would play a key part in that movement by showcasing American art and improving the quality of American design and industry.[15]

Many museums mounted exhibitions aimed at designers, showing off both historical material, for inspiration, and contemporary material, for emulation. They worked with artists, designers, and industrialists to improve the quality of manufactured products. At the Met, Richard Bach, appointed in 1918 as "associate for industrial relations," argued for the economic value of good design. He invited manufacturers to the museum to study the collections and to see special exhibitions that showed off the best of contemporary design.[16] Even the American Museum of Natural History participated in this kind of work, creating special displays of "the decorative arts of primitive peoples" to inspire students of design.[17] Museums reached beyond those professions to work with buyers and salespeople in shops and department stores. "So long as the salesman and saleswoman remain ignorant of artistic principles," wrote *Museum News* in 1918, manufactures won't be able to sell the sorts of goods that require a "higher appreciation of the beautiful."[18]

By the mid-twentieth century, the relationship between artists and museums had begun to sour. Many artists and art teachers challenged the academy and its focus on copying and learning from old masters. Instead, they argued, students should learn from nature and from their own ideas. Radical artists rejected the very idea of the museum. Art and art education became ahistorical, even antihistorical.

Museums had changed, too. With the rise of the field of art history and the professionalization of curatorial training, artists became less common on museum staffs. A group of art museum educators, reviewing their field in 1978, found that many museums had turned their backs on artists. They charged museums with a neglect "that sometimes borders on antipathy."[19] Museums had turned to a new audience for their educational work: the general public, and, more particularly, schoolchildren.

Object Lessons

In the late nineteenth century many museums began to focus on programs for students and to work with schools in educational programs designed to teach with their collections. The Brooklyn Museum, one of the leaders in this area, argued in 1913 that "Teachers in their work require text books from which to teach, but how much greater value is an object about which to teach? How much more lasting in the mind of an individual and in children in particular, is an object lesson, rather than a mental picture?"[20] Museum staff proselytized for the educational value of art and artifact. "The Museum is THE LATEST PHASE OF EDUCATIONAL EVOLUTION," George Sherwood, Curator of Public Education at the American Museum of Natural History (AMNH), wrote in 1920. He explained why:

> It stimulates the imagination; it exercises all the special senses—hearing, taste, sight, smell, and touch; it is especially suited to developing independent judgment, the correctness of which can be tested. In all instruction from books, the child is dependent upon the authority of others. Nature study develops in the child a keen sympathy for all living things; and, most important of all, there is no subject so well suited to training the powers of accurate observation.[21]

The AMNH worked with New York City schools in many ways, starting with lectures for teachers, and later providing materials for teachers to use with their own students. In the 1880s the museum installed in each of New York's 160 schools "a teaching collection of the rocks of Manhattan Island, dried plants, and many specimens of corals, shells, crustaceans, and insects."[22] In 1902, perhaps because of this collaboration, nature study was added to the schools' curriculum. The Museum took on the responsibility of providing "the specimens called for in the Syllabus"—birds, minerals, wood samples, and small mammals. These were circulating collections, each "put up in a small cabinet about the size of a suit-case and . . . accompanied by a leaflet giving simple facts," lent to the schools for a few weeks. The objects could be removed from the cases for closer study. In 1913, some 500 schools—over 1,200,000 students—took advantage of the materials, nearly 10,000 specimens. The museum would also deliver slides from a library of some 30,000 images. "Step by step," the museum declared, "a great system of cooperation has been built up between the regular course work in the schools and the visual instruction in the Museum, until the City of New York now affords the most brilliant example in the world of extension to the school system of all the resources of a great museum."[23]

Schools weren't the only recipients of artifacts. Museums also lent exhibits to factories, army bases, and libraries. The AMNH installed exhibitions in children's rooms in New York City libraries, displaying ethnographic artifacts, industrial models, and taxidermied animals in order to illustrate books on travel, geography, nature study, art, and current events. Some museums even made collections available on loan to individual children.[24] Not everyone approved of this focus on schools and education, with some curators finding it a distraction. A reorganization of the Los Angeles Natural History Museum in 1941, wrote a scientist there, threatened the scientific work of the museum, turning it into "a museum of exhibitions, art instruction and 'education' . . . an instructing agency for the schools, and for circulating study materials."[25]

Nature study seemed most suited to this kind of outreach, but some history museums also believed in this kind of object-learning outreach. The Brooklyn Children's Museum lent out materials for social studies classes: "dioramas showing life in different parts of the world," "dolls representative of almost every civilized country," and "displays which tell the stories of great industries."[26] Edward Page, a professor at the Illinois State Normal School, pushed the idea about as far as it could go. The ideal history museum, he wrote in 1915, was a room spacious enough for all of a collection, but empty; everything had been lent out. "Earlier day relics," he insisted, "should be made to teach." Try on the shoes. Make fire with different devices. Yes, some things may be damaged or lost, but "it is better to lose some things while in use rather than preserve them by 'cold storage.'"[27]

Art museums embraced school work too. Many museums took on the task of Americanization, establishing programs for immigrants, and especially for immigrant children. Boston's MFA, for example, provided free transportation from the city's North End Italian neighborhoods for special programs taught in Italian. It hosted weekend storytelling programs for "slum children." In 1919, some 10,000 children participated at the museum, and some 4,000 in extension programs. The Toledo Museum of Art, which in 1903 had declared itself the first museum in the world to be "child centered," had extensive programs for children, including traveling exhibitions of copies of paintings for sixth, seventh, and eighth grade classrooms.[28] The Met showed off some of its collection in high schools and libraries, reaching some 474,000 visitors in 1937.[29]

Museum outreach to schools would continue, but by the second half of the century it had become secondary to another kind of teaching, gallery teaching. How might museums work with visitors on site, in the museum galleries?

Gallery as Classroom

In the second half of the nineteenth century many natural history museums moved from displays of artifacts designed to be useful to scientists to displays aimed at a more general audience. Art museums began to put only the best art on display, so that visitors might learn from it. History museums began to tell stories with their artifacts. But exhibitions proved a difficult medium for education. Museums began to experiment with new ways of teaching with collections.

In 1877 a writer for the London *Standard* lamented the difficulty of learning from objects in museums. "It is not enough that objects should be exhibited," he wrote. "They must be so explained that visitors can derive some benefit from what they see."[30] Museums answered with more words but also, toward the end of the nineteenth century, a new kind of guided tour. Museums began to engage instructors or, as they came to be called, docents.

Benjamin Ives Gilman, who coined the term docent, had strong feelings about their role, and, as with so many other aspects of the twentieth-century art museum, he set the tone. Docents might work in a variety of ways—some gave talks, some answered questions, some gave tours—but they were not teachers. Their goal was not to impart knowledge. Teaching art history was better done in a lecture, using photographs. Rather, the docent provided "personal companionship" to the visitor, to help him or her understand what the artist intended the work to say. "The purpose of the docent," Gilman wrote, "is to lead his disciples on to enjoyment," and the way to do that was by an intimate connection with art. The docent— indeed, all of the "didactic machinery" of the museum—serves the art. It "ushers the visitor into a royal presence."[31]

The docent's job was to introduce objects to the visitor, and to speak for them. "The essential office of the docent is to get the object thoroughly perceived by the disciple," Gilman wrote. "Those who have already made friends with works of art may profitably be asked to devote a fraction of their time in introducing others to the same friendship." Docents were people who loved, appreciated, and understood museum objects. They were "enthusiasts, experts, . . . each speaking on his own particular hobby, showing off to all who follow him his particular museum pets." Louis Earle Rowe, one of the first docents at the MFA, described his job as striving "to arouse the latent sense of appreciation and criticism which the visitor may possess." The key was individual attention. Success, for Rowe, was when the visitor responded with his or her own critique.[32]

Not everyone agreed with the Gilman style of docent, with objects as "friends" and "pets." Perhaps because teaching with objects was so difficult, museum education quickly devolved into a variety of schools. Some museum teachers turned to formal descriptions of art, and saw their job as teaching the visitor the rules: in 1917 the Met gave lectures teaching visitors "how to recognize good color, good line, and the other qualities that give value to art." Art appreciation classes would allow a visitor to take full advantage of the visit to the art museum—he or she just needed to be taught how to look.[33]

Some museums thought visitors needed more information, especially art history, and offered illustrated lectures. Theodore Low, surveying the field in the early 1940s, found that most museum instructors were trained in art history, and that they taught a watered-down version of university art history.[34]

Other museum teachers thought that the most important purpose of the work of art was as a way to understand the artist's vision. The Barnes Foundation taught that "art must present to the observer an aspect of life that the artist himself has experienced and it must be presented in such a form that it communicates the feelings of that experience to the observer."[35] Still others wanted to discuss art as social and cultural document. During World War II, many museum docents used tours and lectures to reflect on the war, considering both "the examination of values in historical perspective and in contemporary application" and offering solace to those suffering from psychological pressures.[36]

In 1947 Charles Slatkin, who worked at the Met as liaison between New York's public schools and the city's museums, summed up the quandary of museum teaching in a series of questions that showed the range of possibility of art museum teaching. "How much," he asked, "should one lecture; how much discuss; query? Should one educate or inform; elicit information or submerge the listener in a flow of words? Shall one aim for a moment's escape, a vision of man's unfettered genius, a sermon on mortality, the mysteries of the creative process, the enduringness of art, the elements of connoisseurship?"[37]

Museum educators responded with a range of experiments, while at the same time continuing the standard lectures and lecture-tours about art history and art methods. Rika Burnham and Eliott Kai-Kee's history of teaching in art museums notes the appearance of a new set of words in the 1940s and 1950s: game, treasure hunt, informality, freedom, discovery, Socratic method. One educator wrote in 1956 that the goal was "to allow the pupil to draw his own conclusions and to find his own delight." In some museums, the docent became a discussion leader. Her job (most docents were women) was to encourage emotional responses to the art.[38]

Many art museums held art classes, usually separate from the galleries— a return, for some of them, to their roots as art schools. Some focused on children. In the 1940s MoMA, for example, hosted a wide range of hands-on activities for children and families. The "Children's Holiday Circus of Modern Art" gave children the opportunity to "draw and paint . . . , make toys and constructions, put together jigsaw puzzles cut from reproductions of modern paintings, and manipulate the zoo of animals designed by modern artists."[39]

Science teaching in museums also changed radically after World War II. Some natural history museums, once content to teach people about nature, now took on science more generally, and began to focus on scientific principles. They experimented with new, hands-on kinds of teaching. Their audience changed, too. They had always been kid-friendly, but now they were focused almost entirely on children. "Science Centers," as they called themselves, reinvented museum teaching. The Boston Society of Natural History changed its name to the Museum of Science in 1939 and moved to a new building in 1951. It brought some taxidermy with it, but focused on new push-button interactives. The director insisted: "Our exhibits demand the *active participation of the visitor.*"[40]

In the 1960s curiosity, self-expression, creativity, and social consciousness became new buzzwords. Victor d'Amico, head of education at MoMA, combined these when he declared in 1960 that "when people know how to create, they respect others' creativity."[41] Perhaps talking was the wrong approach: students should respond to art with art, drawing, or dancing, or improvisational theater, rather than discussion. Some educators questioned the use of art in art museum teaching: would looking at great art inhibit children's creativity? Would they copy, rather than create on their own? Others insisted that it would make them feel connected: In 1968 the Met's "Old Masters—New Apprentices" program sent students into the gallery to find artists they liked, hoping that this would strengthen their own identities as artists, as well as giving them the security of connecting to a tradition.[42]

Some museum teaching emphasized not the art, but the process of observation. Many museum programs, especially for schoolchildren, concentrated on developing visual perception. Students were to learn how to look: the buzzwords were "focused looking" and "active searching." This might be based on formal elements like line, space, and color, or on mood, or story; the hope was that children would find this useful not just in the museum, but outside its walls, too, in real life. In the 1960s and 1970s, art history, and sometimes even the art itself, lost out to sensitivity, creativity, and sensory and perceptual skills.

The Exploratorium, founded in San Francisco in 1969, applied many of these same principles to science. It lured children with interactive science exhibits that taught observation and the principles of science, and used objects as teaching props. Frank Oppenheimer, the founder, wanted visitors to become experimenters, to both learn and teach. "Explaining science and technology without props," he wrote, "can resemble an attempt to tell what it is like to swim without ever letting a person near the water."[43]

A new way of teaching about art became popular toward the end of the twentieth century, one that had its roots in Gilman and Rowe's work from the early part of the century. Patterson Williams, director of education at the Denver Art Museum, proposed what she called "object-oriented learning." Her goal was "to help visitors have *personally significant experiences with art objects.*" The way to do this was for the visitor to actively engage with art objects in four main ways: Visitors should slow down and look closely at art. They should value their own reactions, their associations, memories, and feelings. They should consider an artwork's cultural context. Finally, they should make judgments about the artwork.[44]

Increasingly, the role of the museum educator was to assist the visitor, mostly by asking questions. A good museum educator, a "master teacher," was a good listener, empathetic, enthusiastic, and flexible. She or he was perceptive and knowledgeable about art, articulate, creative in communication, and skillful in research. Educators debated the proper balance of these skills. Most museum educators were trained as art historians, not as educators, and saw that as the core discipline of the field.

But they weren't teaching art history. Rather, they were teaching—to use the new buzzword of the era—meaning-making. The visitor made meaning in the museum, based on his or her background. Museums embraced constructivism, the idea that knowledge wasn't simply transmitted, but created anew by each learner. Inez Wolins at New York's Bank Street College wrote in 1990 that "museum visitors are active inquirers who construct meaning, and therefore knowledge, about museum objects in relationship to themselves and their views of the world."[45] The educator's job, wrote E. Louis Lankford of the St. Louis Art Museum, is "to invite and motivate visitors to form their own interpretations, ask and pursue their own questions, and find personal relevance in the museum's exhibits and programs."[46]

One of the most popular art museum education approaches, called Visual Thinking Strategies (VTS), focuses on three open-ended questions: What's going on in this picture? What do you see that makes you say that? What more can we find? The facilitator uses three techniques: paraphrasing comments neutrally, pointing to the area being discussed, and linking and framing student comments. VTS and other constructivist approaches teach active looking, playing up the interests of the visitor and playing down the expertise of the curator or docent. This is art for art's sake, the object speaking to the viewer. The docent's job is to help the viewer probe more deeply.[47]

While some criticize VTS and similar meaning-making approaches—what if the visitor was wrong? Wasn't the museum responsible for correcting misapprehensions about the artwork, explaining its meaning and symbolism and historical significance?—others point to success in related areas of development. Students who attended a one-hour school tour at the Crystal Bridges Museum of American Art with a museum facilitator not only remembered "a great deal of factual information," but also "demonstrated stronger critical thinking skills, displayed higher levels of tolerance, had more historical empathy and developed a taste for being a cultural consumer in the future," according to a study of almost 11,000 students.[48]

The idea that the museum's role is to help the visitor use art and artifact in the way he or she finds best—not necessarily to learn about them—has resulted in a range of new approaches. What might the museum do to bring art and artifacts to life, to make them available as things worth learning about, as things through which one learns about oneself, and as tools to learn about something larger, about culture, community, history, or aesthetics?

The result has been an explosion of experimentation.

History museums work with history teachers who want to teach "historical craftsmanship," how to analyze primary sources and to weigh evidence. "You can think of the artifact as another kind of document—one that is sometimes hard to read, but which can tell you a new deeper, more interesting kind of story," the Smithsonian suggests on a site aimed at high school history teachers and their students.[49] Historic house museums, after decades of protecting their artifacts with velvet ropes, began to use them in a more experiential way. A recent manifesto arguing for rethinking the visitor experience in these museums notes that "obviously, guests cannot be allowed to manhandle a priceless object, but usable artifacts could be placed in each room that could be sat on, opened, and engaged."[50]

Museums are finding new ways to let visitors interact with artifacts. Starting in the early 2000s curators at the National Museum of American History organized occasional "Objects Out of Storage" events. Curators would choose objects on a theme, bring them to a public place in the museum, and talk about them with visitors. "With over 3 million objects in our vast collections, the museum is only able to exhibit a fraction at any given time," and so, "a group of seldom-seen objects," showcased on tables, allows visitors a close look—and a chance to learn from the real thing by interacting with an expert curator.[51]

Art museums are also exploring new ways to teach with objects. Many had long made prints, drawings, and photographs available in print rooms. The Harvard Art Museums, reopening in a new building in 2015 with an emphasis on transparency and accessibility, proudly announced that it would make available, in its Art Study Center, any of its artworks, for anyone who asked. The museum's stated goal was "facilitating self-directed teaching and learning from works in all media."[52]

The presentation of art in study centers provides, according to the specialists at Project Zero at the Harvard Graduate School of Education, "exceptional learning environments—rare places where visitors can immerse themselves in prolonged, intimate, and often profound experiences with original works of art." The language visitors used in describing the study center and their time there is revealing: "luxury," "indulgence," "absolutely heady," "treat," "treasure," "rarity," "connoisseur's experience," "like a private museum." Exhibitions might offer free-choice learning; a study center, by letting the visitor choose what he or she wanted to see from the entire collection, would take it to the next level.[53]

Art and artifact online present new possibilities. Online exhibits offer many points of entry and ways of organizing, and online learning sites can add to those many guides. The Smithsonian Learning Lab makes the institution's collections accessible by offering multiple points of access mediated by educators, content experts, and other users. Choose a tour guide, or become a tour guide yourself. Anyone can create a personalized collection of Smithsonian artifacts, not only shaping a collection but also sharing an approach to understanding it and using it for teaching. The Learning Lab's goal is to change online users "from passive recipients of prescribed content into active creators of digital resources personalized for learning in their own classrooms."[54]

And in museum galleries, thousands bloom. A few recent examples show the diversity:

- New York City police officers visited the Met in 2016 to learn to look. Amy Herman, their teacher, told them, "I'm not teaching you about art today. . . . I'm using art as a new set of data, to help you clear the slate and use the skills you use on the job." For Herman, a painting is useful because it captures a scene in carefully

observed detail. "It's extremely evocative and perfect for critical in-
quiry. What am I seeing here? How do I attach a narrative to it?"[55]

- Rachel Shipps, an educator at the Queens Museum who works
  with autistic youth, tries to activate museum objects as social
  objects, to encourage discussion around them. She hopes to create
  social interactions, visitors explaining things to each other.

- The Order of the Third Bird, a group of artists and scholars, offers
  "experimental protocols of Practical Aesthesis and methods of Sustained Attention"—prolonged observation of a single art work.
  Visitors who come across the group in a museum are given a card
  that reads: "The Order of the Third Bird is currently engaged in a
  silent practice of Sustained Attention to Made Things. You are
  welcome to stand with members of the Order and join in giving
  your generous attention to the work."

- Museum Hack, a for-profit museum guide firm, promises "a
  highly interactive, subversive, fun, non-traditional museum tour."
  Their motto adds a dash of humor: "This isn't your Grandma's
  tour."

- The Italian Center for Contemporary Art, in New York, arranges
  small tours led by postdoctoral fellows at the center who are researching and writing about the artists on display. They can answer
  almost any question, in as much depth as the visitor might like.

- The Brooklyn Museum found that visitors long for a personal connection with museum staff and want to ask questions about the art
  on view. The museum responded with ASK, a smartphone app
  that connects visitors to educators via text messages.

So many ways to engage with museum objects! The Peabody Essex Museum sums up the open-ended approaches that characterize twenty-first-
century art museum teaching: "Our goal is not to hang art on the walls
and then tell you what to think." Rather, the museum's collections allow
all visitors to explore the world and to make their own comparisons. Artworks are for the visitors to use as they like: "The connections you make,
because of your own experiences, inspire a journey as important as the
artworks themselves."[56]

Museums have come a long way and yet, in some measure
their roots. In the nineteenth and early twentieth centuries, m
played art to make people better—better citizens, artists, ar
better consumers. Professor Jenks wanted his visitors to undei
place in the world. Later museum educators wanted visitors to connect with
history or their own creativity. In the twenty-first century, at the Peabody
Essex and many other museums, the museum experience is about you,
"your journey." The collections are there to be useful, in old ways and new.

THE PROMISE OF MUSEUMS

JOHN WHIPPLE POTTER JENKS WAS a taxidermist and a museum curator, two professions devoted to freezing time, to making the ephemeral live forever. He cared about the future and what it would think of him. He wrote an autobiography of almost half a million words and gave it to the Brown University library for safekeeping. His son gave Brown a portrait, the frame tagged with the way his father wanted to be remembered: "Founder of These Museums." His tombstone is explicitly about legacy: "This museum, the fruit of his labor, will be his abiding monument."

Museums are supposed to survive forever, but the Jenks Museum barely outlived its founder. Almost all of its collections are lost. The museum is mostly forgotten. It never became his "abiding monument." The portrait hangs in the building where Jenks once worked, but the words "these museums" has lost its meaning.

Yet something of the museum abides, outlasting the locking of the museum's doors and the dispersal of its collections. Jenks used the artifacts and specimens in the museum to reach a wide audience, including students and Providence residents. They learned to look at things deeply, to understand that there is an order to things, to appreciate the world around them in a new way. Some went on to teach others, spreading Jenks's influence. The changed lives of the museum's visitors were his legacy.

In 1817 John Cotton Dana outlined the goal of a museum simply and boldly: Its "one and obvious task," he wrote, is "adding to the happiness, wisdom, and comfort of the members of its community."[1] Museums do this

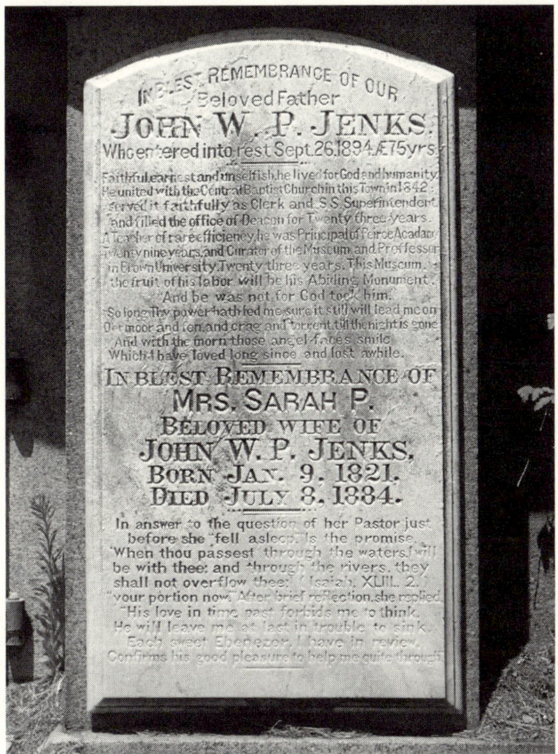

John Whipple Potter Jenks's tombstone, which reads in part, "This Museum, the fruit of his labor, will be his Abiding Monument." Central Cemetery, Middleboro, Massachusetts.

in many ways, and the great diversity of museums presents us with diverse approaches to happiness, wisdom, and comfort. They change lives, build communities, and improve the world. That's the hope and the promise of museums.

Museums Change Lives

The novelist Henry James often wrote about museums, most dramatically in his memories of his visits to the Louvre as a small boy. Those visits were "educative, formative, fertilising," more so than any other experience he remembered. His first visit, on his first morning in Paris, he describes thus: "I gape at Géricault's 'Radeau de la Méduse,' *the* sensation, for splendour and terror of interest." He returned to the Louvre often, where he "looked at pictures, looked and looked again," gaining an aesthetic sensibility that never left him.[2]

A museum also shaped the life and work of the naturalist Edward Wilson. He writes of spending hours as a child "wandering through the halls of the National Museum of Natural History, absorbed by the unending variety of plants and animals on display there, pulling out trays of butterflies and other insects, lost in dreams of distant jungles and savannas." He knew there were curators there, and that gave him "the conception of science as a desirable life goal. I could not imagine any activity more elevating than to acquire their kind of knowledge, to be a steward of animals and plants, and to put the expertise to public service." At age sixteen Wilson wrote to the curator at the museum explaining that he intended to devote himself to the study of the ants of Alabama. The encouraging reply he received set him on the path to a career in science.[3]

Oliver Sacks, who would go on to write many finely observed books on mind and behavior, remembered visits to the Science Museum and the Museum of Natural History as highlights of his London childhood. Sacks had no use for school, with its regimen and routine. But "the museums," he wrote, "allowed me to wander in my own way, at leisure, going from one cabinet to another, one exhibit to another." There was no curriculum to follow, no exam to take. The museum was "a microcosm of the world—compact, attractive, a distillation of the experience of innumerable collectors and explorers, their material treasures, their reflections and thoughts."[4]

Sacks felt at home in the museum. Seeing the real thing mattered to him. His grandfather was the inventor of the Landau lamp, a miner's lamp. And there it was in the Science Museum, next to the famous Davy miner's lamp. Sacks saw John Dalton's wooden models of atoms, and they allowed him to imagine that "atoms actually existed." He loved the neat rows of minerals displayed in the Geology Museum. Museums, he wrote, made him "want to go out in the world and explore for myself."[5]

More than any other exhibit, it was the giant periodic table of the elements at the Science Museum that excited the young Sacks: "I could scarcely sleep for excitement the night after seeing the periodic table. . . . Seeing the table, 'getting' it, altered my life. I took to visiting it as often as I could." Sacks wasn't the only scientist who traced his enthusiasm for science to that exhibit. Freeman Dyson, who would become one of the most important theoretical physicists of the twentieth century, wrote, "As a boy, I stood in front of that display for hours."[6]

Artists, too, remember visits to museums as life-changing. The Museum of Modern Art's oral history program asked artists about the role that museums played in their artistic development, and almost everyone had a story. Robert Mangold remembers his visit to the 1957 Carnegie International at the Carnegie Museum as "the big thing that kind of changed my whole perspective on everything. . . . It was like something that confronted you and had this physical and emotional presence, suddenly."[7] Bruce Nauman remembers a de Kooning painting at the Art Institute of Chicago that he visited many times. But on one visit, "I saw the painting for the first time, I mean, I really understood what was going on, or thought I did. So it was the first time I really had a really deep art experience."[8] James Rosenquist describes hitchhiking from Minnesota to Chicago to visit the Art Institute in order to see real paintings, not reproductions: "I was so surprised that they were so sloppy, vigorous, messy, but beautiful and correct."[9]

Museums shape social and political consciousness, too. Phyllis R. Dixon, an African American novelist, writes movingly of her visit to the National Civil Rights Museum in Memphis. It reveals, she writes, "how important the actions of 'regular people' can be. . . . The exhibits generate a sense of pride. I am proud to be part of a legacy of people who endured so much, yet survived and thrived." And the museum encourages action: "After each visit to the museum, I wake up from complacency and recommit myself not to take things such as voting, public education or even my job for granted. I recommit myself not just to remembering the past, but trying to improve the present and the future."[10]

This is the holy grail for museums: experiences that change lives. The museum worked its magic on these scientists and artists and writers. Some had a deep experience, spending hours with a single object. Others were struck by juxtapositions and connections. Many were drawn in by authenticity, the presence of the real thing: Dalton's wooden models, trays of butterflies, de Kooning's *Excavation* with its layers and layers of paint, Martin Luther King Jr.'s room at the Lorraine Motel. Objects spoke to them.

A certain orderliness was also important. The way the museum arranged things revealed a hidden structure to the world and offered a new way of seeing: cabinets of rocks, vitrines of birds, timelines of history. That periodic table at the Science Museum. An exhibition of New York School paintings that let Robert Mangold know that a new kind of art was possible. The orderliness of the display, the rush of history, was what Henry James remembered most vividly from his childhood visits to the Louvre many years later. In "the most appalling yet most admirable nightmare of his life" he found himself trapped in the Louvre's Galerie d'Apollon, pursued down a "prodigious tube or tunnel through which I inhaled little by little, that is again and again, a general sense of glory." He rushed down the "long perspective, the tremendous, glorious hall" full of its "great line of priceless vitrines."[11] The museum encased him in history, surrounding him with both glory and terror.

For some visitors, it is not the objects or the displays but rather the museum experience itself that is important. They feel comfortable in the museum, roaming free, being able to pick out what interests them, and engaging deeply with it, in their own way, or spending time exploring with family and friends. Critic Adam Gopnik notes how John Updike's 1962 *New Yorker* story "Museum and Women" captures "the eroticism of museums, their atmosphere of mystery." Updike's museum "held an air of silence and a promise of a kind of secret communion that would take place between the individual observer and the work of art or the object of the past." For Gopnik, visiting museums in the 1970s, the museum was something different, "a place where you went to be transformed in another way—not a place where you went to commune with the past, but a place where you went to learn how to be modern."[12] For Australian writer Tim Winton, museums were places where people could imagine what they might be and do "when they saw past the everyday."[13] The museum— open ended, versatile, free-choice—offers each visitor what he or she needs. For some, that is information, entertainment, a place to connect with the past, or to strengthen connections in the present. For others, it is the opportunity to move beyond, to imagine more. A 2003 study that asked "What do you want to get out of your visits to historic sites or museums?" found that many visitors want to escape to another time, to experience a kind of authenticity different from their own, to feel "the aura" of a different period of history. These visitors yearn for an experience of deep engagement, empathy, even awe. John Falk, who studies museum visitors, suggests a range of motivations connected to individual identities, the way we think about ourselves as we consider a museum visit. Some want to explore, to satisfy curiosity. Some want a place to visit with friends and family. Some see it as a place to encounter excellence or to learn new things. Some seek a spiritual retreat, a place to get away.[14]

Museums would like to satisfy all of these motivations with experiences that are meaningful, personal, social, empathetic, maybe life-changing, but certainly engaging, a "flow" experience. The museum provides objects, order, stories, space. Visitors come alone, or with friends and family, and either way make connections that hold new meaning and expanded contexts. Emotion and learning, curiosity and connection, resonance and wonder, art and science and history and the everyday intersect in unknowable ways, and shape us in ways unknown.

Museums Reconnect Communities

The main goal of Professor Jenks's 1874 collecting expedition to Lake Okeechobee was natural history collecting, but he also took the opportunity to research an object that was already in the museum, "an Indian relic constructed of a dozen box-tortoise shells, bound together by deer skin thongs, each one partially filled with wild beans." The relic had been given to the museum by a physician in Providence who had owned it for fifty years but knew nothing about it other than "that it came from the Seminoles." Jenks was eager to learn more, so he visited a Seminole lodge, saw one like it, and was told that it was a leg rattle, used by women as part of their green corn dances. A Seminole woman showed him the dance, or, as he put it, "gave me a specimen of a Seminole reel."[15] He bought the leg rattle, and gave it to the Smithsonian. Two decades later, Thomas Wilson, Smithsonian curator of prehistoric archaeology, would describe and illustrate the rattle in his *Prehistoric Art*.

For Jenks, this was research, fieldwork that helped label and explain a museum artifact. He was collecting information so that his museum would know what it had and to provide information for scholars. That was the way museums worked: they acquired things, learned what they could about them—many artifacts came with no information, or wrong information—organized the objects to fit their theories, and provided information about them to scholars and visitors. Knowledge moved from communities to curators to audiences.

In recent years, a new kind of connection between museums and the communities represented in their collections has become important. Today's anthropologists share collected artifacts with communities in a two-way exchange of knowledge. Starting in the 1980s, many museums began to make contact with those communities in a new way—not to gather more artifacts, or more information, but to converse. Collections may be derived from colonial exploitation, but they contain other stories as well, and conversations between indigenous communities and museums can serve both groups.

Museums gain from this. While some museum artifacts were collected by careful anthropologists who kept good records, many were not. Director Jette Sandahl of Stockholm's Museum of World Culture notes that "our knowledge is thin and faulty. We have torn these objects out of their contexts. We have to reconnect to the knowledge that is in the objects and to the living memories that they still carry."[16] The objects in the museum are brought back to life as part of a living culture.

The community also gains from the exchange. Indeed, it can be an act of reparative justice. Tristram Besterman, former director of the Manchester Museum, writes that "the moral authority to forgive and to move on from past misappropriation lies not with the curator but with the inheritors of dispossession. And it should be on terms agreed with them, not dictated by the museum." He suggests a form of restitution "in which narrative control is re-explored, challenged, widened, and shared." Doing this, he says, not only empowers the community but also enriches the museum.[17]

Working with communities to restore meaning to their artifacts is part of a larger change in museum work. Anthropologist Christina Kreps calls this "curatorship as social practice," a way of considering objects not merely as objects but as part of a web of community relationships. That implies a responsibility for the museum to move beyond things to connections. It means acknowledging that objects mean different things to different people and that there are different ways of caring for objects.[18] Richard Kurin of the Smithsonian explains that "rather than curate dead or captured specimens of a culture, curators are increasingly concerned with the living larger whole." Museums, Kurin suggests, need to do their work in a way that "shapes cultural continuity, rather than destruction."[19]

Considered in this light, objects are there not for their own sake but for the relationships they allow—to the people they represent, first, and then to researchers and visitors. Museum collections, reconnected to their source community, serve both old and new purposes. This happens in many ways. Objects can be repatriated, returned permanently to the communities that once owned them. They can be lent back for ceremonies. They can be shared. Or they might stay in the museum while information about them is made more widely available, an arrangement called e-patriation. The stories told about them in the museum can be changed to serve both the museum's audience and the communities the objects came from, as at the National Museum of the American Indian or the multiversity galleries at the University of British Columbia's Museum of Anthropology.

The Denver Museum of Nature and Science's Indigenous Inclusiveness Initiative provides a model of the ways that collections can connect back to communities. It reimagines anthropology as "an inclusive and dynamic scientific pursuit" by welcoming Native American perspectives in both research and public programs. The initiative strives "to build a museum and a discipline that combines indigenous voices with the best of scientific research so that cultural treasures can be appreciated from every perspective—aesthetic, historical, scholarly, and cultural."[20] A key part of that work is inviting indigenous people to work with their collections. Artists come to learn from and be inspired by artwork. Elders reconnect with family objects. The museum gains from the dialogue and mutual respect of this work, too. "Native participation," report the curators, "brings new ideas, creative energy, and authentic experiences to the museum." The goal is "not merely ensuring that we are not doing harm to the communities we serve but rather actively pursuing some good." That fits with a larger change in the museum relationship with communities. "For much of the twentieth century," the Denver curators write, "the guiding principles of anthropologists in natural history museums have been discovery, salvage, and preservation. These values, however, are being transformed into new ones, such as respect, reciprocity, and dialogue."[21]

The Smithsonian's Recovering Voices program is another example. It works with communities from which collections were gathered "not just to understand the collections, but also to document and revitalize language and knowledge traditions." Letting communities engage with collections, writes Joshua Bell, curator of globalization at the National Museum of Natural History, not only makes collections more useful to the museum but also to the communities. It can be a vital part of heritage revitalization.[22]

He offers some examples of "community research." Samoan *tapa* (bark-cloth) makers work with curators and scientists to reinvent lost techniques. Visitors from the Purari Delta of Papua New Guinea use archives to revitalize language and museum collections to revive art—and are delighted to see photographs of their ancestors. Tangible objects and intangible knowledge survive in different ways, in different places, writes Bell, and together can be much more than the sum of their parts. Museum work, he says, should be "collaborative and symmetrical," and collections, he suggests, can be catalysts. A member of the St. Lawrence Island Yupik people who worked with the Smithsonian put it eloquently: "There is a lot of sleeping information within each material piece—language, memories, and cultural meanings. When elder tribal members visited the Museum's collections, long dormant words and recollections came to them almost like dreams. It is contact with the actual objects and discussion among community members that will awaken the information inside."[23]

Some museums have lent objects back to the communities where they were made to provide a descendant community with a connection to its past. At New Zealand's Te Papa museum, artifacts are loaned to Maori communities for ceremonies. At the Buffalo Museum of Science, Iroquois communities borrow masks for ritual use, returning them to the museum for storage. In 2016 Harvard's Peabody Museum of Archaeology and Ethnology lent an oiled sealskin kayak made around 1850 by Alutiiq artisans to the Alutiiq Museum in Kodiak, Alaska. One of the Alutiiq's museum's missions is to "reawaken cultural traditions." "We believe," the museum notes, "that accurate knowledge of the past is essential to the health of modern communities. History is a resource that can help people confront difficult issues, engage in discussion, and consider many perspectives."

Amy Steffian, the Alutiiq's chief curator, made a personal connection: "It ties you to the power of ancestors."[24]

Museums and indigenous artists working together can help traditions survive by adapting them to the worlds of commerce and art—though always with the risk of changing them in the process. In the 1950s museums in British Columbia became patrons of native artists, hiring them to do work for the museum, buying their work and showing it in exhibitions, and supporting the sale of their art in other ways. Michael Ames, longtime director of University of British Columbia's Museum of Anthropology, supported this work, but also pointed out a problem: It meant that anthropologists were defining ethnic identity. "Anthropologists sit in judgement about what constitutes a proper artefact, a proper price, a proper potlach, and by implication, a proper Indian."[25]

That challenge still holds, but today's relationship is, in general, more equal. Nicholas Thomas, director of the Museum of Archaeology and Anthropology at the University of Cambridge, argues that contemporary museums are institutions that generate knowledge collaboratively. "In many institutions," he writes, "it is anticipated that originating communities are consulted around exhibition or research projects, and they are indeed increasingly full collaborators."[26]

Museums take things out of context; collaboration moves toward putting them back. Community-connected museum work doesn't restore, but it does reconnect. And perhaps most interesting and important, it opens up new possibility for empathetic connections between and beyond communities.

Museums Improve the World

The titles of three of the most important essays on museums in the past half century suggest the transition from museums as places of art and artifact to their new role as places where people connect to something bigger than themselves. Duncan Cameron proposes that museums should be not only a "temple" but also a "forum," a "place where battles are fought." Steven Weil demands that museums be not "about something" but "for somebody," that they shift their "principal focus outward to concentrate on providing a variety of primarily educational services to the public."

Nina Simon calls for a "participatory museum," a place that connects with the public, and insists that museums "demonstrate their value and relevance in contemporary life."[27]

Many contemporary museum missions reflect these new paradigms. They want not merely to display and explain but also to improve the world. The Baltimore Museum of Art wants to serve "as a creative catalyst for the city."[28] The Missouri History Museum seeks to "strengthen the bonds of the community . . . and facilitate solutions to common problems."[29] The National Museum of American History wants to "shape a more humane future."[30] The California Academy of Sciences is dedicated to "sustaining life on Earth."[31] All argue for usefulness, for serving public needs.

Once museums thought in terms of what they did: strategies and activities, that is, the *output* of their work. But output doesn't mean success, so many organizations now look for *outcomes;* how did their work change the world? Or, more generally, *impact:* what effect did it have? Instead of starting with what you do, start with the change you want to make in the world, and work backward. What will you do to create that change and achieve those goals? Advisors to nonprofits have a wonderful phrase they use in getting the organizations they work with to think about their work in a most general way: "theory of change." How does the work you do change individuals, or communities, or the world? They want you to spell it out. How do your activities create outcomes, and what impact do those outcomes have?

The Santa Cruz Museum of Art and History provides an example of a theory of change in action. It undertakes to make the museum a welcoming gathering place and to "find, spark, share, and preserve stories and ideas." This leads to a range of personal outcomes: participants enjoy shared experiences, explore art and history, feel involved and included, build awareness and respect for diverse cultures and peoples, and are inspired to learn more. In turn, these outcomes make a difference in people's lives in three ways: they strengthen the bonds between people, help people build bridges across differences, and empower people to share their creative and civic voices. These individual changes improve the community: it grows stronger and more connected.[32]

Not everyone agrees with the paradigm shift. James Cuno, writing when he was director of the Harvard Art Museums, argues for a vision of the museum not too different from that of Benjamin Ives Gilman a century earlier, doubling down on the power of museums to make a quiet difference to individual lives. After the terrorist attacks of September 11, 2001, Cuno writes, museums once again revealed their true purpose: "People just wanted to be there, in the company of others looking at rare and beautiful things" and appreciating "the varied mysteries of art." Museums, he believes, should be "sanctuaries, places of retreat, sites for spiritual and emotional nourishment and renewal." He contrasts that with what he sees as the wrong path museums took in the 1990s. They had become "an agency of social therapy." Museums had devoted themselves to "after-school programs for at-risk kids and about teaching all kinds of things, from analytical and computational skills to ways of building better self-esteem and self-discipline." They had become "contested sites, where ideas and social identities were in contest." Cuno urges museums to "put works of art front and center and to trust that art is what the museum-going public wants most."[33]

Others blast museums from the other direction: they aren't doing enough. After the 2014 Ferguson, Missouri, shooting of Michael Brown and the unrest that followed, many in the museum world called for museums to respond. #MuseumsRespondtoFerguson trended on Twitter, and a group of "museum bloggers and colleagues" issued a report. "As mediators of culture," they declare, "all museums should commit to identifying how they can connect to relevant contemporary issues irrespective of collection, focus, or mission." Museums must listen and respond to "those we serve and those we wish to serve." They urge museums not only to "respond" but also to "invest" in conversations and partnerships that call out inequity and racism and commit to positive change.[34] Bob Beatty, the president of the American Association for State and Local History, encourages its members to commit "to the cause of interpreting difficult history and using that power to make a difference in the present and the future."[35] Lonnie Bunch, the director of the National Museum of African American History and Culture, calls for museums to concern themselves with social justice. Museums, he writes, need to ask themselves how they are of use not only in the traditional sense—"preserving artifacts, making history and culture accessible, inspiring new generations"—but also in a larger sense. "To me," he says, "the real question is how does a museum make its community, its region, its country better?"[36]

Museums are for individuals, to give people what they want and need, as well as for the public, to make communities stronger. They can do both at the same time. Nicholas Thomas, of the Cambridge Museum of Archaeology and Anthropology, believes that the very heterogeneity of artifacts, individuals, and communities co-existing in museums makes them a space that "elicits and fosters empathy."[37] They bring together artifacts from many times and places. They build on curiosity, sociality, on a willingness to consider the other—and others—in their own terms. Objects are open to interpretation, and museums build on that openness. As flexible institutions, eager to be useful, they have the means and opportunities to help repair the world: wellbeing, education, sustainability, equity, and social justice.

And museums have done, and can do, all of these things. They have a long history of being useful in many ways. In the 1910s the American Museum of Natural History established a department of public health, mounting exhibits that encouraged healthy living; in 2011, Michelle Obama asked museums to participate in her "Let's Move! Museums & Gardens" program. In 1898, Jane Addam's Hull-House settlement house built a museum for workers, showing off not only collections from the Field Museum but also the skilled work produced by immigrant women; today, the Jane Addams Hull-House Museum offers a range of projects highlighting community voices and concerns. In the early twentieth century John Cotton Dana's Newark Museum provided citizenship training. So too, in the early twenty-first century, did the Harvard Art Museums.

All this presents each museum with new challenges. It must consider where, among all of these possibilities, it should focus its efforts, and how it should use its collections and expertise, its space and staff, its wealth, power, and community connections. It must serve not only individuals but also its diverse communities, if possible bridging the gaps between communities. It must look at what it can do that other organizations cannot and at the opportunities its unique nature suggest.

A museum may need to decide whether to focus on teaching children or aim its displays at hobbyists and enthusiasts or experts, whether it should teach facts or emotion, observational skills or creativity. Should history museums focus solely on the past or try to shape the future? Should art museums provide a serene place for reflection, or a lively place for school programs? Should natural history museums teach taxonomy, biodiversity, or environmental justice? Some museums may choose to be activist, encouraging social change and social justice, others to simply be places where art and artifacts are available for visitors to use, however they want.

In 1930 philosopher Georges Bataille wrote that the galleries and collections of a museum are "but a container." The true content of the museum, he said, is its visitors: "The paintings are but dead surfaces, and it is within the crowd that the streaming play of lights and of radiance . . . is produced."[38] It is in those visitors, as individuals, families, communities, even nations, as viewers and users and researchers, that the meaning and value and effect of the museum is felt. Art and artifact are means to an end, and museums serve that end when they put the collections and knowledge they can uniquely offer at the service of their visitors. That happens when museums are both temple and forum, both *about* something and *for* somebody, and participatory, connecting collections, staff, visitors, and communities.

Professor Jenks wanted his visitors to know nature, and through nature, to know God. He wanted his visitors to look, to learn, and to appreciate the wonders of the world. He loved collecting and preserving and displaying his artifacts, but he did it for a purpose. He wanted to change his visitors, to help them understand the world, to make them better people. At some level, that's true of all museum work. Museums are of value because they are useful in that larger project. That's the promise of museums: that they improve people, strengthen communities, and repair the world.

CODA

Critique

The Jenks Museum never recovered from Professor Jenks's death in 1894. The museum became less useful as teaching and research in biology changed. It shrank as artifacts were used up or thrown away, and it finally closed in 1915.

The Jenks Society for Lost Museums revived the museum a century later for a second life as an art project. Mark Dion, an artist whose work is about museums, created the society to reimagine the old museum on the centenary of its closing. The installation slid easily back into its completely renovated former building. It seemed to be both of the past and of the present, both real and fantasy. It was uncanny: in the right building but in the wrong time, accurate but not authentic, true to history but at the same time out of history. It acknowledged the rules of the museum but played with them. Categories were off-kilter, authenticity was uncertain, time and space distorted. The project partook of the seriousness of museums but leavened it with uncertainty and humor. Art served as a way to understand the Jenks Museum.

Over the last few decades, many artists have made art of museums—of museum history, the idea of the museum, the work of the museum, even the act of visiting a museum. They have considered the colonial roots of many museums, the ways that collecting and cataloging shape our view of art, nature, and history, the ways that museums encourage a certain way of looking. Art historians call this kind of work "institutional critique": artists exploring—exposing—the structures and workings of

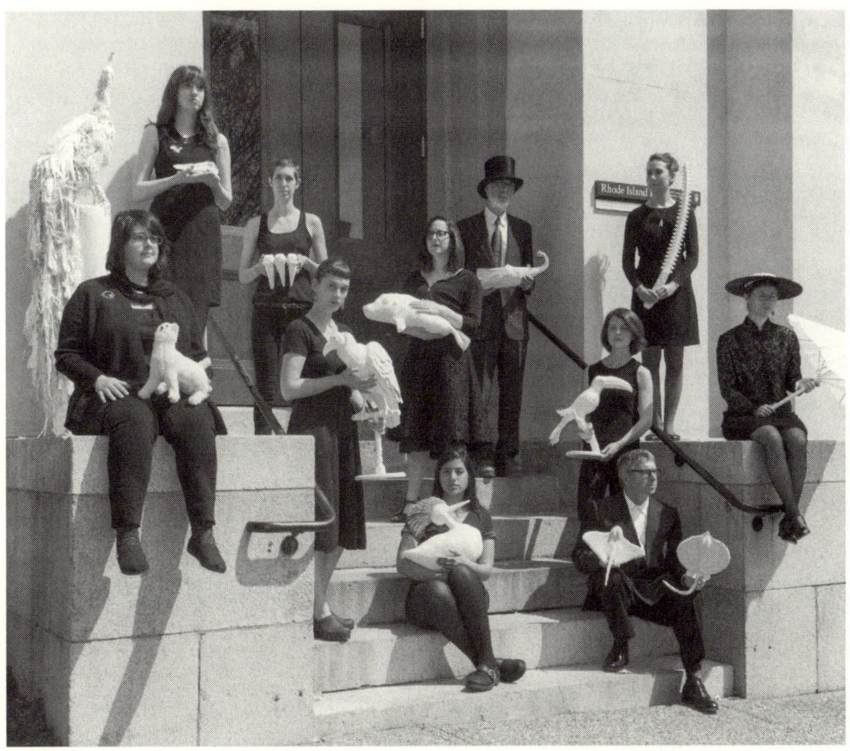

The Jenks Society for Lost Museums, 2014.

the museum, revealing the museum's ideological underpinnings and opening it up to new understandings. Mark Dion notes that there are two camps of artists doing institutional critique. Some do it as criticism, as a way of undermining the museum. Others do it out of love for the institution, to make it, in Dion's words, "more interesting and effective."[1]

This Coda revisits each part of this book—collect, preserve, display, use—through the work of contemporary artists. Artists offer not only a new perspective but also a question: why do museums do things the way they do? Looking at the museum through artists' eyes helps us understand them better.

Collect

What belongs in a museum? Who gets to decide? The triumph of professional curators in the early twentieth century meant that scholarly standards set by the academy drove many collecting decisions. In the late twentieth century, as museums redefined how they might be useful, other groups, both inside and outside the museum, began to have a larger say. Artists, unsurprisingly, have strong feelings about what's collected and why, and who gets to decide. Some have made art from the process. Artist-critics, artist-scholars, and artist-curators practicing institutional critique use the museum's tools, if not to dismantle then to reveal the collecting structures of the museum. They ask what's included, and why or why not; and then they make art and meaning, exploiting and disclosing hidden cultural rules.

Allan McCollum called attention to the choices made in collecting—and the elite group that made them—in his *Collection of Forty Plaster Surrogates* (1982–1984). The work consisted of many plaster frames with a blank black center where the art should be, arranged as though hung in a gallery. "I am very much aware," he wrote, "of the fact that I had nothing to do with choosing what got in there. . . . Museums are filled with objects that were commissioned by, or owned by, a privileged class of people who have assumed and presumed that these objects were important to the culture at large—and who have made sure that they are important to the culture at large." Leaving the frames empty denaturalized them, prompting the viewer to ask what should be there and who gets to decide.[2]

Some artists adopt the role of curator to display things that aren't traditionally shown in museums. They ask, why are these things not usually chosen? What makes something worthy of collecting?

Andrea Fraser, one of the most thoughtful practitioners of institutional critique, considered these questions in *Aren't They Lovely*, an installation she curated at Berkeley's University Art Museum (UAM) in 1992. The exhibit featured a bequest the museum had received from a Berkeley alumna that comprised both art, which the museum accessioned, and other objects—coins, eyeglasses, souvenirs—that the museum did not. Fraser showed them together to examine how the museum makes culture. What's museum-worthy, what's not?

Museums, Fraser wrote in the exhibit brochure, "seduce their public into aspiring toward the essentially class-defined competencies and dispositions—ways of knowing and being—demanded by the culture museums present." She showed how that process worked by telling the story of "the social struggles and subjective aspirations" that the transition of private objects into museum collections reveals. Her exhibit labels disclosed museum decision-making processes, quoting directly from curators' memoranda: "While somewhat less attractive and, as I recall, somewhat abraded in the original drawing, this is a work of significance for the UAM. It is a work of quality, historical interest and useful teaching purposes within the UAM's drawing collection." Fraser hoped that the exhibition would de-idealize "what museums publicize as legitimate culture by returning it to the context of the culture of everyday life." What better way to do that than to show off the parts of the gift the museum rejected as not suitable for the collections and reveal the way curators made their decisions? Juxtaposition demands explanation.[3]

Richard Wentworth, a British artist, raised related questions in his *Questions of Taste* installation at the British Museum in 1997. He juxtaposed ancient Egyptian drinking vessels with modern cans and bottles found discarded near the museum. He described both groups similarly, offering the location of discovery, materials, and manufacturing process for both groups. What made the ancient artifacts valuable, the present-day artifacts trash? Was it rarity, age, material, or method of production? Why were some museum-worthy and others not?

Rosamond Purcell's *Two Rooms* (2003) also juxtaposed museum and not-museum things, but in a very different way. One of the two rooms offered a hauntingly beautiful display of found objects—old books, rusted metal, broken toys—from an Owls Head, Maine, junkyard. The other room was a re-creation of Ole Worm's seventeenth-century museum. The detritus from the Maine junkyard had been saved for its potential sale and reuse value. In Purcell's installation, it was gorgeous: though not "museum-quality," arranged with an eye to pattern, texture, and color. The re-created Worm museum was filled with artifacts of natural history and ethnography (some original to the museum, some not), there because Worm thought them worthy of study and examination. Its display offered a different kind of order: not aesthetic but rather according to the categories that Worm thought would raise questions about nature and culture. Both rooms displayed things that their creators deemed worthy of saving, but for different reasons. Though neither room was a modern museum, the roots of the museum were in both rooms.[4]

Fred Wilson's *Mining the Museum* at the Maryland Historical Society (1992–1993) considered the ways that the museum had neglected the history of African Americans in the state. One section showed what the museum had not collected. Wilson displayed the busts of three famous white men not from Maryland on pedestals. Next to them he placed three empty pedestals labeled with the names of important African American Marylanders whose busts were not represented in the museum's collection. In another room, he projected the names of "enslaved persons who rebelled," for whom no artifacts existed, on top of paintings and artifacts of white abolitionists. Wilson explained: "Museums are afraid of what they will bring up to the surface and how people will feel about certain issues that are long buried. They keep it buried, as if it doesn't exist, as though people aren't feeling these things anyway, instead of opening that sore and cleaning it out so it can heal." He wanted to root out that denial, and to do that he needed to call attention to the stories the museum had decided not to collect.[5]

What's not there, and why, raises historical questions about how museums make collecting decisions. Some artists become historians to explore those questions, creating art from the historical processes of collecting. Scholar-artists draw attention to the forces shaping collections in a more visceral way than scholars whose work shows up only in books and articles.

Mark Dion explored the history of natural history collecting in a series of installations. In *Collectors Collected* (first shown in 1994 at the Reina Sofia in Madrid) he displayed the men responsible for the Expedición al Pacífico, a mid-nineteenth-century natural history expedition, much as they might have exhibited the people they encountered. The explorers would have been surprised to find themselves—their clothing, tools, camping equipment, and portraits—inside the case. In his *Clark Expedition* at the Explorers Club in New York (2012) Dion took over a room usually filled with explorers' trophies and displayed, instead, what he called "the sun-bleached bones of expeditions past"—paper mâché copies of the provisions and equipment that the explorers took with them. Dion's installations call attention to the fact that specimens don't just appear in a museum. To make sense of objects, we must understand the people and processes that made them part of the collections.[6]

Institutional critique artists enjoy exposing the workings of the art markets and the story behind donations to art museums. Hans Haacke, one of the founders of institutional critique, focused on the provenance of one painting in his *Manet-PROJEKT '74*. It took as its subject Édouard Manet's *Bunch of Asparagus* (1880), which had been donated to the Wallraf-Richartz Museum in 1968. Haacke's installation consisted of ten panels tracing the ownership history of the painting. The last panel revealed that the donor to the museum had Nazi connections and was using the donation to cleanse his past. Another Haacke project tracked the provenance of Seurat's *Les Poseuses (Small Version)*. Originally a gift to an anarchist friend of the artist, it was later owned by wealthy lawyers and one of the first art-investment firms. Haacke's installation broke the ultimate museum taboo: he included the price of the artwork.

Preserve

Museums preserve art and artifact by keeping it secure, keeping track of it, and taking care of it, thus making it accessible and useful. Artists call attention to preservation, storage, even cataloging—to the way that art, artifacts, and specimens become museum objects—to raise questions about both art and museums. Mark Dion captured the spirit of this fascination with the behind-the-scenes of the museum when he wrote, "So I say freeze the museum's front rooms as a time capsule and open up the laboratories and storerooms to reveal art and science as the dynamic processes that they are."[7] Opening up the work of museum collections reveals museums as dynamic processes, and a worthy subject of art.

In his seminal 1970 *Raid the Icebox 1* installation at the Museum of the Rhode Island School of Design (RISD), Andy Warhol captured the essence of the storeroom: its combination of treasure and trash, its veneer of organization and spasms of disorder. Collectors Jean and Dominique de Menil had visited the museum's storerooms the previous year—the director was hoping they would fund a conservation project—and, astonished at the riches they found there, proposed that the museum ask an artist to explore the collections and make them visible. They suggested Warhol.

Warhol took on the assignment. But instead of acting as curator, choosing objects that appealed to him, and making sense of them in an exhibition—what the de Menils and the museum's director had hoped for—he chose objects that the museum had not imagined putting on display, and then exhibited them more or less as they had been in storage. Warhol packed a gallery with Windsor chairs kept for spare parts. He crammed hundreds of shoes and parasols into open cabinets. He showed hatboxes, not hats. Paintings, some of them torn, were hung at random, leaned up against the wall, or just shown on sliding metal racks pulled from storage. Shelves were cluttered with bits of sculpture and decorative arts. Furniture and building elements were shoved together. Warhol refused to make selections or organize things in a way that made sense of them, beyond storage sense.

The museum's director noted that "there were exasperating moments when we felt that Andy Warhol was exhibiting 'storage' rather than works of art, and that a series of labels could mean as much to him as the paintings to which they refer." And indeed, the installation looked suspiciously like the museum's storage area. Even the catalog for the show captured the feeling of raw collections. Warhol demanded that each item be fully described, and so the museum's chief curator "simply had the texts from the catalogue cards in the Registrar's Office typed onto lists without any further research." *Raid the Icebox 1* offered museum storage as a place where objects wait. His catalog did similar work, exposing the information sitting in the registrar's files. Warhol's project raised questions about what value curatorial work added, the differences between objects in storage and objects in exhibitions, between the registrar's files and the curator's writing.[8]

Anticipating the fiftieth anniversary of *Raid the Icebox 1*, in 2015 the RISD Museum organized a new behind-the-scenes artist intervention, this one entitled *Raid the Database 1*. It gave Natalja Kent, a Providence artist, access not to the storeroom, like Warhol, but rather to the museum's database of photographs of work at the museum. Her exhibition exists, appropriately, on the museum's website, and offers a new perspective on the museum: "I noticed that by 'peeking behind the curtain' of the Museum's collections, I was able to find images that described the unseen lives of artworks and aspects of their context within the collection. By focusing on touch, I found a break from the most common ways of interacting with art—viewing it, on guarded display, from a distance." Her exhibition included categories like "packing supplies," "white gloves," "blue gloves," and "tools." Like Warhol's *Raid the Icebox*, Kent's *Raid the Database* offers a view of the museum at work, of art objects as physical things.[9]

Kent followed not only in the tradition of Warhol, but also of a long line of artists whose work considers the art in art museum collections as things, rather than as art. One of the founders of institutional critique, Marcel Broodthaers, imagined himself as the director of a fictional museum he called *Musée d'art moderne* (1968–1971). The first version of the museum was in his apartment. He displayed, instead of actual artwork, an eccentric collection of postcards and projected slides of artworks, interspersed with art shipping crates, nicely bifurcating art into image and object, meaning and method—and leaving out the actual art. A critic described a visit—actually, a twenty-four-hour immersion—in one of Broodthaers' later reimagined museums: "I stroll through the storerooms. Wooden packing-cases, shelves, filing cupboards, piles of printed catalogues stacked on pallets. It is as I suspected: the art world is a succession of containers, in which art is clearly in a minority. A museum is 95% storage and 5% art, 99.9% words and 0.1% works."[10]

Vik Muniz, a Brazilian artist, emphasized the physicality of art works, as well as the tools and information required for intellectual control in his *Verso* (2016). He exposed the unseen workings of the museum by flipping paintings over on their back—not literally, but by making precise copies of the backs of paintings. His reproductions of the backs of paintings like the *Mona Lisa, Starry Night,* and *The Girl with a Pearl Earring* revealed evidence of conservation work, exposed registrarial practice, and even disclosed electronic monitoring systems. Emilie Gordenker, director of the Mauritshuis Museum in the Hague and curator of the Muniz exhibition there, appreciated how *Verso* called attention to the physicality of art. "You realise these things are actual physical objects. They are things: they get hung up, they get written on and pasted on."[11]

Preserving art and artifact is only the first part of a museum's preservation mission. Almost as important is preserving information about it. Muniz's *Verso* captured this as well: Labels on the backs of paintings disclose previous owners and borrowers and reveal the work of registrars, collections managers, and collections information systems. There are barcodes and "this side up" stickers. The front of a painting, says Muniz, is eternal;

the back, shows it as continually changing: "You know where they've been; you know who they've been with, how they got hurt."[12]

Steven Prina's project, *Exquisite Corpse: The Complete Paintings of Manet* (begun in 1988 and ongoing) also calls attention to the details of cataloging, abstracting art into metadata and metadata into art. It offers a view of Manet's works by showing not the paintings themselves, but rather their size, shape, name, and date. Prina produced a graphic representation of the size and shape of each of Manet's paintings, to scale, on a single sheet of paper: 556 small boxes showing in chronological order Manet's 556 works, each depicted as a small brownish box. On exhibit, this page is displayed next to an ink wash, also in a monochrome sepia, the same size and shape of one of the Manet originals. The Prina "copy" label gives the name of the Manet work, its date, and the museum it was located in when the list was made. Critic Pedro de Llano notes that the project calls attention not to the actual work but to its history. The appropriation does not "consist in a mere visual quotation from other works of art," he writes, but rather "in a historical analysis of the conditions of their reception." Prina's *Exquisite Corpse* offers insight into the physicality of the art, the way it is described, not what it means.[13]

Museums are responsible for keeping their collections safe and secure, and artists have made art of their occasional failings. Sophie Calle's 1991 *Last Seen . . .* offered museum staff memories of the paintings stolen from the Isabella Stewart Gardner Museum displayed next to photographs of the vacant space where the paintings had once been. Russian artist Ilya Kabakov not only created entire fictional museums, but also the disasters that can befall them. His 1992 *Incident at the Museum or Water Music* offered a museum disaster in action: plastic sheeting covering paintings and sculptures, with plastic buckets filled with water, and a sound installation of dripping water. His intent was not ironic but rather to call attention to the sanctity of the museum.

Artists play with deaccessioning. Michael Asher's contribution to the Museum of Modern Art's 1999 "The Museum as Muse: Artists Reflect" was a printed inventory of that museum's deaccessioned works. Robert Fontenot went further. For his *Recycle LACMA* (2009), he purchased at auction textiles and garments deaccessioned from the Los Angeles County Museum of Art. He carefully deconstructed each one and reconstructed it into new and useful artifacts—and embroidered the original accession number prominently on the object's new incarnation. A 1967 James Galanos long coat became a car seat cover and golf club cover sporting the catalog number as if it were a name. Guatemalan trousers became a pair of stuffed bears. Rita Gonzalez, curator of contemporary art at LACMA, described the project as "a joyful but biting call to all collecting museums to think more radically about recirculating these objects back into the creative community." One might also see it as a physical manifestation of the usually more abstract ways that artists make new art by building on the art in museums.[14]

Display

There's a long history of artists making art of museum exhibitions. Sometimes this is literal: paintings and photographs of the galleries. But some artists go beyond that, remaking the exhibition as their art. They turn the tables on the museum and put the museum itself on display, revealing the choices made, objects included and not included, history told or silenced. They expose the museum's mechanisms. Institutional critique reveals the people, ideologies, and interests behind the scenes.

There are roots of this artist revisionism in the nineteenth century and before. When Samuel F. B. Morse painted his *Gallery of the Louvre* in 1831–1833, he replaced the paintings that actually hung in the Salon Carré—romantic works he didn't approve of—with what he believed were the European masterpieces appropriate for American viewers, mostly historical and religious paintings. In his painting, many of the visitors to the gallery are American artists and art students, demonstrating through their study and copying the right way to appreciate art. Morse intended the painting not only to introduce Americans to great European art but also to argue that the ideals of European art should shape America.[15]

More recently, some artists have gone beyond painting their revision of the museum by actually changing the exhibits. Their art consists of exhibitions that expose the social and cultural constructs that the museum hides, riffing on museums and history. Museums combine objects and stories to make exhibitions; artists do that, too. Fred Wilson describes his work as art using the museum: "I use the museum as my palette.

Curators, whether they think about it or not, really create how you are to view and think about these objects, so I figured, 'If they can do it, I can do it too.'"[16]

Artists respond to existing exhibits in museums, intervening in them, parodying them. They create new exhibits, showing the boundaries by pushing further than museum curators focused on authenticity and constrained to tell a narrow story for a broad audience generally feel comfortable doing. Dion notes that "artists have tools that scientists don't have . . . humor, irony, metaphor. . . . These are the bread and butter of the artist."[17] Curators too lack these tools, or, more accurately, don't feel they can use them in their curatorial work. The artist brings freedom and choice to exhibitions and, sometimes, a breath of fresh air.

Many artists call attention to the colonial roots of the museum, making public what museums often hide. Wilson suggests adding history to objects acquired by conquest and theft: "I like to bring history to the museum, because I feel that the aesthetic anesthetizes the historic and keeps this imperial view within the museum and continues the dislocation of what these objects are about." In one small intervention, he simply added a label: "Stolen from the Zonge tribe, 1899. Private collection."[18]

Performance artists Guillermo Gómez-Peña and Coco Fusco went the opposite direction, toward parody, in *Two undiscovered Amerindians visit . . .*, an artwork that toured many museums in 1992–1993, the five-hundredth anniversary of Columbus's arrival in the Americas. Gómez-Peña and Fusco visited art museums and anthropology museums costumed as members of the previously undiscovered "Guatinau" tribe, from an island in the Gulf of Mexico. They were shown in a cage, performing "traditional" dances, telling "traditional" stories, posing for photographs with visitors. (Their traditions included weightlifting and rap music, their costume sunglasses. Their tribe's name was pronounced "what now.") Two "guards" answered visitors' questions. A label explained the history of the exhibition of indigenous peoples in museums. A fake *Encyclopedia Britannica* article showed the island of Guatinau. "Our original intent," writes Fusco, "was to create a satirical commentary on Western concepts of the exotic, primitive Other." They would remind visitors of the long history of displaying indigenous peoples in museums and zoos, "colonial fantasies" that taught Europeans and Americans about the inferiority of people of color. Fusco and Gómez-Peña would turn it around, taking back control of that tradition.[19]

But visitors didn't see it that way. Many thought it was a real exhibit—it was in a museum, after all, and had some of the elements that make exhibitions seem truthful: labels, "authentic" objects, even something of an environmental setting. Some acted just as previous generations of museum and exposition visitors had, staring at the exotic other. Indeed, there was no appropriate way to respond; the performance offered a truth not only about colonial roots of the museum, but also about how being in a museum shapes perceptions, and the difficulties of reshaping both the museum and those perceptions.

Fred Wilson's 1990 *Rooms with a View: The Struggle between Cultural Content and the Context of Art* called attention to the ways that exhibit setting shaped how visitors understood artifacts. The exhibit was set in three rooms: a white-walled contemporary gallery, one that looked like an ethnographic museum, and one that looked like a salon from the late nineteenth century. The works of thirty artists appeared in the white cube, as well as in one of the other two rooms. "The environment really changed the work," Wilson remembers.[20] Another artist, Judith Barry, contrasted the art museum with other kinds of related spaces—the department store and the exposition—in *Damaged Goods* (1986) and *From Receiver to Remote Control* (1990). Retail settings, she writes, are the "culmination of exhibition design." Her goal was for the exhibition to become "the set for a play with objects describing various possible subject positions and making the viewer spatially as well as visually aware."[21] Visitors come to museums with certain expectations, and undermining those expectations can stimulate a new understanding of the work exhibitions do.

Some artists offer not new displays but interventions in existing ones. Period rooms at art museums appeal to artists as sets for work that casts a new light on an old museum form.

Yinka Shonibare MBE, visiting the Brooklyn Museum to help plan an exhibition of his art, discovered the period rooms there and was enchanted with them. He decided to create a site-specific work for several of them. His installation, *Mother and Father Worked Hard So I Can Play* (2009), adds a mannequin of a child, headless, dressed in Victorian costume made from the African-inspired fabric common in Shonibare's installations, playing in several of the rooms. A child is hiding under the elegantly set table in the Cane Acres Plantation House dining room. She's jumping rope in the John D. Rockefeller House Moorish Smoking Room. She's sitting quietly in the bedroom of the Trippe House, from an eighteenth-century Maryland plantation. All of these rooms refer to Africa, or slavery, in occluded ways; the child makes that connection clear. The title raises many questions: Just who had worked hard so that she can play? Where did the money that built these elegant rooms come from? Is play the right response to that history? And also, perhaps: why are these rooms in this museum?

Mark Dion created a new period room—an imagined period room for an imagined person, but one that spoke to real history. An installation for the Minneapolis Institute of Art's 2012 "More Real? Art in the Age of Truthiness" revealed the long-hidden office of curator Barton Kestle, the museum's first curator of modern art. The installation had two labels, on opposite walls. One identified the room as a period room, c. 1954. The other called it a Mark Dion site-specific installation, *Curator's Office*. Visitors could take it as a period room or as a site-specific installation. Either worked. But a visitor who read both labels had to wonder just what kind of art and history the museum offered, and might then look at other exhibitions with more than the usual degree of questioning. Like many of Dion's installations, the curator's office revealed the people behind the scenes of the museum. It was, Dion says, reminiscent of a frozen crime scene, with a noir back story.[22]

Exhibitions offer context and suggest objectivity and importance by providing a "museum" setting—white walls, frames, vitrines, guards, a certain kind of label using a certain kind of language—and artists have played with those elements, reminding us of the ways that museums shape our knowledge. Some explore the nature of museum display by showing empty cases, pedestals, and frames. Wilson's *In the Course of Arrangement* at the British Museum (1997) showed off display mounts and labels to reveal the hidden supports for art, and the ways that the form of labels—handwritten or printed—shape visitors' perceptions. Kate Ericson and Mel Ziegler's *MoMA Whites* (1990)—a row of eight jars on a shelf, each filled with a different white pigment, each labeled with the name of the curator who preferred that particular white—called into question the neutrality of that cornerstone of the modern art gallery, the white wall.

Some artists have taken on the exhibit script itself. Meleko Mokgosi's *Modern Art: The Root of African Savages (2012–2014)* offers a script as an exhibition—the script of the Met's 2012 exhibition "African Art, New York, and the Avant-Garde." Mokgosi posts the script of the Met's show, some forty typewritten pages, on the wall of the gallery. The original exhibition explained how artists and art dealers used African art in the early twentieth century; Mokgosi looks deeper. Each page is annotated with dozens of corrections "systematically deconstructing," Mokgosi writes, "the power dynamics and cultural biases that underpin these presumably neutral, educational descriptors." In the Met show, these labels seemed authoritative. They told museum visitors what the art meant. Here, they're on display as cultural artifacts themselves, and Mokgosi's notes call attention both to the story they told and the way they told it, raising questions about the accuracy and objectivity of the museum.[23]

Use

The museum is more than the sum of its parts: more than the collections, the color of the walls, the art and artifacts on those walls, and the labels next to them. Museums make science and culture, change lives, shape communities. Artists critique that mission and extend it. They ask, just what is the museum useful for?

Andrea Fraser's 1989 performance at the Philadelphia Museum of Art took the form of a docent-led tour. *Museum Highlights: A Gallery Talk* begins with Fraser introducing herself to museum visitors and telling them about the tour she'll lead: "the Museum's famed period rooms: dining rooms, coat rooms, et cetera, rest rooms. . . . We'll also be talking about the visitor reception areas, and various service and support spaces." She calls attention to a drinking fountain—"a work of astonishing economy and monumentality."[24]

But more interesting to Fraser is the opportunity to address the museum's mission as articulated when the building was built in 1922. She focuses on the argument its supporters made for usefulness, and compares the museum to a poorhouse, also a way of shaping behavior and identification. She reads from Philadelphia Department of Public Health documents about the poor in describing paintings, and from descriptions of paintings in describing rest rooms and fire escapes. At an eighteenth-century sculpture, *The Four Seasons: Autumn as Bacchus,* her group is told, "American, mother, three brothers distinctly subnormal, herself mentally deficient, violent, undisciplined and lacking in every qualification of motherhood." Art and arrangement cannot hide the secret purposes of the museum from the institutional critique artist.[25]

A more recent Fraser work also compares the museum to institutions of social control, but to offer a contrast, not a parallel. Her 2016 Whitney installation *Down the River* pipes the sounds of prisons into the otherwise empty fifth floor of the museum to suggest that museums and prisons are "the bookend institutions of our increasingly polarized society—institutions that celebrate freedom, and institutions that revoke that freedom." She contrasts wealth and power with poverty and mass incarceration. Her work both calls attention to the museum spaces as "an idealized space of achievement and possibility" and provides the context of a divided, unequal, society within which it exists.[26]

One of Chilean artist Alfredo Jaar's projects suggests the possibilities for museums, what they might become—how they might be useful—by explicitly connecting them to the world beyond artists and patrons and the usual museum audience. His *Music (Everything I know I learned the day my son was born)* was part of Dallas's Nasher Sculpture Center tenth anniversary, in 2013–2014. Jaar asked, "What is the most generous thing this institution can do?" and answered: celebrating the birth of others. Jaar's installation consisted of a greenish-blue translucent pavilion in the center of the museum's sculpture garden. Inside the pavilion, a quiet, shaded space, were chairs and sounds: recordings of the first cries of babies born at three nearby Dallas hospitals, played each day at the precise time of birth. Over the course of the exhibition, more were added. "These babies," Jaar says, "are entering the museum not as spectators, but as performers, as artists, themselves." The family of each child was given a one-year membership to the museum, and each child who was recorded was given a lifetime membership. Jaar hoped not only to change the demographics of the museum, but also to make it "a space for the community to come together and celebrate life."[27]

Jaar, one of the most political of contemporary artists, calls this work "the most political gesture I've ever made."[28] By inviting people from outside the museum to visit and to feel at home in the museum, he offers a way for the museum to expand its reach, to open up to new audiences and new uses. Jaar offers a hopeful vision for the future of museums as institutions that can connect cultures, heal wounds, improve lives, repair the world.

Coda

The Jenks Society for Lost Museums did not aim to change the world, or even the museum. We wanted to explore what had happened in that lost space, to discover what had happened to those lost objects, to look inside the lost museum. The project was as research-intensive as a historical exhibition. We investigated archives and publications, visited museums where some of the artifacts survived, even took a trip to the banks of the river where, long ago, the university had unceremoniously dumped those ninety-two truckloads of museum specimens. Our exhibition used the techniques of the historian and the curator. We aimed for factual accuracy. We sought out authentic artifacts and displayed them in a way that would reveal something of their past. We wrote labels that would both describe and evoke the history of the museum.

Our goal was not a history exhibition but something between that and an artist's installation. We were artists as well as curators. Mark Dion urged us to push beyond the usual display of objects with didactic labels and the usual rules that limit exhibitions to authentic artifacts and call for a simplifying clarity. Artists reimagined some of those objects that were lost, offering a chance to critique the process by which museums turn objects into information by turning that information back into objects. Jenks's office was a period room that might have been at home in a history museum ("Taxidermist's Workshop, c. 1890") but it was more than that, too. It offered a portrait, a psychological profile, that used artistic license to go beyond the literal. A display of surviving artifacts was as much a critique of the ways that museums categorize and preserve their collections as it was an explication of the history of the Jenks Museum.

The exhibition played with past and present, offering historic settings in the museum's completely renovated original building. The language of the labels and the design of the exhibition cases had an air of the nineteenth century about them. The labels for those period rooms set nineteenth-century names in modern type. A visitor to Rhode Island Hall was led to consider old and new, past and present, authenticity and accuracy. We knew we were successful when visitors thought long and hard about whether the story was fact or fiction.

Mark Dion writes that he critically analyzes the narratives and techniques of museums to discern the ideology embedded in them. But being critical, he adds, "may also be just another way to love these museums."[29] That attitude applied to the Jenks Society, too. We looked to history to understand how museums work and what work they do in the world. We critically analyzed that lost museum not only to understand it, but also to love it.

I have tried to do the same in this book. Museums deserve loving, critical analysis. Artists have taken the lead with institutional critique. They have questioned museum assumptions and reimagined museum history. They have asked why museums are the way they are, and how they might be better. Museum workers, museum visitors, and museum critics should do the same, considering the history of museums not only for its own sake, but also to understand the trajectories and the possibilities. Museums are too important to take for granted. Look inside: there's much to learn from the lost museum.

Notes

INTRODUCTION

1. Walter Lee Munro, *The Old Back Campus at Brown* (Providence, RI: Haley & Sykes, 1929), 23; Dallas Lore Sharp, "A Bed in the Museum," *Atlantic Monthly,* June 1935, 705.

2. The history of the Jenks Museum is told in J. Walter Wilson, "The Jenks Museum of Brown University," *Books at Brown* 22 (1968): 41–58; Lily Benedict, "The Life and Death of the Jenks Museum," in *A Brief Guide to the MUSEUM Created at Brown University in 1871 by J. W. P. JENKS Now Re-Collected and Re-Imagined by the Jenks Society for Lost Museums* (Providence, RI: Jenks Society for Lost Museums, 2014); and Kathrinne Duffy, "The Dead Curator: Education and the Rise of Bureaucratic Authority in Natural History Museums, 1870–1915," *Museum History Journal* 10, no. 1 (January 2017), 29–49.

3. The project is captured in *A Brief Guide to the MUSEUM* and on the project's website at https://jenksmuseum.org/.

4. See Eilean Hooper-Greenhill, *Museums and the Shaping of Knowledge* (London: Routledge, 1992), 9–10. For an overview of the literature, see Randolph Starn, "A Historian's Brief Guide to New Museum Studies," *American Historical Review* 110, no. 1 (2005): 68–98.

5. Neil Harris, "The Divided House of the American Art Museum," *Daedalus* 128, no. 3 (Summer 1999): 37.

6. Andrew Leonard, "William Gibson: Novelist," *Rolling Stone,* November 15, 2007, 162.

7. Stephen E. Weil, "From Being about Something to Being for Somebody: The Ongoing Transformation of the American Museum," *Daedalus* 128, no. 3 (1999): 229–58.

8. George Brown Goode, *The Principles of Museum Administration* (York: Coultas & Volans, 1895), 10.

9. William Henry Flower, *Essays on Museums and Other Subjects Connected with Natural History* (London: Macmillan, 1898), 13.

10. Frank L. McVey, ed., "Banquet Session," *Transactions and Proceedings of the National Association of State Universities in the United States of America* 17 (1919): 84.

11. J. W. P. Jenks, "Autobiography," TS, 1894, 217, Brown University Biographical Files, Brown University Archives.

CHAPTER 1: WHY COLLECT?

1. J. Walter Wilson, "The Jenks Museum of Brown University," *Books at Brown* 22 (1968): 44–45.

2. J. W. P. Jenks, "Autobiography," TS, 1894, 691, Brown University Biographical Files, Brown University Archives.

3. Steven Conn, *Do Museums Still Need Objects?* (Philadelphia: University of Pennsylvania Press, 2010).

4. Steven Conn, *Museums and American Intellectual Life, 1876–1926* (Chicago: University of Chicago Press, 1998), 22–24.

5. American Association of Museums, *Museums for a New Century: A Report of the Commission on Museums for a New Century* (Washington, DC: American Association of Museums, 1984), 33.

6. Quoted in "Connecting Cultures: A World in Brooklyn," Brooklyn Museum, accessed July 1, 2015, https://www.brooklynmuseum.org/opencollection/exhibitions /3259.

7. *Annual Report of the President to the Corporation of Brown University* (Providence, RI: Angell, Hammett, 1880), 36.

8. Frederic A. Lucas, *Fifty Years of Museum Work* (New York: American Museum of Natural History, 1933), 29–30.

9. Smithsonian Institution, *A Report on the Management of Collections in the Museums of the Smithsonian Institution* (Washington, DC: Smithsonian Institution, 1977), 31, 43.

10. "Smithsonian Collections," Smithsonian: Newsdesk, August 1, 2016, http:// newsdesk.si.edu/factsheets/fact-sheet-smithsonian-collections.

11. Quoted in Ann Stone, "Treasures in the Basement? An Analysis of Collection Utilization in Art Museums" (Ph.D. diss., RAND Graduate School, Santa Monica, CA, 2002), 49.

12. John Cotton Dana, *The New Museum* (Woodstock, VT: Elm Tree Press, 1917), 37.

13. Ibid., 11.

14. Henry H. Higgins, "Museums of Natural History," *Literary and Philosophical Society of Liverpool* 38 (1884): 198.

15. Smithsonian Institution, *A Report on the Management of Collections*, 20–25.

16. Lucas, *Fifty Years of Museum Work*, 40.

17. Christopher Turner, "A Lot of Gall," *Cabinet* 25 (Spring 2007), http://cabinet magazine.org/issues/25/turner.php.

18. Alfred Stieglitz, "General Cesnola and the Metropolitan," *Metropolitan Museum of Art Bulletin* 27, no. 7 (March 1969): 334–35.

19. Alfred Stieglitz, "The Boston Museum (1922–23) and the Metropolitan Museum (1926)," *Metropolitan Museum of Art Bulletin* 27, no. 7 (March 1969): 335–36.

20. Calvin Tomkins, *Merchants and Masterpieces: The Story of the Metropolitan Museum of Art* (New York: E.P. Dutton, 1970), 299–300.

21. Grant H. Kester, "Ongoing Negotiations: Afterimage and the Analysis of Activist Art," in *Art, Activism, and Oppositionality: Essays from* Afterimage, ed. Grant H. Kester (Durham, NC: Duke University Press, 1998), 11.

22. United States Department of the Army, *Army Museums, Historical Artifacts, and Art*, Historical Activities, Army Regulation 870–20 (Washington, DC: Department of the Army, 1999), 16.

23. Douglas Crimp, *On the Museum's Ruins* (Cambridge, MA: MIT Press, 1993), 263.

24. "Donate," Institute of Contemporary Art / Boston, accessed December 26, 2016, https://www.icaboston.org/join-give/donate; "About the J. Paul Getty Museum," J. Paul Getty Museum, accessed December 26, 2016, http://www.getty.edu /museum/about.html.

25. G. Brown Goode, "The Relationships and Responsibilities of Museums," *Science 2*, no. 34 (August 23, 1895): 198.

26. "Mission Statement, adopted by the Board of Trustees, February 28, 1991" Museum of Fine Arts, Boston, accessed December 26, 2016, http://www.mfa.org/about /mission-statement; "Scope and Criteria," Chicago History Museum, accessed December 26, 2016, https://www.chicagohistory.org/collections/donations-to-the-collection /collecting-scope-and-criteria/.

27. "Mission & History," National Museum of American History, accessed March 1, 2012, http://americanhistory.si.edu/museum/mission-history.

CHAPTER 2: COLLECTABLE

1. J. W. P. Jenks, *Hunting in Florida in 1874* (Providence, RI: n.p., 1884), 42.

2. Quoted in Lee Rosenbaum, "Gary Tinterow on the Divine Right of Curators," *Lee Rosenbaum's CultureGrrl Blog*, July 21, 2006, http://www.artsjournal.com/culture grrl/2006/07/gary_tinterow_on_the_divine_ri.html.

3. "Collections Management Policy," *Metropolitan Museum of Art*, last revised September 8, 2015, http://www.metmuseum.org/about-the-met/policies-and-documents /collections-management-policy.

4. American Association of Museums, *National Standards and Best Practices* (Washington, DC: American Association of Museums, 2008), 46.

5. James B. Gardner and Elizabeth E. Merritt, *The AAM Guide to Collections Planning*, Professional Education Series (Washington, DC: American Association of Museums, 2004), 5.

6. Stephen E. Weil, "Twenty-One Ways to Buy Art," appendix to *Museum Strategy and Marketing: Designing Missions, Building Audiences, Generating Revenue and Resources*, by Neil G Kotler and Philip Kotler (San Francisco: Jossey-Bass, 1998), 363–67.

7. Augustus Pitt-Rivers, "Typological Museums, as Exemplified by the Pitt-Rivers Museum at Oxford, and His Provincial Museum at Farnham, Dorset," *Journal of the Society of Arts* 40, no. 2039 (December 1891): 115.

8. George Storey, memorandum to Luigi Palma Di Cesnola, New York, June 3, 1889, Office of the Secretary Records, Metropolitan Museum of Art Archives.

9. Susan Page, "Director of Smithsonian's New African-American Museum Discusses the Power of History," *USA Today*, February 18, 2016, http://www.usatoday.com /story/news/politics/2016/02/18/capital-download-lonnie-bunch-african-american -museum/80470184/.

10. John Whipple Potter Jenks to Professor Ward, December 11, 1886, Ward Natural Science Establishment Collection, Department of Rare Books and Special Collections, Rush Rhees Library, University of Rochester.

11. Carlene E. Stephens, "From Little Machines to Big Themes: Thinking about Clocks, Watches, and Time at the National Museum of American History," *Material History Review*, no. 52 (Fall 2000): 44.

12. Nancy Bercaw, "Slavery, History, and the Exhibition of Catastrophe: Re-Thinking Curatorial Practice" (conference presentation, Worldly Compositions: Humans and Nonhumans in the Making of Global Politics, Georgetown University, October 2015).

13. Robert E. Fry, *An Outline of the Aims and Ideals Governing the Department of Paintings* (New York: Metropolitan Museum of Art, 1906), 9.

14. Theodore E. Stebbins Jr., "Collecting American Art for Yale, 1968–1976: A Curatorial Report," *Yale University Art Gallery Bulletin* 36, no. 2 (April 1977): 1.

15. Edmund P. Pillsbury, "Connoisseurship and the Rôle of Collecting," *Yale University Art Gallery Bulletin* 40, no. 1 (April 1987): 86–87.

16. Carlo Ginzburg and Anna Davin, "Morelli, Freud and Sherlock Holmes: Clues and Scientific Method," *History Workshop*, no. 9 (1980): 7.

17. Charles F. Montgomery, "The Connoisseurship of Artifacts," in *Material Culture Studies in America*, ed. Thomas J. Schlereth (Nashville, TN: American Association for State and Local History, 1982), 135–45.

18. Berry B. Tracy et al., "American Wing," in *Metropolitan Museum of Art Notable Acquisitions, 1965–1975* (New York: Metropolitan Museum of Art, 1975), 23–25.

19. Clara Lieu, "Ask the Art Prof: How Do Art Museums Select Visual Artists to Exhibit?," Art Prof: Visual Art Essentials (blog), May 12, 2013, https://claralieu.word press.com/2013/05/11/ask-the-art-professor-what-is-museum-quality-artwork/.

20. Ezra Pound, *ABC of Reading* (New York: New Directions, 1960), 29.

21. Alanna Martinez, "How Do Museums Collect? Brooklyn Museum Director Arnold Lehman Breaks It Down," *Observer,* May 13, 2015, http://observer.com/2015/05 /how-do-museums-collect-brooklyn-museum-director-arnold-lehman-breaks-it -down/.

22. Fry, *An Outline of the Aims and Ideals,* 3.

23. Carol Vogel, "The Met Makes Its Biggest Purchase Ever," *New York Times,* November 10, 2004, http://www.nytimes.com/2004/11/10/arts/design/the-met-makes -its-biggest-purchase-ever.html.

24. Smithsonian Institution, *A Report on the Management of Collections in the Museums of the Smithsonian Institution* (Washington, DC: Smithsonian, 1977), B-17.

25. Roslyn Russell, Kylie Winkworth, and Collections Council of Australia, *Significance 2.0: A Guide to Assessing the Significance of Collections* (Rundle Mall: Collections Council of Australia, 2009), 10, 40, and 2.

26. Quoted in Smithsonian Institution Office of Policy and Analysis, *Concern at the Core: Managing Smithsonian Collections* (Washington, DC: Smithsonian, 2005), 149.

27. Lonn Taylor, ed., *A Common Agenda for History Museums: Conference Proceedings, February 19–20, 1987* (Washington, DC: Smithsonian, 1987), 20–21; Steven Lubar and Peter Liebhold, "What Do We Keep?," *Invention & Technology* 14, no. 4 (April 1999): 28; Judy M. Chelnick, "From Stethoscopes to Artificial Hearts," *Caduceus* 13, no. 3 (Winter 1997): 19; Smithsonian Institution, *Concern at the Core,* 152.

28. Martinez, "How Do Museums Collect?"

29. "Museum History," Peabody Essex Museum, accessed August 26, 2015, http://www .pem.org/about/museum_history.

30. For the laws and detailed sources of information, see John E. Simmons, *Things Great and Small: Collections Management Policies* (Washington, DC: American Association of Museums, 2006), 188–90.

31. Eileen Kinsella, "Rauschenberg Eagle Ruffles Feathers," *ARTnews,* May 1, 2012, http://www.artnews.com/2012/05/01/rauschenberg-eagle-ruffles-feathers/; Hannah Kim, "Diving into Rauschenberg's *Canyon,*" *Inside / Out,* January 24, 2014, http://www .moma.org/explore/inside_out/2014/01/24/diving-into-rauschenbergs-canyon.

32. National Park Service, "Frequently Asked Questions," accessed August 2, 2016, https://www.nps.gov/nagpra/FAQ/INDEX.HTM#How_many.

33. Samuel Rubinstein, "BPR Interviews: Victoria Reed," *Brown Political Review,* March 24, 2015, http://www.brownpoliticalreview.org/2015/03/bpr-interviews-victoria -reed/.

34. *Convention for the Protection of Cultural Property in the Event of Armed Conflict with Regulations for the Execution of the Convention 1954*, the Hague, May 14, 1954, available at: http://portal.unesco.org/en/ev.php-URL_ID=13637&URL_DO=DO_TOPIC& URL_SECTION=201.html.

35. *Convention on the Means of Prohibiting and Preventing the Illicit Import, Export and Transfer of Ownership of Cultural Property 1970*, Paris, November 14, 1970, http:// portal.unesco.org/en/ev.php-URL_ID=13039&URL_DO=DO_TOPIC&URL_SECTION =201.html.

36. Kwame Anthony Appiah, "Whose Culture Is It?," *New York Review of Books*, February 9, 2006, http://www.nybooks.com.revproxy.brown.edu/articles/archives/2006 /feb/09/whose-culture-is-it/.

CHAPTER 3: ACQUISITIONS

1. Reuben Aldridge Guild, *Memorial Address on the Late John Whipple Potter Jenks* (Providence, RI: Casey Brothers, 1895), 26–30.

2. Ann Stone, "Treasures in the Basement? An Analysis of Collection Utilization in Art Museums" (PhD dissertation, RAND Graduate School, Santa Monica, CA, 2002), 110; MoMA statistics based on analysis of data at "MuseumofModernArt / col lection," GitHub, accessed August 10, 2016, https://github.com/MuseumofModernArt /collection.

3. Association of Art Museum Directors, "Art Museums by the Numbers 2014," January 7, 2015, https://aamd.org/our-members/from-the-field/art-museums-by-the -numbers-2014.

4. John Kelly, "This Would-Be Gift to the Smithsonian Didn't Pass the Smell Test," *Washington Post*, August 29, 2016, http://wpo.st/LgOv1.

5. Steven D. Lubar and Kathleen M. Kendrick, *Legacies: Collecting America's History at the Smithsonian* (Washington, DC: Smithsonian Institution Press in association with the National Museum of American History, 2001), 50–51, 152.

6. William Ivins, quoted in Wendy Kaplan, "R. T. H. Halsey: An Ideology of Collecting American Decorative Arts," *Winterthur Portfolio* 17, no. 1 (April 1982): 43.

7. Gary Kulik, "Designing the Past: History-Museum Exhibitions from Peale to the Present," in *History Museums in the United States: A Critical Assessment*, ed. Warren Leon and Roy Rosenzweig (Urbana: University of Illinois Press, 1989), 14–16.

8. Jeffrey Trask, *Things American: Art Museums and Civic Culture in the Progressive Era* (Philadelphia: University of Pennsylvania Press, 2011), 174.

9. Briann Greenfield, *Out of the Attic: Inventing Antiques in Twentieth-Century New England* (Amherst: University of Massachusetts Press, 2009), 167–69.

10. Lubar and Kendrick, *Legacies*, 108–109, 195–97, and 118–19.

11. Quoted in Eric Nystrom, "'Your Name Would Be Conspicuously Present': Curators, Companies, and the Contents of Exhibits at the Smithsonian" (conference presentation, Organization of American Historians / National Council on Public History Annual Meeting, Washington, DC, March 2006), 9.

12. "The Civil War Is Winding Down," *New York Times,* July 6, 2015, http://www .nytimes.com/2015/07/07/opinion/the-civil-war-is-winding-down.html.

13. Stone, "Treasures in the Basement?," 113.

14. Ibid., 119.

15. Ibid., 125.

16. Robert W. de Forest, *Museum Policies and Scope* (New York: Metropolitan Museum of Art, 1930), 11.

17. Calculated from "MuseumofModernArt / collection."

18. Jorge S. Arango, "Great Donations," *Introspective Magazine,* July 20, 2015, http://www .1stdibs.com/introspective-magazine/glenn-gissler-risd-donations/.

19. M. H. Miller, "The Thoroughly Modern Met," *ARTnews,* October 30, 2014, http:// www.artnews.com/2014/10/30/leonard-lauder-cubist-collection-at-the-met/.

20. Lansing Lamont, "Tom Hoving of the Met Builds, Buys and Makes Much of the Art World Bilious," *People,* November 17, 1975, http://www.people.com/people /archive/article/0,,20065857,00.html.

21. Thomas Hoving, *Making the Mummies Dance: Inside the Metropolitan Museum of Art* (New York: Simon & Schuster, 1993), 181–85.

22. Henry D. Brown, "Can't See the Woods for the Trees," *History News* 11, no. 8 (June 1956): 61.

23. Internal Revenue Service, *Publication 526: Charitable Contributions,* January 19, 2017, 10, https://www.irs.gov/pub/irs-pdf/p526.pdf.

24. Marie C. Malaro, *Museum Governance: Mission, Ethics, Policy* (Washington, DC: Smithsonian Institution Press, 1994), 101.

25. *Annual Report of the President to the Corporation of Brown University* (Providence, RI: Hammon, Angell, 1872), 12.

26. John Whipple Potter Jenks to Prof. Ward, December 28, 1874, and December 6, 1886, Ward's Natural Science Establishment papers, Department of Rare Books, Special Collections, and Preservation, River Campus Libraries, University of Rochester.

27. Sarah Thornton, *Seven Days in the Art World* (New York: W.W. Norton, 2008), xix.

28. Brad Phillips, "How Instagram Is Changing the Art World," *VICE,* May 18, 2016, http://www.vice.com/read/how-instagram-is-changing-the-art-world.

29. *Annual Report of the President to the Corporation of Brown University* (1882), 14.

30. *Annual Report of the Board of Regents of the Smithsonian Institution for the Year 1873* (Washington, DC: Government Printing Office, 1874), 43; Jane McLaren Walsh,

"Collections as Currency," in *Anthropology, History, and American Indians: Essays in Honor of William Curtis Sturtevant*, ed. William L. Merrill and Ives Goddard (Washington, DC: Smithsonian Institution Press, 2002), 203.

31. Walsh, "Collections as Currency." See also Catherine A. Nichols, "Exchanging Anthropological Duplicates at the Smithsonian Institution," *Museum Anthropology* 39, no. 2 (September 2016): 130–46.

32. H. L. Madison, "Tentative Code of Museum Ethics," reproduced in Gaynor Kavanagh, ed., *Museum Provision and Professionalism* (London: Routledge, 2005), 260–65.

33. Carol Vogel, "Brooklyn Museum's Costume Treasures Going to the Met," *New York Times*, December 16, 2008, http://www.nytimes.com/2008/12/16/arts/design/16muse.html.

Chapter 4: In the Field

1. J. W. P. Jenks, *Hunting in Florida in 1874* (Providence, RI: n.p., 1884), 1.

2. Dallas Lore Sharp, "A Bed in the Museum," *Atlantic Monthly*, June 1935, 704.

3. Ralph Waldo Emerson, "The Uses of Natural History," in *The Selected Lectures of Ralph Waldo Emerson*, ed. Ronald A. Basco and Joel Myerson (Athens: University of Georgia Press, 2005), 2–3.

4. Bruno Latour, *Science in Action: How to Follow Scientists and Engineers through Society* (Cambridge, MA: Harvard University Press, 1987), 225.

5. John H. Brinton, *Personal Memoirs of John H. Brinton, Major and Surgeon U.S.V., 1861–1865* (New York: Neale, 1914), 187–88.

6. See Joshua A. Bell and Erin L. Hasinoff, eds. *The Anthropology of Expeditions: Travel, Visualities, Afterlives* (New York: Bard Graduate Center, 2015).

7. William C. Sturtevant, "Does anthropology need museums?," *Proceedings of the Biological Society of Washington* 82 (1969): 640.

8. S. Terry Childs and Eileen Corcoran, "Relevant Laws, Regulations, Policies, and Ethics," *Managing Archeological Collections: Technical Assistance*, October 26, 2000, https://www.nps.gov/archeology/collections/laws_01.htm.

9. Edward W. Nolan, "A Passion for Collecting: Edward N. Fuller and the Genesis of the Washington State Historical Society's Research Collection," *Columbia* 9, no. 4 (Winter 1995): 13.

10. Laurence Vail Coleman, *The Museum in America: A Critical Study* (Washington, DC: American Association of Museums, 1939), 2:245.

11. Bodil Axelsson, "Samdok—Collecting and Networking the Nation as It Evolves," vol. 62, Linköping Electronic Conference Proceedings (Norrköping, Sweden: Linköping University Electronic Press, 2011), 175–81, http://www.ep.liu.se/ecp/article.asp?issue=062&article=019.

12. United States Department of the Army, *Military History: Responsibilities, Policies, and Procedures,* Historical Activities, Army Regulation 870–5, (Washington, DC: Headquarters, Department of the Army, 1999), 3–7, 3–11, C–1, and G–11.

13. Bruce Altshuler, "Collecting the New: A Historical Introduction," in *Collecting the New: Museums and Contemporary Art,* ed. Bruce Altshuler (Princeton, NJ: Princeton University Press, 2005).

14. T. A. Frail, "How Do Smithsonian Curators Decide What to Collect at the Political Conventions?," Smithsonian.com, July 27, 2016, http://www.smithsonianmag.com/smithsonian-institution/how-do-curators-decide-what-collect-political-conventions-180959937.

15. Edith Mayo, "Collecting the Twentieth Century," in *Museum Curatorship: Rhetoric vs. Reality: The Proceedings of the Eighth Museum Studies Conference,* ed. Bryant F. Tolles Jr. (Newark: University of Delaware, 1987), 90–94. See also Kylie Message, *Museums and Social Activism: Engaged Protest* (London: Routledge, 2014).

16. Steven Lubar and Peter Liebhold, "What Do We Keep?," *Invention & Technology* 14, no. 4 (April 1999): 35–36.

17. Kevin Rector, "New Smithsonian Museum Curators Pursue the Stories in Baltimore's Protests," *Los Angeles Times,* June 7, 2015, http://www.latimes.com/nation/la-na-baltimore-smithsonian-20150607-story.html.

18. David H. Shayt, "Artifacts of Disaster: Creating the Smithsonian's Katrina Collection," *Technology & Culture* 47, no. 2 (April 2006): 368. The collection is available at http://amhistory.si.edu/september11/.

19. Laura Anderson and Janet Donlin, "Vietnam Veterans Memorial Scope of Collection Statement" (Washington DC: National Park Service National Mall and Memorial Parks, 2016), 6–7.

20. Edward T. Linenthal, *The Unfinished Bombing: Oklahoma City in American Memory* (Oxford: Oxford University Press, 2003), 167.

21. James B. Gardner and Sarah M. Henry, "September 11 and the Mourning After: Reflections on Collecting and Interpreting the History of Tragedy," *Public Historian* 24, no. 3 (Summer 2002): 38.

22. David Shayt, "What We Are Doing Is Building a Collection for All Time Here," *September 11: Bearing Witness to History,* accessed November 23, 2016, http://amhistory.si.edu/september11/collection/transcript.asp?ID=40.

23. "National September 11 Memorial Museum Collections Management Policy" 9/11 Memorial Museum, 2016, 1–6, https://www.911memorial.org/sites/default/files/Collections%20Management%20Policy_2016.pdf.

24. Erika Lee Doss, *Memorial Mania: Public Feeling in America* (Chicago: University of Chicago Press, 2010), 73–74.

Chapter 5: Who Collects?

1. Louis Agassiz, *Contributions to the Natural History of the United States of America*, vol. 1, *First Monograph* (Boston: Little, Brown, 1857), vi, xv.

2. Dallas Lore Sharp, "Turtle Eggs for Agassiz," *Atlantic Monthly*, February 1910, 158, 162.

3. Reuben Aldridge Guild, *Memorial Address on the Late John Whipple Potter Jenks* (Providence, RI: Casey Brothers, 1895), 21, 33–34.

4. Robert Dodsley, *The General Contents of the British Museum: With Remarks. Serving as a Directory in Viewing That Noble Cabinet* (London: R. and J. Dodsley, 1761), vi.

5. G. Brown Goode, "The Relationships and Responsibilities of Museums," *Science* 2, no. 34 (August 1895): 205.

6. Quoted in Vera L. Zolberg, "Conflicting Visions in American Art Museums," *Theory and Society* 10, no. 1 (January 1981): 106.

7. Quoted in Ernest Samuels and Jayne Samuels, *Bernard Berenson, the Making of a Legend* (Cambridge, MA: Harvard University Press, 1987), 438.

8. Calvin Tomkins, *Merchants and Masterpieces: The Story of the Metropolitan Museum of Art* (New York: E.P. Dutton, 1970), 229–30.

9. William Henry Goodyear, "My Own Duties" (memorandum to Luigi Palma di Cesnola, 1885), Office of the Secretary Records, Metropolitan Museum of Art Archives.

10. "Director's Agreement with Curators," June 10, 1886, Employees-Curators file, Office of the Secretary Records, Metropolitan Museum of Art Archives.

11. James W. Randell to Luigi Palma di Cesnola, April 6, 1893; George Storey to di Cesnola, memorandum, April 9, 1893; di Cesnola to James W. Randell, April 10, 1893, Office of the Secretary Records, Metropolitan Museum of Art Archives.

12. "Report of the Curator," December 1892, Office of the Secretary Records, Metropolitan Museum of Art Archives.

13. J. Elliott Cabot to Charles Eliot Norton, May 10, 1875, quoted in Neil Harris, "The Gilded Age Revisited: Boston and the Museum Movement," *American Quarterly* 14, no. 4 (December 1, 1962): 553.

14. Boston Museum of Fine Arts, *Trustees of the Museum of Fine Arts Boston Eighth Annual Report*: *Eighth annual report, for the year ending Dec. 31, 1883* (Boston: Alfred Mudge & Son, 1884), 12.

15. Martin Brimmer, "The Museum of Fine Arts," *American Architect and Building News* 8, no. 253 (October 30, 1880): 205, quoted in Harris, "The Gilded Age Revisited," 544.

16. Charles C. Perkins, "American Art Museums," *North American Review* 111, no. 228 (July 1870): 8, quoted in Harris, "The Gilded Age Revisited," 553–54.

17. Benjamin Ives Gilman to Morris Gray, Esq., January 8, 1921, Benjamin Ives Gilman papers, Museum of Fine Arts Boston, Archives.

18. Benjamin Ives Gilman, "At a Meeting of the Trustees of the Museum of Fine Arts, in Boston," January 1921, Benjamin Ives Gilman papers, Museum of Fine Arts Boston, Archives.

19. William Henry Flower, *Essays on Museums and Other Subjects Connected with Natural History* (London: Macmillan, 1898), 35–36.

20. F. A. Bather, "The Man as Museum-Curator," *Museums Journal* (1902): 188.

21. A. R. Crook, "The Training of Museum Curators," *Proceedings of the Fifth Annual Meeting of the American Association of Museums*, 1910, 59–63.

22. *Proceedings of the Fifth Annual Meeting of the American Association of Museums*, 1910, 64.

23. Paul J. Sachs, "Fogg Art Museum," *Harvard Graduates' Magazine*, March 1916, 424.

24. Quoted in Janet Tassel, "Reverence for the Object: Art Museums in a Changed World," *Harvard Magazine*, October 2002, 52.

25. Ibid., 50; Sybil Gordon Kantor, *Alfred H. Barr, Jr., and the Intellectual Origins of the Museum of Modern Art* (Cambridge, MA: MIT Press, 2002), 73–74.

26. Committee on the Visual Arts at Harvard, *Report* (Cambridge, MA: Harvard University, 1956), 29.

27. Zolberg, "Conflicting Visions," 109.

28. Ibid., 114.

29. Ibid.

30. *United States National Museum Annual Report* (Washington, DC: Smithsonian Institution, 1952), 1, 6.

31. U.S. Office of Personnel Management, "Position Classification Standard for Museum Curator Series, GS-1015," February 1962, https://www.opm.gov/policy-data -oversight/classification-qualifications/classifying-general-schedule-positions/stan dards/1000/gs1015.pdf, accessed February 17, 2017.

32. Spencer Baird, "General Directions for Collecting and Preserving Objects of Natural History" (Smithsonian Institution, 1848), I-Friends-1975–18, Dickinson College Archives & Special Collections, http://archives.dickinson.edu/sites/all/files/files _document/I-Friends-1975-18_0.pdf.

33. G. Stanley Hall, "Museums of Art and Teachers of History," in *Art Museums and Schools: Four Lectures Delivered at the Metropolitan Museum of Art* (New York: Charles Scribner's Sons, 1913), 71.

34. Tomkins, *Merchants and Masterpieces*, 311.

35. "Smithsonian Exhibit," *Baltimore Afro-American*, December 18, 1965, 5.

36. *Smithsonian Year: Annual Report of the Smithsonian Institution for the Year Ended 30 June 1972* (Washington, DC: Smithsonian Institution Press, 1972), 92.

37. "A History of the World," *BBC*, 2014, http://www.bbc.co.uk/ahistoryof theworld/.

38. "Memories are Made in the Kitchen," accessed February 17, 2017, http://cafmuseum.techno-science.ca/en/whats-on/exhibition-memories-made-in-kitchen.php.

39. Anna Adamek, "A Snapshot of Canadian Kitchens: Collecting Contemporary Technologies as Historical Evidence for Future Research" (unpublished paper, 2015); Anna Adamek, e-mail message to author, September 1, 2015. See "Memories are Made in the Kitchen."

40. Thomas Söderqvist, "The Participatory Museum and Distributed Curatorial Expertise," *NTM Zeitschrift Für Geschichte Der Wissenschaften, Technik Und Medizin* 18, no. 1 (May 2010): 69–78.

Chapter 6: Into the Storeroom

1. J. W. P. Jenks, *Hunting in Florida in 1874* (Providence, RI: n.p., 1884), 2–3.

2. Heritage Health Index, "A Public Trust at Risk: The Heritage Health Index Report on the State of America's Collections" (Washington, DC: Heritage Preservation, 2004), 28.

3. Ibid., 2, 79.

4. Peter Walsh, *It's All Too Much: An Easy Plan for Living a Richer Life with Less Stuff* (New York: Free Press, 2007), 134.

5. George Ellis Burcaw, *Introduction to Museum Work*, 3rd ed. (Walnut Creek, CA: AltaMira, 1997), 102.

6. Linda A. Fairstein, *The Bone Vault* (New York: Scribner, 2003), 334, 355.

7. Mark Starr, "Secrets from the Storeroom," *Newsweek*, July 22, 1990, http://www.newsweek.com/secrets-storeroom-206832.

8. William Mullen, "Space to Grow Becoming Thing of the Past at Field," *Chicago Tribune*, January 14, 2001, http://articles.chicagotribune.com/2001-01-14/news/0101140254_1_collections-field-museum-field-expedition.

9. Mary P. Winsor, *Reading the Shape of Nature: Comparative Zoology at the Agassiz Museum* (Chicago: University of Chicago Press, 1991), 37; Lukas Benjamin Rieppel, "Dinosaurs: Assembling an Icon of Science" (PhD dissertation, Harvard University, 2012), 237.

10. "Surrealist Cabinet of Curiosities Bought for Manchester," *Art Fund*, December 2, 2014, https://www.artfund.org/news/2014/12/02/surrealist-cabinet-of-curiosities-bought-for-manchester.

11. G. Brown Goode, "The Relationships and Responsibilities of Museums," *Science* 2, no. 34 (August 1895): 205; Philip S. Doughty, "Research: Geology Collections," in *The Manual of Curatorship: A Guide to Museum Practice*, ed. John M. A. Thompson (London: Butterworths, 1984), 156.

12. Sharon Macdonald, *Behind the Scenes at the Science Museum* (Oxford: Berg, 2002), 65; Hilary Geoghegan and Alison Hess, "Object-Love at the Science Museum: Cultural Geographies of Museum Storerooms," *Cultural Geographies* 22, no. 3 (July 2015): 445–65.

13. J. E. Gray, "On Museums, Their Use and Improvement," *Annals and Magazine of Natural History* 14 (1864): 287.

14. "Organization and Administration of the United States National Museum, No. 16: Plans for the Installation of Collections," in *Proceedings of the United States National Museum* (Washington, DC: Smithsonian Institution, 1882); G. Brown Goode, *The Principles of Museum Administration* (York: Coultas & Volans, 1895), 40.

15. Roger E. Fry, *An Outline of the Aims and Ideals Governing the Department of Paintings* (New York: Metropolitan Museum of Art, 1906), 9.

16. Benjamin Ives Gilman, *Museum Ideals of Purpose and Method* (Cambridge, MA: Museum of Fine Arts, Boston, 1918), 401; M, "The Boston Art Museum," *Nation* 89, no. 2316 (November 18, 1909): 496. See also Neil Harris, "The Gilded Age Revisited: Boston and the Museum Movement," *American Quarterly* 14, no. 4 (December 1962): 545–66.

17. Laurence Vail Coleman, *The Museum in America: A Critical Study* (Washington, DC: American Association of Museums, 1939), 2:253–54.

18. Henry D. Brown, "Can't See the Woods for the Trees," *History News* 11, no. 8 (June 1956): 61.

19. *United States National Museum Annual Report* (Washington, DC: Smithsonian Institution, 1952), 1.

20. Geraldine Fabrikant, "The Good Stuff in the Back Room," *New York Times*, March 12, 2009.

21. Daniel Robbins, "Confessions of a Museum Director," in Stephen E. Ostrow, *Raid the Icebox 1, with Andy Warhol: An Exhibition Selected from the Storage Vaults of the Museum of Art, Rhode Island School of Design* (Providence: Rhode Island School of Design Museum of Art, 1969), 8.

22. Richard A. Fortey, *Dry Store Room No. 1: The Secret Life of the Natural History Museum* (London: Harper Perennial, 2008), 10–11.

23. "Organization and Administration of the United States National Museum," in *Proceedings of the United States National Museum* (Washington, DC: Smithsonian Institution, 1882), 12; Director's Agreement with Curators, June 10, 1886, Employees-Curators file, Office of the Secretary Records, Metropolitan Museum of Art Archives.

24. *United States National Museum Annual Report*, 5–6.

25. Coleman, *Museum in America*, 2:407; Julian Spalding, "Creative Management in Museums," in *Management in Museums*, ed. Kevin D. Moore (New Brunswick, NJ: Athlone, 1999), 32–34.

26. Brooke Hindle, "How Much Is a Piece of the True Cross Worth?," in *Material Culture and the Study of American Life,* ed. Ian M. Quimby (New York: W. W. Norton, 1978), 18.

27. Gail Anderson, ed., *Reinventing the Museum: The Evolving Conversation on the Paradigm Shift,* 2nd ed. (Lanham, MD: AltaMira, 2012), 285; Douglas Greenberg, " 'History Is a Luxury': Mrs. Thatcher, Mr. Disney, and (Public) History," *Reviews in American History* 26, no. 1 (1998): 305–306.

28. Suzanne Keene, *Fragments of the World: Uses of Museum Collections* (Burlington, MA: Elsevier Butterworth-Heinemann, 2005), 121; Eric D. Gyllenhaal, Deborah Perry, and Emily Forland, "Visitor Understandings about Research," *Current Trends in Audience Research and Evaluation* 10 (May 1996): 22–32.

29. John Liffen, "Behind the Scenes: Housing the Collections," in *Science for the Nation: Perspectives on the History of the Science Museum,* ed. Peter J. T. Morris (Basingstoke, UK: Palgrave Macmillan, 2010), 273–93.

30. Campaign for Good Curatorship, "Manifesto," accessed January 5, 2016, http://campaignforgoodcuratorship.org.uk/manifesto/; Timothy A. M. Ewin and Joanne V. Ewin, "In Defence of the Curator: Maximising Museum Impact," *Museum Management and Curatorship* 31, no. 4 (2016): 324.

31. Carla Sinopoli, "Exploring the Collections: An Anthropologist in the University Museum Storeroom" (lecture, Haffenreffer Museum of Anthropology, Providence, RI, October 8, 2015).

Chapter 7: Paperwork

1. *Annual Report of the President to the Corporation of Brown University* (Providence: Snow & Farnham, 1893).

2. G. Brown Goode, *The Principles of Museum Administration* (York: Coultas & Volans, 1895), 54.

3. E. L. Konigsburg, *From the Mixed-Up Files of Mrs. Basil E. Frankweiler* (1967; repr., New York: Simon and Schuster, 1998), 127.

4. Virginia Zimmerman, "The Curating Child: Runaways and Museums in Children's Fiction," *The Lion and the Unicorn* 39, no. 1 (January 2015): 55.

5. A. C. Smith, "Advice to Administrators of Systematic Collections," *Taxon* 15, no. 6 (July 1, 1966): 202.

6. Robert Dodsley, *The General Contents of the British Museum: With Remarks. Serving as a Directory in Viewing That Noble Cabinet* (London: R. and J. Dodsley, 1761), 1.

7. Gary Edson and David Dean, *The Handbook for Museums* (1996; repr., London: Routledge, 2008), 222–29.

8. Hannah Turner, "Decolonizing Ethnographic Documentation: A Critical History of the Early Museum Catalogs at the Smithsonian's National Museum of Natural History," *Cataloging & Classification Quarterly* 53, nos. 5–6 (July 2015): 658–76.

9. "Organization and Administration of the United States National Museum," in *Proceedings of the United States National Museum* (Washington, DC: Smithsonian Institution, 1882), 28.

10. David W. Scott, "Museum Data Bank Research Report: The Yogi and the Registrar," *Library Trends* 37, no. 2 (1988): 130–41.

11. "Getty Vocabularies," Getty Research Institute, accessed November 27, 2016, https://www.getty.edu/research/tools/vocabularies/.

12. "What Is Nomenclature?," *American Association for State and Local History*, accessed November 27, 2016, http://community.aaslh.org/nomenclature-faqs/.

13. Daniel Robbins, "Confessions of a Museum Director," in *Raid the Icebox 1, with Andy Warhol: An Exhibition Selected from the Storage Vaults of the Museum of Art, Rhode Island School of Design,* by Stephen E. Ostrow (Providence: Rhode Island School of Design Museum of Art, 1969), 14.

14. "18621519," *Curatorial Poetry* (Tumblr blog), August 17, 2015, http://curatorial poetry.tumblr.com/post/126918294432/18621519.

15. William George Maton, "Memoir of the Author," in *A General View of the Writings of Linnæus,* 2nd ed., by Richard Pulteney (London: J. Mawman, 1805), 11–12.

16. Curtis M. Hinsley, *Savages and Scientists: The Smithsonian Institution and the Development of American Anthropology* (Washington, DC: Smithsonian Institution Press, 1981); F. A. Bather, "The Museums of Chicago," *Museums Journal* (1902): 291.

17. Quoted in Rebecca A. Buck, *Collection Conundrums: Solving Collections Management Mysteries* (Washington, DC: American Association of Museums, 2007), 5.

18. L. B. Walton, "Contributions to Museum Technique. I. Cataloguing Museum Specimens," *American Naturalist* 41, no. 482 (1907): 77.

19. Lukas Benjamin Rieppel, "Dinosaurs: Assembling an Icon of Science" (PhD dissertation, Harvard University, 2012), 204; Bather, "The Museums of Chicago," 290.

20. Goode, *The Principles of Museum Administration,* 55.

21. Henry W. Kent, "Some Business Methods in the Metropolitan Museum of Art," *Proceedings of the American Association of Museums* 5 (1911): 32.

22. A. B. Meyer, quoted in Bather, "The Museums of Chicago," 291; F. A. Bather, "The Man as Museum-Curator," *Museums Journal* 1 (1902): 187.

23. See Katherine Jones-Garmil, "Laying the Foundation: Three Decades of Computer Technology in Museums," in *The Wired Museum: Emerging Technology and Changing Paradigms,* ed. Katherine Jones-Garmil (Washington, DC: American Association of Museums, 1997).

24. Seb Chan, "The API at the Center of the Museum," *Cooper Hewitt Labs,* accessed January 7, 2016, http://labs.cooperhewitt.org/2014/the-api-at-the-center-of-the -museum/.

25. Robin Sloan, *Mr. Penumbra's 24-Hour Bookstore: A Novel* (New York: Farrar, Straus and Giroux, 2012), 245–46.

26. Sarah Knapton, "Half of World's Museum Specimens Are Wrongly Labelled, Oxford University Finds," *Telegraph,* November 17, 2015, http://www.telegraph.co.uk /news/science/science-news/11998634/Half-of-worlds-museum-specimens-are -wrongly-labelled-Oxford-University-finds.html.

27. Joseph Stromberg, "The Hope Diamond Was Once a Symbol for Louis XIV, the Sun King," *Smithsonian,* January 28, 2014, http://www.smithsonianmag.com/science -nature/hope-diamond-was-once-symbol-louis-xiv-sun-king-180949482/.

28. Anna Adamek, Twitter posts, August 22–26, 2014, https://twitter.com/curatorn atres/status/504332205502787584.

29. "Danish Museum Finds Lost Charles Darwin Treasure," *Copenhagen Post,* August 21, 2014, http://cphpost.dk/news/danish-museum-finds-lost-charles-darwin -treasure.html.

30. Lisa Peet, "Update: BPL's Lost Artwork Recovered Following Ryan's Resignation," *Library Journal,* June 11, 2015, http://lj.libraryjournal.com/2015/06/industry -news/boston-public-library-searches-for-lost-artwork/.

31. Mitchell Whitelaw, "Generous Interfaces for Digital Cultural Collections," *Digital Humanities Quarterly* 9, no. 1 (2015), http://www.digitalhumanities.org/dhq/vol/9 /1/000205/000205.html.

32. Aaron Straup Cope, "Rijkscolors! (or Colorific Promiscuity)," *Cooper Hewitt Labs,* December 11, 2013, http://labs.cooperhewitt.org/2013/rijkscolors-or-colorific-promis cuity/.

33. Wilcomb E. Washburn, "Collecting Information, Not Objects," *Museum News* 62, no. 3 (February 1984): 14–15.

34. John Updike, "Museums and Women," *New Yorker,* November 18, 1967, 59.

CHAPTER 8: THE ETHICS OF OBJECTS

1. Quoted in J. Walter Wilson, "The Jenks Museum of Brown University," *Books at Brown* 22 (1968): 54.

2. Ibid., 41.

3. American Alliance of Museums, "Code of Ethics for Museums," 1991 amended 2000, http://www.aam-us.org/resources/ethics-standards-and-best-practices/code-of -ethics.

4. Michael A. Mares, "The Moral Obligations Incumbent upon Institutions, Administrators and Directors in Maintaining and Caring for Museum Collections," in

Museum Philosophy for the Twenty-first Century, ed. Hugh H. Genoways (Lanham, MD: AltaMira, 2006), 96.

5. Simon Lambert, "The Early History of Preventive Conservation in Great Britain and the United States (1850–1950)," *CeROArt. Conservation, Exposition, Restauration d'Objets d'Art* 9 (January 22, 2014): http://ceroart.revues.org/3765#tocto2n4.

6. Stefan Michalski, "Climate Guidelines for Heritage Collections: Where We Are in 2014 and How We Got Here," in *Proceedings of the Smithsonian Institution Summit on the Museum Preservation Environment,* ed. Sarah Stauderman and William G. Tompkins (Washington, DC: Smithsonian Institution Scholarly Press, 2016), 9–10.

7. American Defense–Harvard Group, "Notes on Safeguarding and Conserving Cultural Material in the Field" (Committee on Protection of Monuments, 1943), 1:41–44, https://www.aaa.si.edu/collections/items/detail/notes-safeguarding-and-conserving-cultural-material-field-part-i-16189.

8. W. G. Constable, "Curators and Conservation," *Studies in Conservation* 1, no. 3 (1954): 97–102.

9. American Association of Museums, *America's Museums: The Belmont Report. A Report to the Federal Council on the Arts and the Humanities* (Washington, DC: American Association of Museums, 1969), 57; American Association of Museums, *Museums for a New Century: A Report of the Commission on Museums for a New Century.* (Washington, DC: American Association of Museums, 1984), 42.

10. "Smithsonian Institution Declaration on the Collections Preservation Environment," in Sarah Stauderman and William G. Tompkins, eds., *Smithsonian Institution Declaration on the Collections Preservation Environment* (Washington, DC: Smithsonian Institution Scholarly Press, 2016), 109.

11. Barbara Appelbaum, *Conservation Treatment Methodology* (Amsterdam: Butterworth-Heinemann, 2007), xix–xxii.

12. Fred Heiler, "Preserving Classics, Wrinkles and All," *New York Times,* October 5, 2012, http://www.nytimes.com/2012/10/07/automobiles/collectibles/preserving-classics-wrinkles-and-all.html.

13. Quoted in "History of the Collections," *National Museum of the American Indian,* accessed January 19, 2016, http://nmai.si.edu/explore/collections/history/.

14. John Roach, "At New American Indian Museum, Artifacts Are 'Alive,'" *National Geographic News,* September 21, 2004, http://news.nationalgeographic.com/news/2004/09/0914_040913_indians_exhibits.html.

15. Ashley Dunn, "A Heritage Reclaimed; From Old Artifacts, American Indians Shape a New Museum," *New York Times,* October 9, 1994, http://www.nytimes.com/1994/10/09/nyregion/a-heritage-reclaimed-from-old-artifacts-american-indians-shape-a-new-museum.html.

16. Amy Drapeau and Leonda Levchuk, "Cultural Resources Center Fact Sheet," Smithsonian National Museum of the American Indian Office of Public Affairs,

October 2005, http://nmai.si.edu/sites/1/files/pdf/press_releases/2005-10_crc_fact _sheet.pdf.

17. Shabnam Honarbakhsh, Heidi Swierenga, and Mauray Toutloff, "When You Don't Cry over Spilt Milk: Collections Access at the UBC Museum of Anthropology during the Renewal Project," *American Institute for Conservation of Historic and Artistic Works Objects Specialty Group Postprints* 18 (2011): 67–82.

18. James Clifford, *Routes: Travel and Translation in the Late Twentieth Century* (Cambridge, MA: Harvard University Press, 1997), 213; Rosemary A. Joyce, "Things in Motion: Itineraries of Ulua Marble Vases," in *Things in Motion: Object Itineraries in Anthropological Practice*, ed. Rosemary A. Joyce and Susan D. Gillespie (Santa Fe, NM: School for Advanced Research Press, 2015), 31.

19. Christina Kreps, "Curatorship as Social Practice," *Curator: The Museum Journal* 46, no. 3 (July 2003): 311–23.

20. J. W. P. Jenks, "Foreword," in Walter Porter Manton, *Taxidermy without a Teacher: Comprising a Complete Manual of Instruction for Preparing and Preserving Birds, Animals and Fishes* (Boston: Lee and Shepard, 1882), 12.

21. Karen L. Haberman, "On the Significance of Small Dead Things," *Journal of Natural History Education and Experience* 9 (2015): 8–12.

22. "A Manifesto for Active History Museum Collections," *Active Collections*, accessed June 3, 2015, http://www.activecollections.org/manifesto/.

23. Sheila Brennan, "Making Room by Letting Go: A Look at the Ephemerality of Collections," *Preservation Leadership Forum* (blog), August 12, 2014, http://forum.saving places.org/blogs/special-contributor/2014/08/12/making-room-by-letting-go-a-look -at-the-ephemerality-of-collections.

24. Allison Marsh and Lizzie Wade, "Collective Forgetting: Inside the Smithsonian's Curatorial Crisis," accessed July 17, 2014, http://thinkwritepublish.org/the-narratives /collective-forgetting/.

25. Haberman, "On the Significance of Small Dead Things," 9.

26. AAM Museum Accreditation Program, "The Accreditation Commission's Expectations Regarding Collections Stewardship" December 17, 2004, http://www.wipo .int/export/sites/www/tk/en/databases/creative_heritage/docs/aam_collections.pdf. For more detail, see American Alliance of Museums, "Developing a Collections Management Policy," 2015, http://www.aam-us.org/docs/default-source/continuum /developing-a-cmp-final.pdf.

27. AAM, "Accreditation Commission's Expectations."

28. Laurence Vail Coleman, *The Museum in America: A Critical Study* (Washington, DC: American Association of Museums, 1939), 2:241.

29. Stephen E. Weil, "Note: Dispositions by Museums," in *Art Works: Law, Policy, Practice,* by Franklin Feldman and Stephen E Weil (New York: Practising Law Institute, 1974), 1115–18.

30. Stephen E. Weil, "Introduction," in *A Deaccession Reader*, ed. Stephen E. Weil (Washington, DC: American Association of Museums, 1997), 9.

31. National Park Service Museum Management Program, *Museum Handbook* (Washington, DC: National Park Service, 2000), 2:6–1, https://www.nps.gov/museum /publications/handbook.html.

32. Smithsonian Institution Office of Policy and Analysis, *Concern at the Core: Managing Smithsonian Collections* (Washington, DC: Smithsonian Institution, 2005), 186; M. Melissa Wolfe, "Integrating Disruption: Acquiring the Philip J. and Suzanne Schiller Collection of American Social Commentary Art, 1930–1970," *Panorama: Journal of the Association of Historians of American Art* 2, no. 2 (Fall 2016), http://journalpanorama .org/m-melissa-wolfe-curator-of-american-art-saint-louis-art-museum/.

33. Lee Rosenbaum, "My Article on Minneapolis in Today's WSJ—Part I," *Lee Rosenbaum's CultureGrrl* (blog), July 18, 2006, http://culturegrrl.blogspot.com/2006 /07/my-article-on-minneapolis-in-todays_18.html.

34. "Indianapolis Museum of Art Launches Searchable Database of Deaccessioned Artworks," March 16, 2009, https://www.imamuseum.org/sites/default/files/Deac cessions%20database%20release-FINAL.pdf; and Indianapolis Museum of Art, "Deaccession Policy," July 6, 2015, http://www.imamuseum.org/sites/default/files/Final _IMA_Deaccession_policy.pdf.

35. "Deaccessioned Artworks," *Indianapolis Museum of Art*, accessed August 11, 2016, http://collection.imamuseum.org/browse/deaccessioned-artworks/?l=20.

36. American Alliance of Museums, "Direct Care of Collections: Ethics, Guidelines and Recommendations," April 2016, http://www.aam-us.org/docs/default-source /default-document-library/direct-care-of-collections-ethics-guidelines-and-recommen dations-pdf.pdf; American Alliance of Museums, "AAM Statement on Delaware Art Museum," March 27, 2014, http://www.aam-us.org/about-us/media-room/2014/delaware -art-museum.

37. Michael O'Hare, "Capitalizing Art Museum Collections: Awkward for Museums but Good for Art and for Society" (working paper, Goldman School of Public Policy, GSPP08–005, November 2005), http://papers.ssrn.com/abstract=1262403. See also Michael O'Hare, "Museums Can Change—Will They?," *Democracy Journal*, no. 16 (Spring 2015), http://democracyjournal.org/magazine/36/museums-can-change will-they/.

Chapter 9: Objects, Stories, and Visitors

1. *Annual Report of the President to the Corporation of Brown University* (Providence, RI: Angell, Hammett, 1880), 36.

2. *Annual Report of the President to the Corporation of Brown University* (Providence, RI: Snow & Farnham, 1893), 42.

3. Anthony McCabe, "The College Buildings in Other Days," in *Memories of Brown; Traditions and Recollections Gathered from Many Sources,* ed. Robert Perkins Brown, Henry Robinson Palmer, Harry Lyman Koopman, and Clarence S. Brigham (Providence, RI: Brown Alumni Magazine, 1909), 332–33.

4. Dallas Lore Sharp, "A Bed in the Museum," *Atlantic Monthly,* June 1935, 707.

5. Igor Stravinsky, *Poetics of Music in the Form of Six Lessons,* trans. Arthur Knodel and Ingolf Dahl (Cambridge, MA: Harvard University Press, 1947), 25–26.

6. James Clifford, *Routes: Travel and Translation in the Late Twentieth Century* (Cambridge, MA: Harvard University Press, 1997), 188–219.

7. Cathy Aldridge, "Exhibit on Everyone's Mind," *New York Amsterdam News,* February 1, 1969, 38, quoted in Bridget R. Cooks, *Exhibiting Blackness: African Americans and the American Art Museum* (Amherst: University of Massachusetts Press, 2011), 64.

8. Daisy Nguyen, "Vietnam War Exhibit Ignites Its Own Protest," *San Diego Union-Tribune,* September 3, 2004, http://legacy.sandiegouniontribune.com/uniontrib /20040903/news_1c3exhibit.html.

9. See Steven C. Dubin, *Displays of Power: Memory and Amnesia in the American Museum* (New York: New York University Press, 1999), chap. 6; and the essays in "History and the Public: What Can We Handle? A Round Table about History after the *Enola Gay* Controversy," *Journal of American History* 82, no. 3 (December 1995): 1029–1144.

10. Holland Cotter, "Review: New Whitney Museum's First Show, 'America Is Hard to See,'" *New York Times,* April 23, 2015, http://www.nytimes.com/2015/04/24 /arts/design/review-new-whitney-museums-first-show-america-is-hard-to-see.html.

11. Margaret A. Lindauer, "From Salad Bars to Vivid Stories: Four Game Plans for Developing 'Educationally Successful' Exhibitions," *Museum Management and Curatorship* 20, no. 1 (March 2005): 41–55.

12. American Alliance of Museums, "Museum Facts," accessed August 29, 2016, http://www.aam-us.org/about-museums/museum-facts; Betty Farrell and Maria Medvedeva, "Demographic Transformation and the Future of Museums" (Washington, DC: American Association of Museums Press, 2010), 12, http://www.aam-us.org/docs /center-for-the-future-of-museums/demotransaam2010.pdf.

13. Andrew J. Pekarik et al., "IPOP: A Theory of Experience Preference," *Curator: The Museum Journal* 57, no. 1 (2014): 5–27.

14. Mihaly Csikszentmihalyi and Kim Hermanson, "Intrinsic Motivation in Museums: Why Does One Want to Learn?," in *Public Institutions for Personal Learning: Establishing a Research Agenda,* ed. John H. Falk and Lynn D. Dierking (Washington, DC: American Association of Museums, 1995), 67–75.

15. Richard Rabinowitz, "Eavesdropping at the Well: Interpretive Media in the 'Slavery in New York' Exhibition," *Public Historian* 35, no. 3 (August 2013): 23.

16. Richard Kohen and Kim Sikoryak, *Interp Guide: The Philosophy and Practice of Connecting People to Heritage,* 3rd ed. (Lakewood, CO: U.S. Department of the Interior, National Park Service, 2005), 9, http://tot.unm.edu/documents/InterpGuide-XP.pdf.

17. Freeman Tilden, *Interpreting Our Heritage: Principles and Practices for Visitor Services in Parks, Museums, and Historic Places* (Chapel Hill: University of North Carolina Press, 1957), 9.

18. Jay Rounds, "Strategies for the Curiosity-Driven Museum Visitor," *Curator: The Museum Journal* 47, no. 4 (October 2004): 389–412.

19. Henry James, *The American* (Boston: Houghton Mifflin, 1877), 5–6.

20. Charlotte Brontë, *Villette* (Edinburgh: J. Grant, 1924), 1:331.

21. Benjamin Ives Gilman, *Museum Ideals of Purpose and Method* (Cambridge, MA: Museum of Fine Arts, Boston, 1918), 251–69; See also Gareth Davey, "What Is Museum Fatigue?," *Visitor Studies Today* 8, no. 3 (2005): 17–21.

22. Nelson Goodman, "The End of the Museum?," *Journal of Aesthetic Education* 19, no. 2 (1985): 57–58.

23. Smithsonian Institution Office of Policy and Analysis, "Art Museums and the Public" (Washington, DC: Smithsonian Institution, 2001), 6, https://www.si.edu /Content/opanda/docs/Rpts2001/01.10.ArtPublic.Final.pdf.

24. Henry H. Higgins, "Museums of Natural History," *Literary and Philosophical Society of Liverpool* 38 (1884): 183–87.

25. See George E. Hein, *Learning in the Museum* (London: Routledge, 1998), 103–104.

26. Zahava D. Doering, "Strangers, Guests, or Clients? Visitor Experiences in Museums," *Curator: The Museum Journal* 42, no. 2 (April 1999): 74–87.

27. Beverly Serrell, *Judging Exhibitions: A Framework for Assessing Excellence* (Walnut Creek, CA: Left Coast, 2006).

28. David Hofstede, *Audrey Hepburn: A Bio-Bibliography* (Westport, CT: Greenwood, 1994), 30.

29. John H. Falk, *Identity and the Museum Visitor Experience* (Walnut Creek, CA: Left Coast, 2009), 133; Paulette M. McManus, "Memories as Indicators of the Impact of Museum Visits," *Museum Management and Curatorship* 12, no. 4 (December 1993): 367–80.

30. Higgins, "Museums of Natural History," 187.

Chapter 10: Objects on Display

1. Roy Rosenzweig and David Thelen, *The Presence of the Past: Popular Uses of History in American Life* (New York: Columbia University Press, 1998), chap. 4; Reach Advisors, "Museums and Trust," *Museums R+D Monthly Memo* 1, no. 8 (June 2015), 3; BBVA Foundation Department of Social Studies and Public Opinion, "BBVA Foundation

International Study on Scientific Culture: General Attitudes to Science," 31, http://www.fbbva.es/TLFU/dat/culturacientificanotadeprensalarga.EN.pdf.

2. Thomas Thiemeyer, "Work, Specimen, Witness: How Different Perspectives on Museum Objects Alter the Way They Are Perceived and the Values Attributed to Them," *Museum and Society* 13, no. 3 (July 2015): 396–412.

3. Stephen Greenblatt, "Resonance and Wonder," in *Exhibiting Cultures: The Poetics and Politics of Museum Display,* ed. Ivan Karp and Steven Lavine (Washington, DC: Smithsonian Institution Press, 1991), 42.

4. Benjamin Ives Gilman, *Museum Ideals of Purpose and Method* (Cambridge, MA: Museum of Fine Arts, Boston, 1918), 81.

5. Greenblatt, "Resonance and Wonder," 42.

6. Mark Twain, *The Adventures of Huckleberry Finn* (New York: Bantam Books, 1981), 247.

7. Margaret T. Jackson, "History and Traditions of Methods of Display," in *Proceedings of the American Association of Museums,* vol. 11 (Charleston, SC: The Association, 1917), 58.

8. George Ellis Burcaw, *Introduction to Museum Work,* 3rd ed. (Walnut Creek, CA: AltaMira, 1997), 193.

9. Rachel P. Maines and James J. Glynn, "Numinous Objects," *Public Historian* 15, no. 1 (January 1993): 10.

10. Julie Stein, "Natural History Museums," *Natural Histories Project,* 2016, accessed March 8, 2017, http://naturalhistoriesproject.org/conversations/natural-history-museums.

11. Monique P. Yazigi, "The Auction: How to Play the Game," *New York Times,* June 4, 1995, http://www.nytimes.com/1995/06/04/nyregion/the-auction-how-to-play-the-game.html.

12. Constance Classen, *The Deepest Sense: A Cultural History of Touch* (Urbana: University of Illinois Press, 2012), 138–40.

13. Maria Zytaruk, "America's First Circulating Museum: The Object Collection of the Library Company of Philadelphia," *Museum History Journal* 10, no. 1 (January 2017): 68–82.

14. Classen, *The Deepest Sense,* 144.

15. Kenneth Hudson, *Museums of Influence* (Cambridge: Cambridge University Press, 1987), 30.

16. [Anna] Jameson, *Companion to the Most Celebrated Private Galleries of Art in London* (London: Saunders and Otley, 1844), xxxiv.

17. Classen, *The Deepest Sense,* 146.

18. John Whipple Potter Jenks to Prof. Ward, October 1874, Ward Natural Science Establishment Collection, Department of Rare Books and Special Collections, Rush Rhees Library, University of Rochester.

19. J. Earle Clauson, "These Plantations," *Providence Evening Bulletin*, March 30, 1934.

20. James Dwight Dana, *Science and Scientific Schools. An Address before the Alumni of Yale College, at the Commencement Anniversary, August, 1856* (New Haven, CT: S. Babcock, 1856), 27.

21. G. Brown Goode, *The Museums of the Future* (Washington, DC: Government Printing Office, 1891), 427.

22. Ken Arnold and Thomas Söderqvist, "Medical Instruments in Museums: Immediate Impressions and Historical Meanings," *Isis* 102, no. 4 (2011): 718–29; Samuel J. M. M. Alberti, "The Museum Affect: Visiting Collections of Anatomy and Natural History," in *Science in the Marketplace: Nineteenth-Century Sites and Experiences*, ed. Bernard V. Lightman and Aileen Fyfe (Chicago: University of Chicago Press, 2007): 371–403.

23. Classen, *The Deepest Sense*, 177; Fiona Candlin, "Museums, Modernity and the Class Politics of Touching Objects," in *Touch in Museums: Policy and Practice in Object Handling*, ed. Helen J. Chatterjee (Oxford: Berg, 2008): 9–20.

24. Constance Classen, *The Book of Touch* (Oxford: Berg, 2005), 282.

25. Susan Stewart, "Prologue: From the Museum of Touch," in *Material Memories: Design and Evocation*, ed. Marius Kwint, Christopher Breward, and Jeremy Aynsley (Oxford: Berg, 1999), 17–37.

26. Donald Preziosi, "Brain of the Earth's Body: Museums and the Framing of Modernity," in *The Rhetoric of the Frame: Essays on the Boundaries of the Artwork*, ed. Paul Duro (Cambridge: Cambridge University Press, 1996), 96–110.

27. Kathleen Curran, *The Invention of the American Art Museum: From Craft to Kulturgeschichte, 1870–1930* (Los Angeles: Getty Research Institute, 2016), 57.

28. Gilman, *Museum Ideals of Purpose and Method*, 238–48.

29. Andrew McClellan, *The Art Museum from Boullée to Bilbao* (Berkeley: University of California Press, 2008), 72.

30. Kate M. Hall, "The Smallest Museum," *Museums Journal* 1–2 (August 1901): 42, quoted in Alison Griffiths, *Shivers Down Your Spine: Cinema, Museums, and the Immersive View* (New York: Columbia University Press, 2012), 179–80.

31. Carolyn Ethel Macdonald, "Museum Resources for Teachers" (EdD dissertation, Teachers College, Columbia University, 1944).

32. Gilman, *Museum Ideals of Purpose and Method*, 393.

33. Anna Billings Gallup, "The Children's Museum as an Educator," *Popular Science Monthly* 72 (April 1908): 371–79.

34. Susan Stewart, "Prologue: From the Museum of Touch," 28–30.

35. Ken Arnold, *Cabinets for the Curious: Looking Back at Early English Museums* (Burlington, VT: Ashgate, 2006), 100–101.

36. Gilman, *Museum Ideals of Purpose and Method*, 268.

37. Edmund de Waal, *The Hare with Amber Eyes: A Hidden Inheritance* (New York: Picador, 2011), 61–66.

38. E. H. Gombrich, *The Story of Art*, 16th ed., rev. (London: Phaidon, 1995), 156–57.

39. Barbara E. Savedoff, "Frames," *The Journal of Aesthetics and Art Criticism* 57, no. 3 (July 1999): 350.

40. David Phillips, *Exhibiting Authenticity* (Manchester: Manchester University Press, 1997), 206.

Chapter 11: Organizations and Juxtapositions

1. G. Brown Goode, *The Museums of the Future* (Washington, DC: Government Printing Office, 1891), 433.

2. Philip Fisher, *Making and Effacing Art: Modern American Art in a Culture of Museums* (Oxford: Oxford University Press, 1991), 8.

3. Ivan Gaskell, *Vermeer's Wager: Speculations on Art History, Theory and Art Museums* (London: Reaktion Books, 2000), 86.

4. For an overview of this literature, see Randolph Starn, "A Historian's Brief Guide to New Museum Studies," *American Historical Review* 110, no. 1 (2005): 73–75.

5. See Gary Kulik, "Designing the Past: History-Museum Exhibitions from Peale to the Present," in *History Museums in the United States: A Critical Assessment*, ed. Warren Leon and Roy Rosenzweig (Urbana: University of Illinois Press, 1989), 2–37.

6. Barbara Maria Stafford, *Artful Science: Enlightenment Entertainment and the Eclipse of Visual Education* (Cambridge, MA: MIT Press, 1996), 238.

7. Andrew McClellan, *The Art Museum from Boullée to Bilbao* (Berkeley: University of California Press, 2008), 119–20.

8. Margaret T. Jackson, "History and Traditions of Methods of Display," *Proceedings of the American Association of Museums*, vol. 11 (Charleston, SC: The Association, 1917), 55.

9. Jonathan Crary, "Unbinding Vision," *October* 68 (1994): 23.

10. David Jenkins, "Object Lessons and Ethnographic Displays: Museum Exhibitions and the Making of American Anthropology," *Comparative Studies in Society and History* 36, no. 2 (April 1994): 242–43; Carlo Ginzburg, "Clues: Roots of an Evidential Paradigm," in *Clues, Myths, and the Historical Method*, trans. John Tedeschi and Anne C. Tedeschi (Baltimore: Johns Hopkins University Press, 2013), 102–108.

11. John E. Taylor, "Collecting and Preserving," *Hardwicke's Science-Gossip*, 1873, 25.

12. Quoted in G. Brown Goode, *The Principles of Museum Administration* (York, UK: Coultas & Volans, 1895), 27.

13. Jenkins, "Object Lessons and Ethnographic Displays," 269.

14. McClellan, *The Art Museum*, 121–22; quote from Casmir Varon, 1794.

15. Ibid., 120.

16. See Julia Noordegraaf, *Strategies of Display: Museum Presentation in Nineteenth- and Twentieth-Century Visual Culture* (Rotterdam: Museum Boijmans Van Beuningen, 2004), 65–66.

17. Benjamin Ives Gilman, *Museum Ideals of Purpose and Method* (Cambridge, MA: Museum of Fine Arts, Boston, 1918), 65.

18. John Cotton Dana, *The New Relations of Museums and Industries* (Newark, NJ: Newark Museum Association, 1919), 17–20.

19. Noordegraaf, *Strategies of Display*, 91.

20. Robert E. Fry, *An Outline of the Aims and Ideals Governing the Department of Paintings* (New York: Metropolitan Museum of Art, 1906), 6–8.

21. Brian O'Doherty, *Inside the White Cube: The Ideology of the Gallery Space*, exp. ed. (Berkeley: University of California Press, 1999), 14–15.

22. Umberto Eco, *Foucault's Pendulum*, trans. William Weaver (San Diego, CA: Harcourt Brace Jovanovich, 1989), 8.

23. Jennifer Wild Czajkowski, "Changing the Rules: Making Space for Interactive Learning in the Galleries of the Detroit Institute of Arts," *Journal of Museum Education* 36, no. 2 (2011): 171–78.

24. Derek Allan, "André Malraux, the Art Museum, and the Digital *Musée Imaginaire*" (paper presented at Imagining Identity symposium, National Portrait Gallery, Canberra, Australia, July 2010), http://home.netspeed.com.au/derek.allan/musee %20imaginaire.htm.

25. Richard Saul Wurman, *Information Anxiety* (New York: Doubleday, 1989).

26. G. Brown Goode, *Museum-History and Museums of History* (New York: Knickerbocker, 1889), 268. Every curator with one object and two stories to tell has had the same dilemma, and it was not just Goode who thought of wheels as a solution. Augustus Pitt-Rivers had the same idea: he acknowledged that his museum "would require constant rearrangement, as knowledge increased, and the cases might, perhaps, be put on wheels to facilitate their readjustment." Augustus Pitt-Rivers, "Typological Museums, as Exemplified by the Pitt-Rivers Museum at Oxford, and His Provincial Museum at Farnham, Dorset," *Journal of the Society of Arts* 40, no. 2039 (December 18, 1891): 117.

27. Susan Stewart, *On Longing: Narratives of the Miniature, the Gigantic, the Souvenir, the Collection* (Durham, NC: Duke University Press, 1993), xi. See also Steven Lubar, "Timelines in Exhibitions," *Curator: The Museum Journal* 56, no. 2 (2013): 169–88.

28. Cyrus Adler, Immanuel Moses Casanowicz, Walter Hough, et al., *Annual Report of the Board of Regents of the Smithsonian Institution: Report of the U.S. National Museum* (Washington, DC: Government Printing Office, 1900), 26.

29. Tracy Teslow, *Constructing Race: The Science of Bodies and Cultures in American Anthropology* (Cambridge: Cambridge University Press, 2014), 89.

30. Samuel Cauman, *The Living Museum: Experiences of an Art Historian and Museum Director: Alexander Dorner* (New York: New York University Press, 1958), 206.

31. Lynn Allyn Young, *Beautiful Dreamer: The Completed Works and Unfulfilled Plans of Sculptor Lorado Taft* (Chicago, IL: Artistic License Limited, 2012).

32. Mary Louise Hinckley, "Modern Biology Descends from the 'Cat Course' of 1896," *Providence Evening Bulletin*, March 2, 1937.

33. Quoted in Karen A. Rader and Victoria E. M. Cain, *Life on Display: Revolutionizing U.S. Museums of Science and Natural History in the Twentieth Century* (Chicago: University of Chicago Press, 2014), 41.

34. Edward P. Alexander, "Functional History in the Small Museum," *Museum News*, October 13, 1936.

35. Holland Cotter, "Making Museums Moral Again," *New York Times*, March 17, 2016, http://www.nytimes.com/2016/03/17/arts/design/making-museums-moral-again.html.

36. James Panero, "The Museum of the Present," *New Criterion*, December 2016, http://www.newcriterion.com/articles.cfm/The-museum-of-the-present-8547.

37. Judy Rand, "The 227-Mile Museum, or a Visitors' Bill of Rights," *Curator: The Museum Journal* 44, no. 1 (January 2001): 12–13.

38. Marty Sklar, "Mickey's 10 Commandments" (presentation to American Association of Museums Annual Meeting, San Francisco, June 1987), http://www.themedattraction.com/mickeys10commandments.htm.

39. Erich Zuern, "Financial Planning," in *Manual of Museum Exhibitions*, 2nd ed., ed. Barry Lord and Maria Piacente (Lanham, MD: Rowman and Littlefield, 2014), 373–78.

Chapter 12: Explanations and Encounters

1. Barbara Maria Stafford, *Artful Science: Enlightenment Entertainment and the Eclipse of Visual Education* (Cambridge, MA: MIT Press, 1996), 238, 251; Tony Bennett, "Pedagogic Objects, Clean Eyes, and Popular Instruction: On Sensory Regimes and Museum Didactics," *Configurations* 6, no. 3 (1998): 348–50.

2. Bennett, "Pedagogic Objects," 351.

3. Ibid., 359. See also Joel J. Orosz, *Curators and Culture: The Museum Movement in America, 1740–1870* (Tuscaloosa: University of Alabama Press, 1990); and Neil Harris, *Humbug: The Art of P. T. Barnum*, Phoenix ed. (Chicago: University of Chicago Press, 1981), chap. 2 and 3. Quotation from William Maclure, "An Epitome of the Improved Pestalozzian System of Education," *American Journal of Science and Arts* 10 (1826): 145.

4. Barnum's American Museum, *Catalogue or Guide Book of Barnum's American Museum* (New York: Printed and published for the proprietor, n.d.), 13, 75, 21.

5. W. H. Fowler, "Inaugural Address," *Nature* 40 (September 1889): 467.

6. G. Brown Goode, *The Principles of Museum Administration* (York, UK: Coultas & Volans, 1895), 40.

7. Bennett, "Pedagogic Objects," 361.

8. Julia Noordegraaf, *Strategies of Display: Museum Presentation in Nineteenth- and Twentieth-Century Visual Culture* (Rotterdam: Museum Boijmans Van Beuningen, 2004), 55.

9. Quoted in Brandon Taylor, *Art for the Nation: Exhibitions and the London Public, 1747–2001* (Manchester: Manchester University Press, 1999), 47.

10. Benjamin Ives Gilman, *Museum Ideals of Purpose and Method* (Cambridge, MA: Museum of Fine Arts, Boston, 1918), 431. For similar statements from curators at other art museums, see Andrew McClellan, *The Art Museum from Boullée to Bilbao* (Berkeley: University of California Press, 2008), 176.

11. "Report of a Survey of Visitors to the Museum in Its First Year," *Philadelphia Museum Bulletin* 25, no. 130 (December 1929): 11.

12. Gilman, *Museum Ideals of Purpose and Method*, 319–23; Benjamin Ives Gilman, "Staff Docket, Item 33," (Typescript, Boston, January 14, 1916), TS, Museum of Fine Arts Boston, Archives.

13. Alfred H. Barr Jr., introduction to *Modern Works of Art; Fifth Anniversary Exhibition, November 20, 1934–January 20, 1935* (New York: Museum of Modern Art, 1934).

14. Menil Foundation, *The Menil Collection: A Selection from the Paleolithic to the Modern Era* (New York: Harry N. Abrams, 1987), 7; Gail Gregg, "Your Labels Make Me Feel Stupid," *ArtNews*, Summer 2010, http://www.artnews.com/2010/07/01/your-labels-make-me-feel-stupid/.

15. Gilman, "Staff Docket, Item 33."

16. Quoted in Karen A. Rader and Victoria E. M. Cain, *Life on Display: Revolutionizing U.S. Museums of Science and Natural History in the Twentieth Century* (Chicago: University of Chicago Press, 2014), 37.

17. Joseph Grinnell, "The Museum Conscience," *Museum Work: Including the Proceedings of the American Association of Museums* 4, no. 2 (October 1921): 62.

18. Margaret Mead, *Anthropologists and What They Do* (New York: F. Watts, 1965), 56.

19. Frederic A. Lucas, "Museum Labels and Labelling," *Proceedings of the American Association of Museums* 5 (1911): 96.

20. Albert Bigelow Paine, "Children's Room in Smithsonian Institution," *Annual Report of the Board of Regents of the Smithsonian Institution*, 1902, 560.

21. See Rader and Cain, *Life on Display*, 34–47.

22. B. D. Sorsby and S. D. Horne, "The Readability of Museum Labels," *Museums Journal* 80, no. 3 (1980): 158, quoted in Dana Fragomeni, "The Evolution of Exhibit Labels," *Faculty of Information Quarterly* 2, no. 1 (March 15, 2010): 5, http://fiq.ischool.utoronto.ca/index.php/fiq/article/download/15408/12436.

23. Beverly Serrell, *Exhibit Labels: An Interpretive Approach* (Walnut Creek, CA: AltaMira, 1996), 233–34.

24. R. Mileham, "Wise words," *Museums Journal* 106, no. 11 (2006), 18–19, quoted in Fragomeni, "The Evolution of Exhibit Labels," 5.

25. Sara Bodinson, Stephanie Pau, and Jackie Armstrong, "Writing Effective Interpretive Labels for Art Exhibitions: A Nuts and Bolts Primer" (presentation at New York City Museum Education Roundtable 2012 Annual Conference, New York, May 2012), http://www.slideshare.net/stephaniepau/writing-effective-interpretive-labels-for-art-exhibitions-a-nuts-and-bolts-primer. Other excellent guides for label writing include Serrell, *Exhibit Labels* and "Gallery Text at the V&A: A Ten Point Guide" (London: Victoria and Albert Museum, 2013), http://www.vam.ac.uk/__data/assets/pdf_file/0009/238077/Gallery-Text-at-the-V-and-A-Ten-Point-Guide-Aug-2013.pdf.

26. Janice Majewski, "Smithsonian Guidelines for Accessible Exhibition Design" (Washington, DC: Smithsonian Accessibility Program, 1996), 17–26.

27. Peter Liebhold and Harry R. Rubenstein, "Bringing Sweatshops into the Museum," in *Sweatshop USA: The American Sweatshop in Historical and Global Perspective*, ed. Daniel E. Bender and Richard A. Greenwald (New York: Routledge, 2003), 57–74.

28. Dr. Ant. Fritsch, "The Museum Question in Europe and America," *Museums Journal* 3, no. 8 (February 1904): 252, quoted in Alison Griffiths, *Shivers Down Your Spine: Cinema, Museums, and the Immersive View* (New York: Columbia University Press, 2012), 235.

29. "A Museum Comes to Life," *Popular Mechanics*, April 1931, 572.

30. Robert Beverly Hale et al., "Reports of the Departments," *Metropolitan Museum of Art Bulletin* 23, no. 2 (1964): 63.

31. John C. Ewers, "Problems and Procedures in Modernizing Ethnological Exhibits," *American Anthropologist* 57, no. 1 (1955): 1–12.

32. Ibid.

33. George Francis Dow, comment on George Francis Kunz's "To Increase the Number of Visitors to Our Museums," *Proceedings of the American Association of Museums* 5 (1911): 90.

34. Waldemar Kaempffert, *From Cave-Man to Engineer: The Museum of Science and Industry* (Chicago: The Museum of Science and Industry, 1933), 4.

35. Rader and Cain, *Life on Display*, 101–107.

36. Kaempffert, *From Cave-Man to Engineer*, 12; "A Museum Comes to Life," 572.

37. Sterling Gleason, "Push-Button Museum Shows Secrets of Science," *Popular Science*, April 1936, 14.

38. "The Museum's First Interactive Exhibit," display in "Museum of Science: Then and Now," Boston Museum of Science, visited December 2016.

39. Edward P. Alexander, "A Fourth Dimension for History Museums," *Curator: The Museum Journal* 11, no. 4 (December 1968): 274.

40. F. A. Bather, "The Museums of New York State," *Museum Journal* 1, no. 3 (1901), 73, quoted in Griffiths, *Shivers Down Your Spine*, 235.

41. Curt Germundson, "Alexander Dorner's Atmosphere Room: The Museum as Experience," *Visual Resources* 21, no. 3 (September 2005): 270.

42. Quoted in Griffiths, *Shivers Down Your Spine*, 169, 171.

43. See ibid., 184, 274–76.

44. George E. Hein, *Progressive Museum Practice: John Dewey and Democracy* (Walnut Creek, CA: Left Coast, 2012), 189–90.

45. Griffiths, *Shivers Down Your Spine,* 190; Lisa Roberts, *From Knowledge to Narrative: Educators and the Changing Museum* (Washington, DC: Smithsonian Institution, 1997), 86.

46. Sandra H. Dudley, "What's in the Drawer? Surprise and Proprioceptivity in the Pitt Rivers Museum," *Senses and Society* 9, no. 3 (November 2014): 296–309. Quote on p. 298.

47. Alexander, "A Fourth Dimension for History Museums," 286.

48. "Brass Candlestick Making," The Henry Ford, accessed March 8, 2017, https://www.thehenryford.org/current-events/calendar/brass-candlestick-making.

49. "To Fire a Flintlock Musket," Colonial Williamsburg.com, accessed March 17, 2016, https://www.colonialwilliamsburg.com/plan/calendar/fire-flintlock-musket /?eventId=15659.

50. "Process Lab," CooperHewitt.org, October 27, 2014, https://www.cooperhewitt .org/events/current-exhibitions/process-lab/.

51. Quoted in Jennifer Czajkowski, "Changing the Rules: Making Space for Interactive Learning in the Galleries of the Detroit Institute of Arts," *Journal of Museum Education* 36, no. 2 (Summer 2011): 176.

Chapter 13: Setting the Scene

1. Barbara Kirshenblatt-Gimblett, *Destination Culture: Tourism, Museums, and Heritage* (Berkeley: University of California Press, 1998), chap. 2; Raymond Corbey, "Ethnographic Showcases, 1870–1930," *Cultural Anthropology* 8, no. 3 (August 1993): 338–69.

2. Frank Baker, "Some Instructive Methods of Bird Installation," *Proceedings of the American Association of Museums* 1 (1907): 55.

3. Frederick Starr, "Anthropology at the World's Fair," *Popular Science Monthly,* 43 (September 1893), 615–16, quoted in David Jenkins, "Object Lessons and Ethnographic Displays: Museum Exhibitions and the Making of American Anthropology," *Comparative Studies in Society and History* 36, no. 2 (April 1994): 246.

4. Quoted in Marjorie Schwarzer and Mary Jo Sutton, "The Diorama Dilemma: A Literature Review and Analysis," November 25, 2009, 10, https://www.academia.edu /6727013/The_Diorama_Dilemma_A_Literature_review_and_Analysis_MS_and _MJS_The_Diorama_Dilemma_A_Literature_Review_and_Analysis.

5. Quoted in Karen A. Rader and Victoria E. M. Cain, *Life on Display: Revolutionizing U.S. Museums of Science and Natural History in the Twentieth Century* (Chicago: University of Chicago Press, 2014), 45.

6. See Corbey, "Ethnographic Showcases"; and Robert W. Rydell, *All the World's a Fair: Visions of Empire at American International Expositions, 1876–1916* (Chicago: University of Chicago Press, 1987).

7. Margaret Mead, *Anthropologists and What They Do* (New York: F. Watts, 1965), 11–12.

8. Susan Harding and Emily Martin, "Anthropology Now and Then in the American Museum of Natural History," *Anthropology Now,* December 13, 2016, http:// anthronow.com/print/anthropology-now-and-then-in-the-american-museum-of -natural-history.

9. Pauline Wakeham, *Taxidermic Signs: Reconstructing Aboriginality* (Minneapolis: University of Minnesota Press, 2008), 4.

10. *Anti-Columbus Day Tour; Decolonize This Museum* (brochure, October 2016), http://decolonizethisplace.org/brochures/161010_AMNH%20map%20final.pdf; "Rename The Day—Remove The Statue—Decolonize This Museum," *Decolonize This Place* (blog), October 10, 2016, http://www.decolonizethisplace.org/post/rename-the -day-remove-the-statue-decolonize-this-museum.

11. Based on Mary Jo Arnoldi, "From the Diorama to the Dialogic: A Century of Exhibiting Africa at the Smithsonian's Museum of Natural History," *Cahiers d'études africaines* 39, no. 155 (1999): 701–26.

12. Ibid., 711–16.

13. Ibid., 722.

14. "Pequot Village," Mashantucket Pequot Museum and Research Center, accessed March 17, 2016, http://www.pequotmuseum.org/PermanentExhibits.aspx.

15. Patricia West, *Domesticating History: The Political Origins of America's House Museums* (Washington, DC: Smithsonian Books, 1999), chap. 1.

16. Quoted in Dianne H. Pilgrim, "Inherited from the Past: The American Period Room," *American Art Journal* 10, no. 1 (May 1, 1978): 7–8.

17. See Mark B. Sandberg, *Living Pictures, Missing Persons: Mannequins, Museums, and Modernity* (Princeton, NJ: Princeton University Press, 2003).

18. Quoted in Pilgrim, "Inherited from the Past," 7–8.

19. Quoted in Jay Anderson, "Living History: Simulating Everyday Life in Living Museums," *American Quarterly* 34, no. 3 (1982): 293.

20. Quoted in Steven Conn, *Museums and American Intellectual Life, 1876–1926* (Chicago: University of Chicago Press, 1998), 156.

21. Quoted in William A. Griswold and Donald W. Linebaugh, eds., *Saugus Iron Works: The Roland W. Robbins Excavations* (Washington, DC: National Park Service, 2010), 362.

22. "Early American Hall Opens," *Smithsonian Torch*, February 1957, 2.

23. Malcom Watkins, Interview 3, 26 February 1992, transcript, p. 32, Smithsonian Institution Archives Record Unit 9586. Quoted in Michele Gates Moresi, "Exhibiting Race, Creating Nation: Representations of Black History and Culture at the Smithsonian Institution, 1895–1976" (PhD dissertation, George Washington University, Washington, DC, 2003), 108. See also William S. Walker, *A Living Exhibition: The Smithsonian and the Transformation of the Universal Museum* (Amherst: University of Massachusetts Press, 2013), chap. 2.

24. Edward P. Alexander, "A Fourth Dimension for History Museums," *Curator: The Museum Journal* 11, no. 4 (December 1968): 279–80.

25. Spencer R. Crew and James E. Sims, "Locating Authenticity: Fragments of a Dialogue," in *Exhibiting Cultures: The Poetics and Politics of Museum Display*, ed. Ivan Karp and Steven D. Lavine (Washington DC: Smithsonian Institution Press, 1991), 159–75.

26. Quoted in Robert A. Gross, review of "After the Revolution: The Smithsonian History of Everyday Life in the Eighteenth Century," *Journal of American History* 76, no. 3 (December 1989): 860.

27. Crew and Sims, "Locating Authenticity," 162, 171.

28. Elena Tajima Creef, *Imaging Japanese America: The Visual Construction of Citizenship, Nation, and the Body* (New York: New York University Press, 2004), chap. 4; and Jennifer Locke Jones, Chair and Curator, Division of Armed Forces History, National Museum of American History, in personal communication with author, March 19, 2016.

29. Leon Wieseltier, "After Memory: Reflections on the Holocaust Memorial Museum," *New Republic*, May 3, 1993, 19.

30. Alison Landsberg, *Prosthetic Memory: The Transformation of American Remembrance in the Age of Mass Culture* (New York: Columbia University Press, 2004), 131–33.

31. Jonathan Blitzer, "Antique Shopping with the Tenement Museum's Curator of Furnishings," *New Yorker*, December 19, 2016, http://www.newyorker.com/magazine /2016/12/19/antique-shopping-with-the-tenement-museums-curator-of-furnishings.

32. Benjamin Filene, "Make Yourself at Home—Welcoming Voices in *Open House: If These Walls Could Talk*," in *Letting Go? Sharing Historical Authority in a User-Generated World*, ed. Bill Adair, Benjamin Filene, and Laura Koloski (Philadelphia: Pew Center for Arts & Heritage, 2011), 138–55.

33. Steven Lubar, "The Making of 'America on the Move' at the National Museum of American History," *Curator: The Museum Journal* 47, no. 1 (January 2004): 19–51.

34. "You Are There," Indiana Historical Society, accessed August 18, 2016, http://www.indianahistory.org/indiana-experience/you-are-there.

35. Quoted in Anderson, "Living History," 297. See also Jeremy Dupertuis Bangs, "The Hypothetical Nature of Architecture in Plimoth Plantation's Pilgrim Village," *Museum History Journal* 8, no. 2 (July 1, 2015): 130.

36. "Follow the North Star," Conner Prairie Interactive History Park, accessed August 18, 2016, http://www.connerprairie.org/Things-To-Do/Events/Follow-the-North-Star. See also Carl R. Weinberg, "The Discomfort Zone: Reenacting Slavery at Connor Prairie," *OAH Magazine of History* 23, no. 2 (April 2009), http://archive.oah.org/magazine-of-history/issues/232/connor_prairie.html; and Olivia Lewis, "Conner Prairie Slavery Re-Enactment Draws Criticism," *Indianapolis Star*, August 6, 2016, http://www.indystar.com/story/news/2016/08/06/conner-prairie-slavery-re-enactment-draws-criticism/82987036/.

37. Victoria Newhouse, *Art and the Power of Placement* (New York: Monacelli, 2005), 271.

38. M. S. Prichard, "Current Theories of the Arrangement of Museums of Art," in *Communications to the Trustees Regarding the New Building*, vol. 1 (Boston: Museum of Fine Arts, Boston, 1904), 18–22.

39. Julia Noordegraaf, *Strategies of Display: Museum Presentation in Nineteenth- and Twentieth-Century Visual Culture* (Rotterdam: Museum Boijmans Van Beuningen, 2004), 64–66.

40. M. S. Prichard, "Current Theories of the Arrangement of Museums of Art," in *Communications to the Trustees*, vol. 1, 20, 27.

41. Benjamin Ives Gilman to Morris Gray, June 14, 1916, Benjamin Ives Gilman papers, Museum of Fine Arts Boston, Archives.

42. Benjamin Ives Gilman, "On the Distinctive Purpose of Museums of Art," in *Communications to the Trustees*, vol. 1 (Boston: Museum of Fine Arts, 1904), 46. For more on the debate at the MFA, see Kathleen Curran, *The Invention of the American Art Museum: From Craft to Kulturgeschichte, 1870–1930* (Los Angeles: Getty Research Institute, 2016), 52–79.

43. R. T. Haines Halsey, *The Homes of Our Ancestors, as Shown in the American Wing of the Metropolitan Museum of Art of New York* (Garden City, NY: Garden City, 1937), xxii. See also Wendy Kaplan, "R. T. H. Halsey: An Ideology of Collecting American Decorative Arts," *Winterthur Portfolio* 17, no. 1 (April 1982): 43–53.

44. "History of the Museum: The Met Cloisters," Metropolitan Museum of Art, accessed August 18, 2016, http://www.metmuseum.org/about-the-met.

45. Sally Anne Duncan, "From Period Rooms to Public Trust: The Authority Debate and Art Museum Leadership in America," *Curator: The Museum Journal* 45, no. 2 (April 1, 2002): 98–99.

46. Robert W. de Forest, *Museum Policies and Scope* (New York: Metropolitan Museum of Art, 1930), 14. More generally, see Michelle Henning, *Museums, Media and Cultural Theory* (Maidenhead, UK: Open University Press, 2006), chap. 1 and 2.

47. "Report of a Survey of Visitors to the Museum in Its First Year," *Philadelphia Museum Bulletin* 25, no. 130 (December 1929): 3–11.

48. Quoted in Neil Harris, "Museums, Merchandising, and Popular Taste: The Struggle for Influences," in *Cultural Excursions: Marketing Appetites and Cultural Tastes in Modern America* (Chicago: University of Chicago Press, 1990), 65.

49. Lee Simonson, "Skyscrapers for Art Museums," *American Mercury*, August 1927, 401, quoted in Neil Harris, "Museums, Merchandising, and Popular Taste," 75.

50. Fred Turner, *The Democratic Surround: Multimedia and American Liberalism from World War II to the Psychedelic Sixties* (Chicago: University of Chicago Press, 2013), 96–97.

51. Quoted in Sandra Karina Löschke, "Material Aesthetics and Agency: Alexander Dorner and the Stage-Managed Museum," *Interstices: Journal of Architecture and Related Arts* 14, no. 1 (January 2013): 28.

52. Curt Germundson, "Alexander Dorner's Atmosphere Room: The Museum as Experience," *Visual Resources* 21, no. 3 (September 2005): 263–73.

53. Elias Svedberg, "Museum Display," *Journal of the Royal Society of Arts* 97, no. 4804 (1949): 859.

54. Neil Harris, *Capital Culture: J. Carter Brown, the National Gallery of Art, and the Reinvention of the Museum Experience* (Chicago: University of Chicago Press, 2013), 371 and 389.

55. Harold Koda and Andrew Bolton, *Dangerous Liaisons: Fashion and Furniture in the Eighteenth Century* (New Haven, CT: Yale University Press, 2006), available at http://www.metmuseum.org/art/metpublications/Dangerous_Liaisons_Fashion_and _Furniture_in_the_Eighteenth_Century.

CHAPTER 14: TURNED INSIDE OUT

1. Amy Boyle, "Proposal for a Re-Imagination of the Haffenreffer Museum of Anthropology," (class presentation, "The Narrative Museum," Rhode Island School of Design, Providence, May 2011).

2. Quoted in Suzanne Keene, *Fragments of the World: Uses of Museum Collections* (Burlington, MA: Elsevier Butterworth-Heinemann, 2005), 116.

3. Michael M. Ames, "Visible Storage and Public Documentation," *Curator: The Museum Journal* 20, no. 1 (March 1977): 65–80.

4. Ibid., 65.

5. Celestine Bohlen, "Museums as Walk-In Closets; Visible Storage Opens Troves to the Public," *New York Times*, May 8, 2001, http://www.nytimes.com/2001/05/08/arts/museums-as-walk-in-closets-visible-storage-opens-troves-to-the-public.html.

6. Sally Yerkovich, review of the Henry Luce III Center for the Study of American Culture, *Public Historian* 24, no. 2 (Spring 2002): 127–31.

7. Karen Kedmey, "These New York Museums Let Visitors Go behind the Scenes, Exploring Their Brimming Storage Facilities," Artsy.com, February 12, 2016, https://www.artsy.net/article/artsy-editorial-new-york-museums-open-their-storage-to-the-public-putting-their-vast-collections-on-display.

8. Ibid.

9. Georgina Bath, "Visible Storage at the Smithsonian American Art Museum," in *The Manual of Museum Management*, 2nd ed., by Gail Dexter Lord and Barry Lord, (Lanham, MD: AltaMira, 2009), 180. See also Rachel M. Allen, "Making Collections Visible: The Luce Foundation Center for American Art," *American Art* 15, no. 1 (2001): 2–3.

10. Jori Finkel, "LACMA, Broad, Other Art Museums Work to Put Storage on Display," *Los Angeles Times*, July 20, 2013, http://articles.latimes.com/2013/jul/20/entertainment/la-et-cm-lacma-broad-museum-storage-20130721.

11. Bohlen, "Museums as Walk-In Closets."

12. Ibid.

13. Carrie Rebora, "Curators' Closet," in *Care of Collections*, ed. Simon J. Knell (London: Routledge, 1994), 223.

14. Smithsonian Institution Office of Policy and Analysis, *Donald W. Reynolds Center Visitor Survey; Study Highlights and Frequency Distributions* (Washington DC: Smithsonian Institution, 2010), http://www.si.edu/content/opanda/docs/Rpts2010/10.05.DWRCVisitor.Final.pdf.

15. Karen Kedmey, "These New York Museums."

16. Abigail Glogower and Margaret Olin, "Between Two Worlds: Ghost Stories under Glass in Vienna and Chicago," in *Visualizing and Exhibiting Jewish Space and History*, ed. Richard I. Cohen (Oxford: Oxford University Press, 2012), 233–34.

17. Ibid., 227–29.

18. Edward Rothstein, "The Problem with Jewish Museums," *Mosaic*, February 2016, http://mosaicmagazine.com/essay/2016/02/the-problem-with-jewish-museums/.

19. Michael M. Ames, *Cannibal Tours and Glass Boxes: The Anthropology of Museums* (Vancouver: University of British Columbia Press, 1992), 95–96.

20. "Tales from the Crypt: Museum Storage and Meaning" (paper delivered at The Indian-European Advanced Research Network, London, October 2014), 3–4, https://enfilade18thc.files.wordpress.com/2015/03/iearn-museum-storage-workshop-2014-report-copy.pdf.

21. Jennifer Kramer, "Möbius Museology: Curating and Critiquing the Multiversity Galleries at the Museum of Anthropology at the University of British Columbia," in *The International Handbooks of Museum Studies*, vol. 4, *Museum Transformations*, ed. Annie E. Coombes and Ruth B. Phillips (Chichester, West Sussex: Wiley-Blackwell, 2015), 490.

22. Ibid., 490–91.

23. Shabnam Honarbakhsh, Heidi Swierenga, and Mauray Toutloff, "When You Don't Cry over Spilt Milk: Collections Access at the UBC Museum of Anthropology during the Renewal Project," *AIC Objects Specialty Group Postprints* 18 (2011): 67–82, http://resources.conservation-us.org/osg-postprints/postprints/v18/honar bakhsh/.

24. Winifred Kehl, "Turning the Museum Inside Out: Opening Collections, Engaging Audiences," *Museum,* September / October 2015, http://aam-us.org/docs /default-source/museum/burkemuseum.pdf.

25. Finkel, "LACMA, Broad."

26. Steven Lubar and Emily Stokes-Rees, "Getting Everyone to Think with Things: New Approaches to Teaching and Learning at the Haffenreffer Museum of Anthropology, Brown University," in *A Handbook for Academic Museums*, vol. 2, *Beyond Exhibitions and Education,* 2nd ed., ed. Stefanie S. Jandl and Mark S. Gold (Edinburgh: MuseumsEtc, 2015), 69–100.

Chapter 15: What Use Is a Museum?

1. *Annual Report of the President to the Corporation of Brown University* (Providence, RI: The Providence Press, 1892), 8.

2. David M. Robinson, "The Place of Archaeology in the Teaching of the Classics," *Classical Weekly* 10, no. 1 (1916): 2.

3. *Annual Report of the President to the Corporation of Brown University* (Providence, RI: Brown University, 1907), 24.

4. Rush Hawkins, "Announcement: To the Editor of the Providence Daily Journal," *Providence Daily Journal,* July 1, 1907, on display at the Annmary Brown Memorial. See also Rebecca Soules, " 'Nothing Must Be Changed': Rush Hawkins' Lost Memorial Museum," *Museum History Journal* 10, no. 1 (January 2017): 15–28.

5. Amos Perry, *The Library and Cabinet of the Rhode Island Historical Society: Their Origin and Leading Features Together with a Classified Summary of Their Contents* (Providence, RI: Printed for the Society by Snow & Farnham, 1892), 21–22.

6. Robert P. Emlen, "Picturing the Worthies: An Introduction to the Brown Portrait Collection," *Brown University Office of the Curator,* August 2003, http://library.brown .edu/cds/portraits/intro.html.

7. Steven Conn, *Museums and American Intellectual Life, 1876–1926* (Chicago: University of Chicago Press, 1998), 4.

8. J. Walter Wilson, "The Jenks Museum of Brown University," *Books at Brown* 22 (1968): 54–55.

9. Thomas Beck, *The Age of Frivolity: A Poem; Addressed to the Fashionable, the Busy, and the Religious World,* 3rd ed., rev. and enl. (London: Maxwell and Wilson, 1809), 80–81.

10. Samuel Quiccheberg, *The First Treatise on Museums: Samuel Quiccheberg's Inscriptiones, 1565,* trans. Mark A. Meadow and Bruce Robertson (Los Angeles: Getty Research Institute, 2013).

11. John Cotton Dana, *The New Museum* (Woodstock, VT: Elm Tree Press, 1917), 12, 36, 9. See also Carol Duncan, *A Matter of Class: John Cotton Dana, Progressive Reform, and the Newark Museum* (Pittsburgh: Periscope, 2009) and Nicolas Maffei, "John Cotton Dana and the Politics of Exhibiting Industrial Art in the US, 1909–1929," *Journal of Design History* 13, no. 4 (2000): 301–17.

12. John Cotton Dana, *The Gloom of the Museum* (Woodstock, VT: Elm Tree Press, 1917), 4.

13. Rossiter Howard, "Changing Ideals of the Art Museum," *Scribner's Magazine* 71, no. 1 (January 1922): 125–28.

14. Benjamin Ives Gilman, *Museum Ideals of Purpose and Method* (Cambridge, MA: Museum of Fine Arts, Boston, 1918), 55, 118, xi, 94.

15. Thomas R. Adam, *The Civic Value of Museums* (New York: American Association for Adult Education, 1937), 11, 20.

16. Theodore L. Low, *The Museum as a Social Instrument* (New York: Metropolitan Museum of Art, 1942), reprinted in Gail Anderson, ed., *Reinventing the Museum: The Evolving Conversation on the Paradigm Shift,* 2nd ed. (Lanham, MD: AltaMira, 2012), 42.

17. Blanche R. Brown, review of *The Museum as a Social Instrument,* by Theodore L. Low, *College Art Journal* 2, no. 4 (1943): 131.

18. "John Hay Whitney Announces Museum of Modern Art Will Serve as Weapon of National Defense," February 28, 1941, Museum of Modern Art, Press Release Archives online, http://www.moma.org/momaorg/shared/pdfs/docs/press_archives /676/releases/MOMA_1941_0015_1941-02-28_41228-14.pdf. See also Clarissa J. Ceglio, "A Cultural Arsenal for Democracy: The War Work of U.S. Museums, 1939–1946" (PhD dissertation, Brown University, 2015), chap. 4, https://repository.library.brown .edu/studio/search/?q=ceglio. Quote from p. 237.

19. Pamela M. Henson, "The Smithsonian Goes to War," in *Science and the Pacific War: Science and Survival in the Pacific, 1939–1945,* ed. Roy M. MacLeod (Boston: Kluwer Academic, 1999), 27–50.

20. American Association of Museums, *America's Museums: The Belmont Report. A Report to the Federal Council on the Arts and the Humanities* (Washington, DC: American Association of Museums, 1969), 19, 21, 43.

21. American Association of Museums, *Museums for a New Century: A Report of the Commission on Museums for a New Century* (Washington, DC: American Association of Museums, 1984), 17–19.

22. American Association of Museums, *Excellence and Equity: Education and the Public Dimension of Museums* (Washington, DC: American Association of Museums, 1992), 3–8.

23. American Alliance of Museums, "Museum Facts," accessed August 29, 2016, http://www.aam-us.org/about-museums/museum-facts.

24. Museum of Fine Arts, "The Economic and Community Impacts of the Museum of Fine Arts, Boston," http://www.mfa.org/about/economic-impact-report.

25. Metropolitan Museum of Art, "Three Metropolitan Museum Exhibitions Stimulate $742 Million 2013 Economic Impact for New York," October 7, 2013, http://www.metmuseum.org/press/news/2013/economic-impact-release-october-2013.

26. Qatar Museums Authority, "Museums & Galleries," accessed December 22, 2016, http://www.qm.org.qa/en/area/museums-galleries; Qatar Museums Authority, "Exclusively from Our Chairperson," 2014, http://www.qm.org.qa/en/blog/exclusively -our-chairperson.

27. "Mad about Museums," *Economist*, December 21, 2013, http://www.economist .com/news/special-report/21591710-china-building-thousands-new-museums-how -will-it-fill-them-mad-about-museums.

28. Initiative for a Competitive Inner City, "MASS MoCA: Rethinking an Industrial Complex as a New Museum and Urban Anchor," September 29, 2014, http://icic .org/mass-moca-rethinking-industrial-complex-new-museum-urban-anchor/.

29. Holland Cotter, "Making Museums Moral Again," *New York Times*, March 17, 2016, http://www.nytimes.com/2016/03/17/arts/design/making-museums-moral-again .html.

CHAPTER 16: MUSEUMS MAKE COMMUNITIES

1. James Montgomery and David J. Jeremy, *Technology and Power in the Early American Cotton Industry: James Montgomery, the Second Edition of His "Cotton Manufacture" (1840), and the "Justitia" Controversy about Relative Power Costs* (Philadelphia: American Philosophical Society, 1990), 158.

2. *Annual Report of the President to the Corporation of Brown University* (Providence: Angell, Hammett, 1879), 29. See also Montgomery and Jeremy, *Technology and Power*, 238n2.

3. Frederick L. Lewton, "Samuel Slater and the Oldest Cotton Machinery in America," in *Annual Report of the Board of Regents of the Smithsonian Institution 1926* (Washington, DC: Smithsonian Institution, 1926), 505–11.

4. Samuel Quiccheberg, *The First Treatise on Museums: Samuel Quiccheberg's Inscriptiones, 1565,* trans. Mark A. Meadow and Bruce Robertson (Los Angeles: Getty Research Institute, 2013), 63, 80.

5. Francis Bacon, *The Works of Lord Bacon: Philosophical Works* (London: Henry G. Bohn, 1854), 1:269.

6. On technology at the Peale museum, see Gary Kulik, "Designing the Past: History-Museum Exhibitions from Peale to the Present," in *History Museums in the United States: A Critical Assessment,* ed. Warren Leon and Roy Rosenzweig (Urbana: University of Illinois Press, 1989), 2–37; Laura Rigal, *The American Manufactory: Art, Labor, and the World of Things in the Early Republic* (Princeton, NJ: Princeton University Press, 1998).

7. On the patent office displays, see Douglas E. Evelyn, "Exhibiting America: The Patent Office as Cultural Artifact," *Smithsonian Studies in American Art* 3, no. 3 (Summer 1989): 24–37. Quote from p. 31.

8. J. Elfreth Watkins, *The Camden and Amboy Railroad: Origin and Early History* (Washington, DC: Gedney & Roberts, 1892), 61.

9. "Report of the Secretary," in *Annual Report of the Board of Regents of the Smithsonian Institution,* 1924, 3.

10. Arthur P. Molella, "The Museum That Might Have Been: The Smithsonian's National Museum of Engineering and Industry," *Technology and Culture* 32, no. 2 (April 1991): 254.

11. Jacqueline Trescott, "The Motherlode of Invention: High-Tech Innovator Gives $10.4 Million for Smithsonian Center," *Washington Post,* June 1, 1995.

12. Lemelson Center for the Study of Invention and Innovation, "Exhibitions," accessed August 19, 2016, http://invention.si.edu/about/exhibitions.

13. "Brown Museum," *Providence Journal,* November 26, 1896.

14. John Adams to Abigail Adams," May 1780, Massachusetts Historical Society, Adams Family Papers, http://www.masshist.org/digitaladams/archive/doc?id =L178004xxja.

15. Jeremy Belknap, *Circular Letter, of the Historical Society* (Boston: Printed by Belknap and Young, 1791), http://www.masshist.org/database/viewer.php?item _id=65.

16. Quoted in Erin R. Lawrimore, " 'Let Us Hasten to Redeem the Time That Is Lost': J. G. M. Ramsey's Role in the Collection and Promotion of Tennessee History," *Libraries & the Cultural Record* 41, no. 4 (2006): 425.

17. Evan J. Friss, "From University Heights to Cooperstown: Halls of Fame and American Memory," *Journal of Archival Organization* 3, no. 4 (July 18, 2006): 87–104.

18. Ibid.; James A. Vlasich, *A Legend for the Legendary: The Origin of the Baseball Hall of Fame* (Bowling Green, OH: Popular Press, 1990).

19. Rhode Island Heritage Hall of Fame, "Future Home of the Rhode Island Heritage Hall of Fame," accessed August 19, 2016, http://www.riheritagehalloffame.org/future.cfm.

20. Jeanne Zeidler, "The Hampton University Museum Collections," *International Review of African American Art* 11, no. 4 (January 1994): 46.

21. Andrea A. Burns, *From Storefront to Monument: Tracing the Public History of the Black Museum Movement* (Amherst: University of Massachusetts Press, 2013).

22. Grace Glueck, "A Very Own Thing in Harlem," *New York Times*, September 15, 1968, http://query.nytimes.com/gst/abstract.html?res=9A03E2DE1030EE3BBC4D52DFBF668383679EDE.

23. Arlene M. Dávila, *Latino Spin: Public Image and the Whitewashing of Race* (New York: New York University Press, 2008), 124.

24. John Kuo Wei Tchen, "Creating a Dialogic Museum: The Chinatown History Museum Experiment," in *Exhibiting Cultures: The Poetics and Politics of Museum Display,* ed. Ivan Karp and Steven D. Lavine (Washington, DC: Smithsonian Institution Press, 1991), 286.

25. Ron Chew, "The Wing Luke Asian Museum: Gathering Asian American Stories," in *Chinese America: History and Perspectives 2000* (Brisbane, CA: Chinese Historical Society, 2000), 64.

26. See Elena L. Gonzales, "Museums Working for Social Justice: Resonance and Wonder" (PhD dissertation, Brown University, 2015), chap. 4; and Karen Mary Davalos, *Exhibiting Mestizaje: Mexican (American) Museums in the Diaspora* (Albuquerque: University of New Mexico Press, 2001).

27. Peggy Levitt, *Artifacts and Allegiances: How Museums Put the Nation and the World on Display* (Oakland: University of California Press, 2015), 78–79.

28. Amanda J. Cobb, "The National Museum of the American Indian: Sharing the Gift," *American Indian Quarterly* 29, nos. 3–4 (2005). Quote p. 362.

29. Amy Lonetree, "Missed Opportunities: Reflections on the NMAI," *American Indian Quarterly* 30, nos. 3–4 (2006). Quote p. 637.

30. Edward Rothstein, "Museum with an American Indian Voice," *New York Times*, September 21, 2004, http://www.nytimes.com/2004/09/21/arts/design/museum-with-an-american-indian-voice.html.

31. Edward Rothstein, "To Each His Own Museum, as Identity Goes on Display," *New York Times*, December 28, 2010, http://www.nytimes.com/2010/12/29/arts/design/29identity.html; Edward Rothstein, "Reopened Museum Tells Chinese-American Stories," *New York Times*, September 21, 2009, http://www.nytimes.com/2009/09/22/arts/design/22museum.html.

32. National Museum of African American History and Culture, "About the Museum," accessed January 4, 2016, https://nmaahc.si.edu/about/museum.

33. Ellen Lekka, "Communities of South Sudan Build Their National Museum," United Nations Educational, Scientific and Cultural Organization, June 19, 2014, http://www.unesco.org/new/en/media-services/single-view/news/communities_of_south_sudan_build_their_national_museum.

34. Stefan Berger, "National Museums in between Nationalism, Imperialism and Regionalism, 1750–1914," in *National Museums and Nation-Building in Europe, 1750–2010*, ed. Peter Aronsson and Gabriella Elgenius (New York: Routledge, 2014), 13.

35. Quoted in ibid., 14.

36. Lyndon B. Johnson, "Remarks at the Dedication of the Smithsonian Institution's Museum of History and Technology," January 22, 1964, The American Presidency Project, http://www.presidency.ucsb.edu/ws/index.php?pid=26023.

37. National Park Service, "Manzanar National Historic Site: Long-Range Interpretive Plan," August 2007, 4, https://www.nps.gov/manz/learn/management/upload/manz-lrip-2007.pdf.

38. National Park Service, "World War II Valor in the Pacific: Purpose of the Monument," accessed May 30, 2016, https://www.nps.gov/valr/learn/park-purpose.htm.

39. National September 11 Memorial & Museum, "Mission Statement," accessed August 20, 2016, http://www.911memorial.org/mission-statements-0.

40. Benedict R. Anderson, *Imagined Communities: Reflections on the Origin and Spread of Nationalism* (London: Verso, 1983) chap. 10, quotes on pp. 186 and 189.

41. James B. Cuno, *Whose Culture? The Promise of Museums and the Debate over Antiquities* (Princeton, NJ: Princeton University Press, 2009), 28–29.

42. Kwame Anthony Appiah, "Whose Culture Is It?," *New York Review of Books*, February 9, 2006, http://www.nybooks.com/articles/archives/2006/feb/09/whose-culture-is-it/.

43. Orhan Pamuk, "Orhan Pamuk's Manifesto for Museums," *Art Newspaper*, June 7, 2016, http://theartnewspaper.com/comment/orhan-pamuk-s-manifesto-for-museums/.

44. United States Holocaust Memorial Museum, "Mission and History," accessed August 20, 2016, https://www.ushmm.org/information/about-the-museum/mission-and-history.

45. UN Live Museum for Humanity, "A Brief Introduction," 2015, 4, http://www.unlivemuseum.org/wp-content/uploads/2015/12/UN_Live_A_Brief_Introduction.pdf.

46. UN Live Museum for Humanity, "Pre-Feasibility Report," 2015, 7, http://www.unlivemuseum.org/Pre-feasibility_Report.pdf.

CHAPTER 17: LEARNING FROM THINGS

1. Spencer Fullerton Baird, *Directions for Collecting, Preserving, and Transporting Specimens of Natural History* (Washington: Smithsonian Institution, 1852), 22.

2. J. W. P. Jenks, "Autobiography," TS, 1894, 199, Brown University Biographical Files, Brown University Archives.

3. *Tenth Annual Report of the Board of Regents of the Smithsonian Institution* (Washington, DC: A.O.P. Nicholson, 1856), 48, 59.

4. Spencer Fullerton Baird, *The Mammals of North America: The Descriptions of Species Based Chiefly on the Collections in the Museum of the Smithsonian Institution* (Philadelphia: J. B. Lippincott, 1857), 461.

5. J. A. Allen, *Catalogue of the Mammals of Massachusetts: With a Critical Revision of the Species* (Cambridge, MA: University Press, 1869), 145.

6. Ibid., 26, 28; Elliott Coues and J. A. Allen, *Monographs of North American Rodentia* (Washington, DC: Government Printing Office, 1877), iii–iv. See also Mary P. Winsor, *Reading the Shape of Nature: Comparative Zoology at the Agassiz Museum* (Chicago: University of Chicago Press, 1991), 77.

7. Baird, *Directions for Collecting*, 13.

8. Nicole L. Bedford and Hopi E. Hoekstra, "Peromyscus Mice as a Model for Studying Natural Variation," *eLife* 4 (2015): Table 1, accessed August 13, 2015, http://www.ncbi.nlm.nih.gov/pmc/articles/PMC4470249.

9. Wilfred Hudson Osgood, *Revision of the Mice of the American Genus Peromyscus* (Washington, DC: Government Printing Office, 1909), 120.

10. Bedford and Hoekstra, "Peromyscus Mice."

11. *United States National Museum Annual Report for the year ended June 30, 1953* (Washington, DC: Smithsonian Institution, 1954), 1–2.

12. Margaret Mead, *Anthropologists and What They Do* (New York: F. Watts, 1965), 13–18.

13. Richard Fortey, *Dry Store Room No. 1: The Secret Life of the Natural History Museum* (London: Harper Perennial, 2008), 32.

14. Lucile H. Brockway, *Science and Colonial Expansion: The Role of the British Royal Botanic Gardens* (New Haven, CT: Yale University Press, 2002), 6, 75.

15. Laura Dassow Walls, *Emerson's Life in Science: The Culture of Truth* (Ithaca, NY: Cornell University Press, 2003), 84–85.

16. Bruno J. Strasser, "Collecting Nature: Practices, Styles, and Narratives," *Osiris* 27, no. 1 (January 2012): 322.

17. William James, *Louis Agassiz: Words Spoken by Professor William James at the Reception of the American Society of Naturalists by the President and Fellows of Harvard*

College, at Cambridge, on December 30, 1896 (Cambridge: Printed for the University, 1897), 9.

18. Edward O. Wilson, *Naturalist,* rev. ed. (Washington, DC: Island Press, 2013), 203–4.

19. Ibid., 227.

20. Ibid., 231.

21. Strasser, "Collecting Nature."

22. J. Grinnell, "The Uses and Methods of a Research Museum," *Popular Science Monthly,* August 1910, quoted in James R. Griesemer and Elihu M. Gerson, "Collaboration in the Museum of Vertebrate Zoology," *Journal of the History of Biology* 26, no. 2 (Summer 1993): 197.

23. Steve Beissinger, "Grinnell Resurvey Project," The Grinnell Survey Project, Museum of Vertebrate Zoology, University of California, Berkeley, accessed April 18, 2016, http://mvz.berkeley.edu/Grinnell/.

24. President's Committee of Advisors on Science and Technology, Biodiversity and Ecosystems Panel, *Teaming with Life: Investing in Science to Understand and Use America's Living Capital* (Washington, DC: PCAST Biodiversity and Ecosystems Panel, 1998), 29–32.

25. E. McClung Fleming, "Artifact Study: A Proposed Model," *Winterthur Portfolio* 9 (1974): 153–73; Jules David Prown, "Mind in Matter: An Introduction to Material Culture Theory and Method," *Winterthur Portfolio* 17, no. 1 (1982): 1–19; James Deetz, *In Small Things Forgotten: The Archaeology of Early American Life* (Garden City, NY: Anchor / Doubleday, 1977).

26. Nicholas Thomas, "The Museum as Method," *Museum Anthropology* 33, no. 1 (Spring 2010): 7.

27. Leora Auslander, "Beyond Words," *American Historical Review* 110, no. 4 (2005): 1017–18.

28. Laurel Thatcher Ulrich, *The Age of Homespun: Objects and Stories in the Creation of an American Myth* (New York: Alfred A. Knopf, 2001), 41–74.

29. David J. Jeremy, "Innovation in American Textile Technology during the Early 19th Century," *Technology and Culture* 14, no. 1 (January 1973): 71.

30. Edwin Battison, "Eli Whitney and the Milling Machine," *Smithsonian Journal of History* 1, no. 2 (Summer 1966): 9–34.

31. William L. Withuhn, "Testing the John Bull," in John H. White, *The John Bull, 150 Years a Locomotive* (Washington, DC: Smithsonian Institution Press, 1981), 114, 121.

32. Corinna da Fonseca-Wollheim, "More Musicians Are Trying Period Instruments," *New York Times,* August 30, 2013, http://www.nytimes.com/2013/09/01/arts/music/more-musicians-are-trying-period-instruments.html.

33. Pravina Shukla, *Costume: Performing Identities through Dress* (Bloomington: Indiana University Press, 2015), 190.

34. Deborah Arenz, "Why Every Curator Should Enter a Plowing Match," *ALHFAM*, February 4, 2016, https://alhfam.wordpress.com/2016/02/04/why-every-curator-should-enter-a-plowing-match-2/.

35. Robert B. Gordon, "The 'Kelly' Converter," *Technology and Culture* 33, no. 4 (1992): 769–79.

36. Arthur Caswell Parker, *A Manual for History Museums* (New York: Columbia University Press, 1935), 108.

37. Cary Carson, "Doing History with Material Culture," in *Material Culture and the Study of American Life,* ed. Ian M. Quimby (New York: W. W. Norton, 1978), 42.

38. Alexander Fenton, "Collections Research: Local, National and International Perspectives," in *Manual of Curatorship: A Guide to Museum Practice,* ed. John M. A. Thompson, 2nd ed. (1992; repr., London: Routledge, 2012), 493.

CHAPTER 18: TEACHING WITH THINGS

1. Hermon Carey Bumpus, *Hermon Carey Bumpus: Yankee Naturalist* (Minneapolis: University of Minnesota Press, 1947), 36.

2. Ibid., 36–40; Kathrinne Duffy, "The Dead Curator: Education and the Rise of Bureaucratic Authority in Natural History Museums, 1870–1915," *Museum History Journal* 10, no. 1 (January 2017): 29–49.

3. Quoted in Kenneth A. Yellis, "Museum Education," in *The Museum: A Reference Guide,* ed. Michael Steven Shapiro and Louis Ward Kemp (New York: Greenwood, 1990), 168.

4. S. S. Conant, "The First Century of the Republic—Eighteenth Paper—Progress of the Fine Arts," *Harper's New Monthly Magazine,* April 1876.

5. Nikolaus Pevsner, *Academies of Art, Past and Present* (Cambridge: Cambridge University Press, 1940), 172–73.

6. "Students and Copyists in the Louvre Gallery Paris," *Harper's Weekly,* January 11, 1868, 25.

7. *Catalogue of the Rhode Island School of Design* (Providence, RI: Dart and Bigelow, 1903), 10.

8. Isaac Edwards Clarke, *Art and Industry: Industrial and Technical Training in Voluntary Associations and Endowed Institutions* (Washington, DC: U.S. Government Printing Office, 1897), 260.

9. *Report of the Commissioner of Education for the Year Ended June 30, 1915,* vol. 1 (Washington, DC: U.S. Government Printing Office, 1915), 552.

10. John Shapley, "The Art Museum as a College Laboratory," *Museum Work* 1 (January 1918): 127.

11. Louis Earle Rowe, "Docent Service at the Museum of Fine Arts," in *Proceedings of the American Association of Museums* 5 (1911): 10.

12. Quoted in Kathryn Brush, *Vastly More than Brick & Mortar: Reinventing the Fogg Art Museum in the 1920s* (Cambridge, MA; Harvard University Art Museums, 2003), 60–61, 84.

13. Clarke, *Art and Industry*, 761–65.

14. Ibid., 148.

15. Jeffrey Trask, *Things American: Art Museums and Civic Culture in the Progressive Era* (Philadelphia: University of Pennsylvania Press, 2011), 1.

16. Ibid., 140–41.

17. George H. Sherwood, *Free Education by the American Museum of Natural History in Public Schools and Colleges* (New York: American Museum of Natural History, 1918), 15.

18. "Museum Extension Work," *Museum News Letter* 1, no. 8 (April 1918): 3, https://books.google.com/books?id=85UjAQAAMAAJ&pg=PA26#.

19. Barbara Y. Newsom and Adele Z. Silver, eds., *The Art Museum as Educator: A Collection of Studies as Guides to Practice and Policy* (Berkeley: University of California Press, 1978), 568.

20. William Henry Fox, *Museums of the Brooklyn Institute of Arts and Sciences Report for the Year 1913* (Brooklyn: Printed for the Museums, 1913), 12.

21. Henry Fairfield Osborn and George H. Sherwood, *The Museum and Nature Study in the Public Schools* (New York: American Museum of Natural History, 1913), 4.

22. Sherwood, *Free Education*, 7–8.

23. Osborn and Sherwood, *Museum and Nature Study*, 2–4.

24. *The Brooklyn Children's Museum* (Brooklyn: Brooklyn Institute of Arts and Sciences, 1936), 8.

25. A. W. Bell, "Reorganization at the Los Angeles Museum," *Science* 94, no. 2437 (1941): 255–56.

26. *Brooklyn Children's Museum*, 8.

27. Edward Carlton Page, "How the Working Museum of History Works," *History Teacher's Magazine* 6 (1915): 307–11.

28. Quoted in Rika Burnham and Elliott Kai-Kee, *Teaching in the Art Museum: Interpretation as Experience* (Los Angeles: J. Paul Getty Museum, 2011), 26; "Lending Reproductions to Schools," *Museum News Letter* 1, no. 6 (February 1918): 2, https://books.google.com/books?id=85UjAQAAMAAJ&pg=PA17.

29. Richard F. Bach, "Neighborhood Exhibitions of the Metropolitan Museum of Art," *Museum News* 16 (September 1938): 7, quoted in Neil Harris, "The Divided House of the American Art Museum," *Daedalus* 128, no. 3 (1999): 37–38.

30. Quoted in Alison Griffiths, *Shivers Down Your Spine: Cinema, Museums, and the Immersive View* (New York: Columbia University Press, 2012), 185.

31. Benjamin Ives Gilman, *Museum Ideals of Purpose and Method* (Cambridge, MA: Museum of Fine Arts, Boston, 1918), 287; Benjamin Ives Gilman, "On the Distinctive

Purpose of Museums of Art," in *Communications to the Trustees*, vol. 1 (Boston: Museum of Fine Arts, 1904), 50. More generally, see George E. Hein, *Progressive Museum Practice: John Dewey and Democracy* (Walnut Creek, CA: Left Coast, 2012), 126.

32. Gilman, *Museum Ideals*, 306–15; Hein, *Progressive Museum Practice*, 127; Rowe, "Docent Service at the Museum of Fine Arts," 14.

33. Quoted in Burnham and Kai-Kee, *Teaching in the Art Museum*, 24.

34. Quoted in ibid., 29.

35. Mary Mullen, *An Approach to Art* (Merion, PA: Barnes Foundation, 1923), 24, quoted in Burnham and Kai-Kee, *Teaching in the Art Museum*, 25.

36. Burnham and Kai-Kee, *Teaching in the Art Museum*, 32.

37. Charles E. Slatkin, "Aims and Methods in Museum Education," *College Art Journal* 7, no. 1 (Autumn 1947): 30, quoted in Burnham and Kai-Kee, *Teaching in the Art Museum*, 31.

38. Burnham and Kai-Kee, *Teaching in the Art Museum*, 31, 97–98.

39. Sarah Newmeyer, "Children's Holiday Circus Opens at Museum of Modern Art" (press release, Museum of Modern Art, December 1, 1943), http://www.moma.org/momaorg/shared/pdfs/docs/press_archives/914/releases/MOMA_1943_0066_1943-12-06_431206-62.pdf?2010.

40. Quoted in Karen A. Rader and Victoria E. M. Cain, *Life on Display: Revolutionizing U.S. Museums of Science and Natural History in the Twentieth Century* (Chicago: University of Chicago Press, 2014), 160.

41. Victor D'Amico, *Experiments in Creative Arts Teaching: A Progress Report on the Department of Education* (New York: Museum of Modern Art, 1960), quoted in Burnham and Kai-Kee, *Teaching in the Art Museum*, 36.

42. Blythe Bohnen, "Old Masters—New Apprentices," *Bulletin of the Metropolitan Museum of Art* 27, no. 4 (December 1968): 230–36, quoted in Burnham and Kai-Kee, *Teaching in the Art Museum*, 37.

43. Frank Oppenheimer, "Rationale for a Science Museum," *Curator: The Museum Journal* 11, no. 3 (September 1968): 206.

44. Quoted in Burnham and Kai-Kee, *Teaching in the Art Museum*, 39. Emphasis in original.

45. Inez Wolins, "Teaching the Teachers," *Museum News* 69, no. 3 (May/June 1990): 72, quoted in Burnham and Kai-Kee, *Teaching in the Art Museum*, 45.

46. E. Louis Lankford, "Aesthetic Experience in Constructivist Museums," *Journal of Aesthetic Education* 36, no. 2 (2002): 145.

47. Burnham and Kai-Kee, *Teaching in the Art Museum*, 47–48, 102–104. See also "What Is VTS?," Visual Thinking Strategies, accessed August 25, 2016, http://www.vtshome.org/what-is-vts.

48. Veronica Bagwell, "Crystal Bridges Museum of American Art & University of Arkansas Department of Education Reform Announce Results of a Study on

Culturally Enriching School Field Trips" (press release, Crystal Bridges Museum of American Art, September 17, 2013), http://crystalbridges.org/blog/crystal-bridges-museum-of-american-art-university-of-arkansas-department-of-education-reform-announce-results-of-a-study-on-culturally-enriching-school-field-trips/.

49. Steven Lubar and Kathleen Kendrick, "Looking at Artifacts, Thinking about History," *Artifact & Analysis: A Teacher's Guide to Interpreting Objects and Writing History*, accessed August 24, 2016, http://www.smithsonianeducation.org/idealabs/ap/essays/looking7.htm.

50. Franklin D. Vagnone and Deborah E. Ryan, *Anarchist's Guide to Historic House Museums* (Walnut Creek, CA: Left Coast, 2016), 115.

51. National Museum of American History, "Out of Storage," *Oh Say Can You See?* (blog), June 29, 2011, http://americanhistory.si.edu/blog/2011/06/out-of-storage.html.

52. Harvard Art Museums, "Art Study Center," accessed December 31, 2015, http://www.harvardartmuseums.org/teaching-and-research/art-study-center. See also Sebastian Smee, "Art Study Center at Harvard Allows Close-up Access to Genius," *Boston Globe,* February 8, 2015, http://www.bostonglobe.com/arts/theater-art/2015/02/08/art-study-center-harvard-provides-close-encounters-with-masters/9vVsPmnVxcwrE27xaixmLP/story.html.

53. Shari Tishman, Alythea McKinney, and Celka Straughn, "Study Center Learning: An Investigation of the Educational Power and Potential of the Harvard Art Museum's Study Centers" (Cambridge, MA: A collaborative project between the Harvard Art Museum and Project Zero at the Harvard Graduate School of Education, September 2007), 1, 17, http://www.pz.harvard.edu/sites/default/files/Study%20Center%20report%20Tishman%20et%20al.pdf.

54. Darren Milligan, "What We Have Learned So Far: The Digital Learning Resources Project," Smithsonian Learning Lab, October 14, 2014, https://learninglab.si.edu/news/what-we-have-learned-so-far-the-digital-learning-resources-project.

55. Sarah Lyall, "Off the Beat and Into a Museum: Art Helps Police Officers Learn to Look," *New York Times,* April 26, 2016, http://www.nytimes.com/2016/04/27/arts/design/art-helps-police-officers-learn-to-look.html.

56. Peabody Essex Museum, "Collections," accessed December 30, 2016, http://www.pem.org/collections/.

Chapter 19: The Promise of Museums

1. John Cotton Dana, *The New Museum* (Woodstock, VT: Elm Tree Press, 1917), 36.

2. Henry James, *A Small Boy and Others: A Critical Edition,* ed. Peter Collister (Charlottesville: University of Virginia Press, 2011), 278.

3. Edward O. Wilson, *Naturalist,* rev. ed. (Washington, DC: Island Press, 2013), 95–96.

4. Oliver Sacks, *Uncle Tungsten: Memories of a Chemical Boyhood*, repr. (New York: Vintage, 2002), 57–59.

5. Ibid., 117, 151, 57–59.

6. Ibid., 187–94, 203.

7. Robert Mangold, interview by Claire Dienes and Lilian Tone, November 18, 1999, 4, transcript, Museum of Modern Art Oral History Program, https://www .moma.org/momaorg/shared/pdfs/docs/learn/archives/transcript_mangold.pdf.

8. Bruce Nauman, interview by Kathy Halbreich, January 9, 2012, 18, transcript, Museum of Modern Art Oral History Program, http://www.moma.org/momaorg /shared/pdfs/docs/learn/archives/transcript_nauman.pdf.

9. James Rosenquist, interview by Leah Dickerman, April 17, 2012, 3, transcript, Museum of Modern Art Oral History Program, http://www.moma.org/momaorg /shared/pdfs/docs/learn/archives/transcript_rosenquist.pdf.

10. Phyllis R. Dixon, "More than a Dream," in *Chicken Soup for the African American Woman's Soul: Laughter, Love and Memories to Honor the Legacy of Sisterhood*, ed. Jack Canfield, Mark Victor Hansen, and Lisa Nichols (New York: Backlist, 2012), available at http://www.chickensoup.com/book-story/43748/more-than-a-dream.

11. James, *A Small Boy and Others*, 276–77.

12. Adam Gopnik, "The Mindful Museum," *The Walrus*, September 12, 2012, https://thewalrus.ca/the-mindful-museum/.

13. Tim Winton, "Spurned No Longer," in *Treasure Palaces: Great Writers Visit Great Museums*, ed. Maggie Fergusson (London: Economist Books, 2016), 65.

14. Catherine M. Cameron and John B. Gatewood, "Seeking Numinous Experiences in the Unremembered Past," *Ethnology* 42, no. 1 (2003): 55–71; John H. Falk, *Identity and the Museum Visitor Experience* (Walnut Creek, CA: Left Coast, 2009), 62–65.

15. J. W. P. Jenks, *Hunting in Florida in 1874* (Providence, RI: n.p., 1884), 34.

16. Quoted in Peggy Levitt, *Artifacts and Allegiances: How Museums Put the Nation and the World on Display* (Oakland: University of California Press, 2015), 141.

17. Tristram Besterman, "Crossing the Line: Restitution and Cultural Equity," in *Museums and Restitution: New Practices, New Approaches*, ed. Louise Tythacott and Kostas Arvanitis (Farnham, Surrey, UK: Ashgate, 2014), 28.

18. Christina Kreps, "Curatorship as Social Practice," *Curator: The Museum Journal* 46, no. 3 (July 2003): 311–23.

19. Richard Kurin, *Reflections of a Culture Broker: A View from the Smithsonian* (Washington, DC: Smithsonian Institution Press, 1997), 93.

20. Chip Colwell-Chanthaphonh, Steven R. Holen, and Stephen E. Nash, *Crossroads of Culture: Anthropology Collections at the Denver Museum of Nature & Science* (Boulder: University Press of Colorado, 2010), 39.

21. Ibid., 10.

22. Joshua A. Bell, "A View from the Smithsonian: Connecting Communities to Collections," *Practicing Anthropology* 37, no. 3 (December 2014): 14–16.

23. Smithsonian Institution, "Smithsonian Collections," *Recovering Voices,* 2016, http://recoveringvoices.si.edu/resourcesandgrants/sicollections.htm.

24. Eve M. Kahn, "A Preserved Kayak Floats an Exhibition in Alaska," *New York Times,* April 14, 2016, http://www.nytimes.com/2016/04/15/arts/design/a-preserved -kayak-floats-an-exhibition-in-alaska.html; Alutiiq Museum, "Our Philosophy," accessed August 26, 2016, https://alutiiqmuseum.org/explore/programs.

25. Michael M. Ames, *Cannibal Tours and Glass Boxes: The Anthropology of Museums* (Vancouver: University of British Columbia Press, 1992), 57.

26. Nicholas Thomas, "The Museum as Method," *Museum Anthropology* 33, no. 1 (Spring 2010): 7.

27. Duncan F. Cameron, "The Museum, a Temple or the Forum," *Curator: The Museum Journal* 14, no. 1 (March 1971): 11–24; Stephen E. Weil, "From Being about Something to Being for Somebody: The Ongoing Transformation of the American Museum," *Daedalus* 128, no. 3 (1999): 229; Nina Simon, *The Participatory Museum* (Santa Cruz, CA: Museum 2.0, 2010), i.

28. Malcolm Gay, "Rose Art Museum Director Leaving for Baltimore," *Boston Globe,* May 2, 2016, https://www.bostonglobe.com/arts/theater-art/2016/05/02/rose -art-museum-director-bedford-bound-for-new-baltimore-appointment/8tPUBQ1 WBzQXcGXdZksoTI/story.html.

29. Missouri History Museum, "Museum Moving Forward," accessed February 24, 2017, http://mohistory.org/forward.

30. National Museum of American History, "Mission & History," accessed August 8, 2015, http://americanhistory.si.edu/museum/mission-history.

31. California Academy of Sciences, "About," accessed August 26, 2016, https://www .calacademy.org/about-us.

32. Santa Cruz Museum of Art & History, "About—Santa Cruz Museum of Art & History," accessed December 25, 2016, https://santacruzmah.org/about/.

33. James Cuno, "A World Changed? Art Museums after September 11," *Bulletin of the American Academy of Arts and Sciences* 55, no. 4 (Summer 2002): 33–35.

34. "Joint Statement from Museum Bloggers and Colleagues on Ferguson and Related Events December 11, 2014," *Museums & Social Issues* 10, no. 2 (October 1, 2015): 106–8. See also Gretchen Jennings, "The #museumsrespondtoFerguson Initiative, a Necessary Conversation," *Museums & Social Issues* 10, no. 2 (October 2015): 97–105; Alyssa Boge and Katherine Rieck, "A Guide to Discussions Surrounding Ferguson," *Museums & Social Issues* 10, no. 2 (October 2015): 113–16.

35. American Association for State and Local History, "History Organizations Positioned to Be Powerful Participants in Dialogue on Ferguson and Related Events," *AASLH*

(blog), December 16, 2014, http://blogs.aaslh.org/history-organizations-positioned
-to-be-powerful-participants-in-dialogue-on-ferguson-and-related-events/.

36. Adrianne Russell, "The Media Needs a History Lesson When Addressing Civic
Unrest, Says the Director of the African American History Museum," *Smithsonian.com*,
May 1, 2015, http://www.smithsonianmag.com/smithsonian-institution/media-needs
-history-lesson-addressing-civic-unrest-director-african-american-history-museum
-180955140/.

37. Nicholas Thomas, *The Return of Curiosity: What Museums Are Good for in the
21st Century* (London: Reaktion Books, 2016), 60–63.

38. Georges Bataille and Annette Michelson, "Museum," *October* 36 (Spring 1986): 24.

Coda

1. Quoted in Marion Endt, "Beyond Institutional Critique: Mark Dion's Surrealist
Wunderkammer at the Manchester Museum," *Museum and Society* 5, no. 1 (2007): 2.

2. Kynaston McShine, *The Museum as Muse: Artists Reflect* (New York: Museum of
Modern Art, 1999), 145.

3. Andrea Fraser, "Aren't They Lovely? An Introduction," in *Museum Highlights:
The Writings of Andrea Fraser,* ed. Alexander Alberro (Cambridge, MA: MIT Press,
2005), 141–45.

4. Rosamond Wolff Purcell and Lisa Melandri, *Rosamond Purcell: Two Rooms*
(Santa Monica, CA: Santa Monica Museum of Art, 2003); Rosamond Purcell, *Owls
Head: On the Nature of Lost Things* (New York: Quantuck Lane, 2007).

5. Leslie King-Hammond, "A Conversation with Fred Wilson," in *Mining the Museum:
An Installation by Fred Wilson,* ed. Lisa G. Corrin (New York: W. W. Norton, 1994), 34.

6. Lisa G. Corrin, Miwon Kwon, and Norman Bryson, *Mark Dion* (London: Phaidon,
1997), 80; Allison Meier, "The Ambition and Arrogance of Exploration," *Hyperallergic,*
June 12, 2012, http://hyperallergic.com/52761/mark-dion-explorers-club/.

7. Endt, "Beyond Institutional Critique," 1.

8. Daniel Robbins, "Confessions of a Museum Director," in *Raid the Icebox 1, with
Andy Warhol: An Exhibition Selected from the Storage Vaults of the Museum of Art,
Rhode Island School of Design,* by Stephen E. Ostrow, (Providence: Rhode Island
School of Design Museum of Art, 1969), 15; Deborah Bright, "Shopping the Leftovers:
Warhol's Collecting Strategies in *Raid the Icebox I,*" *Art History* 24, no. 2 (April 2001): 286.

9. A. Will Brown, "Raid the Database 1 with Natalja Kent," *Manual / A Resource
about Art and Its Making,* accessed July 2, 2015, http://risdmuseum.org/manual/340
_raid_the_database_1_with_natalja_kent.

10. Guillaume Désanges, "Infinite Infoldings of Emptiness upon Emptiness: Account of a Night with Marcel Broodthaers," accessed February 24, 2017, http://
guillaumedesanges.com/spip.php?article106.

11. Senay Boztas, "Revealed: The Unseen Flip-Sides of the World's Most Famous Paintings," *The Guardian,* June 16, 2016, https://www.theguardian.com/artanddesign /2016/jun/16/revealed-worlds-famous-paintings-exhibition-vik-muniz-verso -mauritshuis.

12. VikMunizStudio, *This Is Verso Part 1 (of 2),* 2008, https://www.youtube.com /watch?v=sThr7kyozgc.

13. Pedro de Llano, "Displacement and Translation in the Work of Stephen Prina," *Afterall,* July 9, 2009, http://www.afterall.org/online/displacement.and.translation .in.the.work.of.stephen.prina/#.V2ln1pMrLdQ.

14. Candace Jackson, "An Artist Transforms a Museum's Leftovers," *Wall Street Journal,* September 4, 2009, http://on.wsj.com/1w7qxex.

15. See Peter Bacon Hales, "Surveying the Field," *Art Journal* 54, no. 3 (Autumn 1995): 35–41.

16. Ivan Karp and Fred Wilson, "Constructing the Spectacle of Culture in Museums," in *Thinking about Exhibitions,* ed. Reesa Greenberg, Bruce W. Ferguson, and Sandy Nairne (London: Routledge, 2005), 253.

17. Quoted in Bárbara C. Cruz and Noel Smith, "Mark Dion's Troubleshooting: Empowering Students to Create and Act," *Art Education* 66, no. 3 (May 1, 2013): 29–38.

18. Karp and Wilson, "Constructing the Spectacle," 253.

19. Coco Fusco, "The Other History of Intercultural Performance," *TDR* 38, no. 1 (1994): 143.

20. Karp and Wilson, "Constructing the Spectacle," 252.

21. Judith Barry, "Dissenting Spaces," in Greenberg, Ferguson, and Nairne, eds., *Thinking About Exhibitions,* 218–21.

22. Robin Cembalest, "The Curator Vanishes: Period Room as Crime Scene," *ARTnews,* February 28, 2013, http://www.artnews.com/2013/02/28/mark-dion-curator -office-in-minneapolis/.

23. Meleko Mokgosi, "Text Installations," accessed June 27, 2016, http://www .melekomokgosi.com/text-installations/.

24. Fraser, "Museum Highlights: A Gallery Talk," in *Museum Highlights,* 96, 104.

25. Ibid., 102.

26. Charlotte Burns, "Andrea Fraser: The Artist Turning the Whitney into a Prison," *Guardian,* February 24, 2016, https://www.theguardian.com/artanddesign/2016/feb /24/andrea-fraser-artist-turning-whitney-museum-prison.

27. Nasher Sculpture Center, *Nasher XChange—Alfredo Jaar,* accessed August 27, 2016, https://www.youtube.com/watch?v=bPEjTfnMVo8.

28. Carol Strickland, "Deep in the Art of Texas," *Art in America,* October 30, 2013, http://www.artinamericamagazine.com/news-features/news/deep-in-the-art-of-texas/.

29. McShine, *The Museum as Muse,* 98.

Acknowledgments

One of the delights of museum work is the opportunity to participate in teams with a wide range of expertise. Writing this book brought back pleasant memories of working with and learning from colleagues and students. Michael Folsom, founder of the Charles River Museum of Industry, offered me my first real museum job and introduced me to the challenges and pleasures of the museum world. My former colleagues at the Smithsonian Institution taught me how museums work, among them Wendy Abel-Weiss, David Allison, Richard Barden, Jeanne Benas, Nancy Bercaw, Nigel Briggs, Lonnie Bunch, Lynn Chase, Judy Chelnick, Harold Closter, Spencer Crew, Tom Crouch, Janet Davidson, Nanci Edwards, Jon Eklund, Doug Evelyn, Steven Fisher, John Fleckner, Robert Friedel, Jim Gardner, Karen Garlick, Anne Golovin, Judy Gradwohl, Margarite Grandine, Rayna Green, Laura Hansen, Elizabeth Harris, Kate Henderson, Andy Heymann, Brooke Hindle, Cynthia Hoover, Harry Hunter, James Hutchins, Ray Hutt, David Jaffe, Gretchen Jennings, Paula Johnson, Jennifer Locke Jones, Larry Jones, Katy Kendrick, Roger Kennedy, Peggy Kidwell, Gary Kulik, Richard Kurin, Peter Liebhold, Bonnie Campbell Lilienfeld, Melinda Machado, Edie Mayo, Nancy McCoy, John McDonagh, Charlie McGovern, Joan Mentzer, Art Molella, Martha Morris, Howard Morrison, Susan Myers, Stephanie Norby, Craig Orr, Katherine Ott, Catherine Perge, Bob Post, Constantine Raitzky, Betsy Robinson, Rodris Roth, Harry Rubenstein, Fath Davis Ruffins, Debora Scriber-Miller, Dave Seibert, Bob Selim, Elizabeth Sharpe, David Shayt, Ann Silverman, Jim Sims, Barbara Clark Smith, Katherine Spiess, Carlene Stephens, Shari Stout,

Gary Sturm, Gabrielle Tayac, Lonn Taylor, Susan Tolbert, Deborah Warner, Jack White, Bill Withuhn, Helena Wright, Bill Yeingst, and Marilyn Zoidis.

My thanks also to my students, friends, and colleagues in the Public Humanities program, the Department of American Studies, and the Haffenreffer Museum of Anthropology at Brown University, and at the Rhode Island School of Design and the RISD Museum, who helped me think about museums in new ways. I have also learned much from presentations and discussions at meetings of the American Alliance of Museums, the Consortium for American Material Culture, the National Council for Public History, and the Society for the History of Technology, and at the Beautiful Data Workshop.

Special thanks to members of the Jenks Society for Lost Museums, who did a spectacular job researching and reimagining the Jenks Museum: Raina Belleau, Lily Benedict, Elizabeth Crawford, Ann Daly, Mark Dion, Kathrinne Duffy, Layla Ehsan, Sophia LaCava-Bohanan, Robin Ness, Kristen Orr, Jessica Palinski, Lukas Rieppel, Benjamin Shaykin, Rebecca Soules, and Jamie Topper. Mark Dion's enthusiasm for this project, and for museums in general, allowed me to see them in a new light, for which I am enormously grateful. Mark's art, and his approach to museums and to museum work, inspired this book, and I am delighted to have had the opportunity to work with him and with all of the members of the Jenks Society.

For help with research, my thanks to Jim Moske at the Archives of the Metropolitan Museum of Art; Andrew Witkin, at Barbara Krakow Gallery, for his enthusiastic descriptions of modern and contemporary art; Jennifer Betts and Raymond Butti at the Brown University Archives; the interlibrary loan staff at Brown's Rockefeller Library; Judy Chupasko and Mark Omura at the Mammalogy Department of the Museum of Comparative Zoology, Harvard University; Paul McAlpine and Maureen Melton of the Archives of the Museum of Fine Arts, Boston; Erin Bilyeu and Suzanne Peurach of the National Museum of Natural History; Dr. David Berson and Russell Racette for assistance in understanding and building the skiascope; and to the community of museum folks on Twitter.

I am grateful to the John Simon Guggenheim Foundation for a fellowship in support of this project, and to Brown University, for support of *The Lost Museum* project and for the sabbatical that made the book possible.

My thanks to the readers of all or part of the manuscript—Matthew Battles, Lily Benedict, Jackie Delamatre, Kate Duffy, Kathy Franz, Elena Gonzales, Matthew Guterl, Debbie Moore, Marjory O'Toole, Malgorzata Rymsza-Pawlowska, and the anonymous referees—for their advice and encouragement, and to editor Janice Audet and copy editor Carol Noble, for making this book much better.

Finally, my deep love and appreciation to my spouse, Lisa Thoerle, who not only read the manuscript several times, improving it every step of the way, but also maintained an unfailing enthusiasm for the project, from the discovery of the Jenks Museum to the platypus cake to the last footnote. This book is dedicated to her.

Illustration Credits

p. 255   Courtesy of the John Hay Library, Brown University Library.

p. 270   Mammalogy Department, Museum of Comparative Zoology, Harvard University. Photograph by Lukas Rieppel.

p. 286   Courtesy of the Brown University Archives.

p. 303   Photograph by the author.

p. 317   Jenks Society for Lost Museums, 2014. Photograph by Jodie Mim Goodnough.

Index